California's Frontier Naturalists

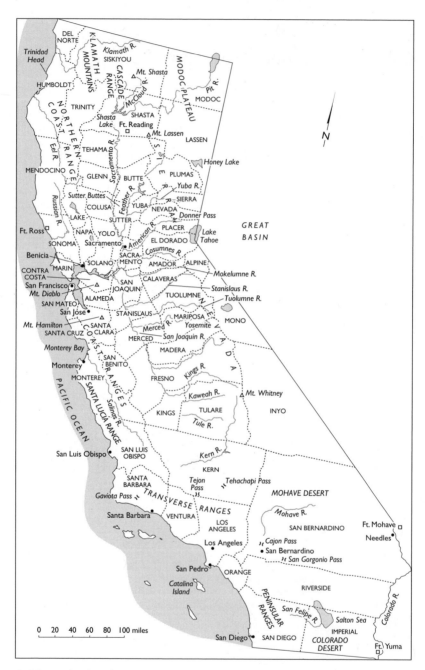

California landmarks.

California's Frontier Naturalists

Richard G. Beidleman

UNIVERSITY OF CALIFORNIA PRESS
Berkeley · Los Angeles · London

University of California Press, one of the most
distinguished university presses in the United States,
enriches lives around the world by advancing
scholarship in the humanities, social sciences, and
natural sciences. Its activities are supported by the UC
Press Foundation and by philanthropic contributions
from individuals and institutions. For more
information, visit www.ucpress.edu.

University of California Press
Berkeley and Los Angeles, California

University of California Press, Ltd.
London, England

Library of Congress Cataloging-in-Publication Data
Beidleman, Richard G.
  California's frontier naturalists / Richard G.
Beidleman.
     p.   cm.
  Includes bibliographical references and index.
  ISBN 0-520-23010-8 (cloth : alk. paper)
    1. Natural history—California.   2. Scientific
expeditions—California.   3. Naturalists—California—
History.   I. Title.
QH105.C2B45   2006
508.794—dc22   2005002729

Manufactured in the United States of America
15   14   13   12   11   10   09   08   07   06
10   9   8   7   6   5   4   3   2   1

This book is printed on Natures Book, containing 50%
post-consumer waste, and meets the minimum
requirements of ANSI/NISO Z39.48–1992 (R 1997)
(Permanence of Paper). ∞

The August and Susan Frugé Endowment Fund in California
Natural History honors the memory of August Frugé, who
served as Director of University of California Press from 1951
to 1977, and his wife, Susan. The fund commemorates the
Frugés' dedication to California's natural environment by
supporting the finest books UC Press publishes in this subject.
We gratefully acknowledge the founding donors whose
generous contributions established this fund:

Benefactors
Harlan and Esther Kessel

Supporters
Ernest Callenbach and Christine Leefeldt
James H. Clark
Grete Cubie
Joanna and Alan Curtis
Phyllis Faber
Donald and Katharine Foley
Loren Furtado
Robert Harlan
Nora Harlow
Harold and Meryle Ireland
Sally Lilienthal
Thomas and Dianne McFarland
Sylvia McLaughlin
Beverly Jarrett Mills
Susan H. Peters
Margaret Philbrook
Bill Platt
Anne Wertheim Rosenfeld
Thomas G. and Lilo Rosenmeyer
Jack Schulman
Jean B. Siri
A. K. Smiley Public Library / Larry E. Burgess, Director
James P. Smith, Jr.
Lori and Robert Torrano
Lynne Withey
William and Janice Wood

Contents

Illustrations

Preface

California's Frontier Naturalists may seem a provincial book, with its consideration of only a single state—except that this state is California! Probably no other terrain of equal expanse in the world possesses such a diversity of natural bounty that attracted such an interesting, stalwart, enthusiastic, and talented array of naturalists over many early decades. What follow herein are vignettes of these frontier naturalists' lives, their demanding fieldwork and often exciting adventures, against a backdrop of California and world history.

Fifty years ago there was little interest in field naturalists who had explored the natural history of frontier North America. Today, the increasing fascination with such natural scientists amounts to a cottage industry, involving books (both factual and fictitious), periodical articles, lectures, symposia, and even special museum exhibits and field excursions. Through 2004 the American Philosophical Society in Philadelphia featured a major exhibition, "Stuffing Birds, Pressing Plants, Shaping Knowledge: Natural History in North America, 1730–1860." And also commencing in 2004, the Jefferson National Expansion Memorial in St. Louis designed a tribute to the Lewis and Clark exploration, one of many diversified tributes during the bicentennial celebration of the United States' first transcontinental expedition of discovery.

Although I am an ecologist by training, research, and profession, I have long been interested, as well, in the history of the field natural sciences: botany, zoology, geology, paleontology, climatology, archaeology, and ethnology. As a university student I experienced a unique history of biology course taught by Joseph Ewan. Meeting in a university library, we delved into original works, from herbals to American scientific expedition reports, and each week we were charged to produce a relevant essay based on library research. In 1954–55 I was fortunate to receive a Ford Foundation grant, supporting a sabbatical leave focusing on the impact of the American frontier on natural science. Fittingly, during that Lewis and Clark sesquicentennial year I researched pertinent library and museum collections, from Wilkes Expedition documents at the Huntington Library to Torrey correspondence at the New York Botanical Garden. In the field, with naturalist narratives in hand, I found it exciting to explore frontier naturalist haunts—Bartram's Garden along the Schuylkill near Philadelphia, Audubon's wilderness log-cabin site overlooking the Ohio River, Lewis and Clark's encampment on a Missouri River island, and Monterey Bay's shoreline, where Nuttall collected shells.

The writing of this book could have been accomplished only with many invaluable years of background research and field travel, assisted and encouraged by my first wife, Reba, who died in 1990, and my present wife, Linda, a biologist and author, whose meticulous editing, computer skills, and support have seen this treatise through to completion. There are many individuals, as well as libraries and other institutions, here and abroad, who have made valuable contributions, including Eric Groves, the authority on Menzies at the British Museum of Natural History; Kirk Beidleman, geologist Dr. Thomas Grose, and the Huntington Library for material on the Beckwith Expedition; Janet and Carol Beidleman and New Gloucester town clerk Barbara Seaver for Sarah Lemmon background; Dr. Thomas Jervis for new Gambel material; and botanist Dr. Barbara Ertter for information on the Brandegees and others. Special thanks go to the University of California, Berkeley, libraries, especially the Bancroft, Bioscience, and Herbarium Library/Archives, as well as to the authors of hundreds of books and articles whose material I drew on; to Doris Kretschmer, Jenny Wapner, Laura Harger, and others at the University of California Press; and to the late Dr. Robert Ornduff, who urged that I undertake this task. But alas and alack, my children, now grown, never again want to visit a dead naturalist's grave in some forsaken spot.

This historical panorama is presented as a narrative rather than a treatise replete with footnotes and ibids. However, interspersed are brief quotations from expedition and individual naturalists' field journals, correspondence, and official publications; and there is an extensive reference list for those who would like to delve further. But more exciting would be to take the book and a good map and head for the Golden State countryside. The naturalist travels are presented in sufficient detail with respect to where they went and what they saw and did, so that you can retrace and in some measure relive the experiences of California's frontier naturalists.

Prelude

CAPTAIN COOK

Let us now proceed to the Voyage.
                    —John Rickman (1781)

Captain Cook's three oceanic expeditions between 1768 and 1778 set a pattern for the scientific exploration of the world. The first expedition was made to observe the transit of Venus (the passage of Venus across the face of the sun) in Tahiti and to search for a possible southern polar landmass to "balance" those in the Northern Hemisphere. This expedition's distinctive significance was its inclusion aboard HMS *Endeavour* of naturalists, financed and headed by an affluent amateur botanist, Joseph Banks. Cook's next expedition was, in part, a continued search for a southern polar continent, while the third sought for the hypothetical Northwest Passage by which ships could cross the North American continent from Pacific to Atlantic. On March 7, 1778, Cook's first West Coast landfall missed Alta California by only sixty miles. Naturalists also accompanied these later expeditions, firmly establishing the pattern that was emulated by many future oceanic and inland explorations. Their observations and collections inevitably stimulated interest in the natural diversity of the earth.

There were limitations on scientific expeditions in the 1700s. Travel definitely was not easy, while natural scientists and published knowledge about the natural world were minimal. Swedish botanist Carl Linnaeus in the 1730s began devising a universal system for "classifying" and naming plants and animals that he called *Systema Naturae,* and its tenth edition (1778) became the worldwide classification bible. As Linnaeus sent disciples to far-flung corners of the globe to collect new flora and fauna (some dying in the quest), others were inspired to do likewise. In the second half of that century the great French biologist Comte Georges Buffon produced his forty-four-volume *Histoire Naturelle,* an optimistic attempt to describe the natural history, especially animals, of the world. Works by Linnaeus and Buffon were avidly perused by savants and were included in the library of many expedition vessels.

During the early days of exploration, botany was of foremost interest. Plants might prove to be important sources of food, fiber, and medicine. And increasingly they became desirable for horticultural purposes, in private gardens and public botanical gardens such as the Royal Botanic Gardens, Kew, near London and the Jardin des plantes in Paris. Indeed, during Victorian times even outdoor and indoor ferneries became the rage. Plants were relatively easy to collect and keep, either alive or preserved, while animals were difficult to obtain—most having to be shot, trapped, hooked, or netted—and when bigger specimens were collected by naturalists, they tended to be eaten by nonnaturalists as well as naturalists themselves. Preserving mammals and birds was time-consuming, usually involving toxic preservatives and producing bulky collections. Live birds and mammals were even more difficult to deal with and keep alive. "Cold-blooded" vertebrates and many invertebrates had to be preserved in containers of drinkable alcohol, the containers difficult to transport and easily broken, while the liquor itself posed a consummate attraction for expedition members.

Institutional and personal herbaria (collections of dried plants) generally developed before collections of preserved animals did. The early private animal collections usually favored seashells and insects, especially beetles and butterflies. Among vertebrates, excepting large game mammals, only birds were favored, especially as mounted specimens. The development of natural history museums such as the British Museum, Bullock's Egyptian Museum, Peale's Philadelphia Museum, the Smithsonian Institution, and the Berlin Museum provided the stimulus for large-scale collecting of animal specimens for research (skins, skulls, bird nests and eggs, etc.), with taxidermy mounts for exhibition. Meanwhile, interest in both public and private menageries and aquaria encouraged collections of the world's diversity of terrestrial and aquatic live vertebrates and some invertebrates. Likewise, the broad spectrum of geological specimens, especially gems and fossils, and many archaeological and anthropological artifacts, came to be sought by both individuals and institutions, while geological phenomena, new land forms, and valuable rocks, minerals, and deposits stimulated pursuits. As expeditions and their naturalists returned with specimens and published their findings, including illustrations, further interest was aroused among naturalists back home as well as among the public at large.

THE CORNUCOPIA OF CALIFORNIA

The California encountered by early scientific expeditions and frontier naturalists was a land—Sir Francis Drake's New Albion—that had long been inhabited by tribes of Native Americans representing more than twenty language groups. Although Spain had been interested in colonization and exploitation of the New World (Latin America) since the time of Columbus, for Alta California the Spanish influence commenced when Juan Cabrillo set foot on its southern coast (at San Diego) in 1542. Settlement here proceeded at a casual pace. Spanish land grants to a chosen few made possible great ranches in early California, especially of cattle, whose main value resided in tallow and hides for leather. Catholic Spain extended its missionary zeal to California's Indians starting in 1769, when Father Junípero Serra initiated a chain of twenty-one Franciscan missions from San Diego to north of San Francisco. For inland travelers, including early naturalists, the missions provided convenient and safe stopping places along a thoroughfare, El Camino Real, outlined by European yellow mustard. Beyond the mission chain for years was little but wilderness. The missions were a mixed blessing, agricultural oases worked by subjugated Indians, with occasional padres who evinced cultural and natural history interests.

During decades of Spanish domination, Alta California (with Monterey as its capital), far removed from the European mother country, suffered from the "tyranny of distance" that affected isolated colonies economically and in many other ways. Visiting foreign expeditions and naturalists quickly recognized this. At Monterey on April 11, 1822, the Spanish flag finally came down. But Mexican Alta California, with Monterey still its capital, remained a colony now controlled by, but far removed from, Mexico City. On August 17, 1833, the Mexican government began secularizing the California missions, which fell into decline (although padres were later permitted to return). Alta California existed as a country of leisure and hospitable living for some, requiring little energetic endeavor. It continued to be a countryside of extensive ranchos populated by thousands of livestock, with only dots of communities along the coast, the remainder of the land still wilderness. The only other foreign presence was a small Russian colony, established in 1812 some leagues north of San Francisco Bay.

The great changes began with the United States' 1848 victory in the Mexican War, the gold rush burgeoning in 1849, and California statehood in 1850. During the decade of the 1850s the population increased from under one hundred thousand to about four hundred thousand, and by the "end of the frontier" in 1890 to well over a million. But still much of the bounty that had attracted naturalists was relatively intact, except for that impaired by hydraulic mining for gold, logging for lumber, and overgrazing by livestock.

California's natural cornucopia embodies a diversified terrain of 160,000 square miles including the highest point (Mount Whitney) in the Lower Forty-eight as well as the lowest point, 282 feet below sea level (Badwater, Death Valley). California's ecological landscape has representatives of all of the world's major ecosystems save the tropical rain forest, and all of western America's altitudinal life zones. Here the diversity includes hot and cool desert, coniferous forest, temperate rain forest, evergreen oak and deciduous woodlands, coastal and foothills chaparral, semiarid brushland, meadows and marshland, steppe, and tundra. There are five thousand great to tiny lakes and thousands of miles of river systems with their riparian corridors, plus nearly a thousand-mile ocean coastline in rocky to sandy settings.

California is a matrix of ancient major tectonic plates and innumerable drifted and collided terranes, almost too complex geologically to assemble into coherent order even today. There are sedimentary and granitic and volcanic mountain ranges, vast semiarid basins, desertscapes, diversified shorelines. There are ores of gold and silver and copper, gems of tourmaline and jade and even tiny diamonds. There are arid lands bereft of rainfall and wintery summits buried beneath the deepest of snow, Mediterranean climate in the southwest and boreal climate in the northwest. This is a land of major earthquake faults, active or quiescent, volcanoes extinct or merely waiting, glaciers active or long disappeared.

California is home to about 1,300 species of vertebrate animals, among them more than 450 species of birds and over 200 species of mammals. Among the innumerable invertebrates are more than 28,000 species of insects. California's vegetation is diverse and distinctive, with nearly 8,000 species of higher plants, including almost 1,500 endemics (species that occur only in the state). Along the northern coast are among the world's tallest trees, the Redwoods, while in the Sierra Nevada are among the world's most massive trees, the Giant Sequoias. At California's edge of the Sonoran Desert is the world's largest cactus, the Saguaro, while the largest and most bizarre yucca (Joshua Tree) thrives in the Mohave Desert. Tall native palms grow in secluded southeastern canyons, and the oldest conifer on the continent, the Western Bristlecone Pine, grows atop the White Mountains. Past flora and fauna are finely represented by fossil remains scattered from among limestone cliffs, shale outcrops, and alluvium to ancient sandstone shorelines and asphaltum seeps. And bordering Alta California on the west is the diversified world of marine life.

This was the natural bounty that excited the frontier naturalists who explored Alta California over the years. They came as members of expeditions, foreigners and Yankees, professional and private collectors, inquisitive visitors, mostly men at first, and eventually some became residents. Though seldom appearing in historical accounts of California, these natural scientists were most frequently with exploration's vanguard. Despite manifold hardships, the naturalists reveled in their adventures and discoveries. Zoologist John Kirk Townsend captured their essence when he wrote in May 1834, after an early morning afield in a western streamside woodland, "None but a naturalist can appreciate a naturalist's feelings, his delight amounting to ecstasy, when a specimen such as he has never before seen, meets his eye, and the sorrow and regret which he must feel when he is forced to tear himself away from a spot abounding with all which he has anxiously and unremittingly sought for."

PART ONE

The Oceanic Expeditions

CALIFORNIA'S FIRST NATURALISTS VISITED AS official members of oceanic expeditions flying the ensigns of France, Spain, England, Russia, and finally the United States. For these men, every landscape, every flower, bird, beetle, and seashell was new and exciting. Back to Europe went specimens of the first roadrunner, Redwood, oak, ground squirrel, boa snake, and California Poppy. Into expedition reports went artist portrayals of California Quail, Sutter Buttes, an Ohlone hunter with bow and arrow, and Zorillo the skunk. The naturalists confronted grizzlies, rattlesnakes, Indians, bad weather, and on occasion naval and Alta California bureaucratic meddling. While most exploration was done near the coast, the American naturalists traveled inland from the Columbia River to San Francisco Bay. Memorable among the oceanic expedition naturalists was Captain Vancouver's Archibald Menzies, whose advice and collections would influence generations of future naturalists.

The French La Pérouse Expedition at Monterey, 1786

Fish to anatomize, quadrupeds to describe; insects to catch,
shells to class . . . and nature to contemplate . . .
                                        —Robert-Paul de Lamanon

It was September 1786. Most of the varied bird species of the Monterey Peninsula had long since ceased singing, the flowers of the California springtime had come and gone, and the wild oats from Europe gave the cattle-grazed slopes among the oaks and pines a sunburned hue. Dimly visible through the late-summer fog shrouding the bay were two specially refitted French frigates, *L'Astrolabe* and *La Boussole,* anchored in twelve fathoms about four hundred feet from Monterey's shoreline.

Aboard was the greatest scientific entourage ever assembled for a worldwide exploring expedition up to that time, under the command of Jean-François de Galaup, Comte de La Pérouse, with orders directly from Louis XVI of France and a memorandum from the French Academy of Sciences. La Pérouse was a distinguished choice as leader. He entered the navy in his early teens, serving in Canada's Hudson Bay toward the end of the Seven Years' War, and during the American Revolution again opposed the British in the northwestern Atlantic. In 1784, honoring his support of the American cause, La Pérouse was inducted into the Society of the Cincinnati, an American patriotic group whose president was George Washington.

Inspiration for this enterprising French scientific expedition came to the French king while he was reading about the late Captain Cook's expeditions, and the king initiated plans for a voyage similar to Cook's Third, across the Pacific to North America. Indeed, two of Cook's dip needles, used in the compass to measure magnetic north, were actually loaned to La Pérouse for navigation, thanks to Joseph Banks. Considering that the British, Spanish, Dutch, and Portuguese had already coursed the Pacific, the French government's official instructions emphasized that "everything seems to invite the French who share the empire of the seas with them, to perfect a work of which, until now, they have done only a small share." There was to be consideration of fur trade possibilities, the extent of Russian and Spanish settlement, and a search for the Northwest Passage. Thomas Jefferson was so intrigued by what he heard of the French expedition that he wrote to John Paul Jones, asking the naval officer to obtain as many details on the scientific expedition as he could.

Seventeen savants—"Engineers, Men of Science, and Artists"—constituted La Pérouse's scientific party. They had their own library—travel chronicles including the three Cook expeditions and Louis de Bougainville, as well as sixty-five volumes on natural history including books by Linnaeus and Buffon—along with necessaries such as fifty reams of brown paper for drying plants, eight butterfly nets, two microscopes, nine boxes of dissecting instruments, thermometers, barometers, and hygrometers, a hydrostatic balance, and a twenty-six-foot-tall linen balloon for observational and experimental purposes. On the two frigates the naturalists were to be billeted in special quarters christened *chambres des savants.* Aboard *L'Astrolabe* (named for the early navigation instrument) were seven natural scientists, including Bossieu de la Martinière, a botanist and medical doctor who would later have a large genus of Australian pea-flowered shrubs, *Bossiaea,* named in his honor; cleric Claude-François-Joseph Receveur, "Collector of Natural Curiosities"; naturalist Dufresne; and the older artist Prévost. The larger group was on *La Boussole* (meaning "compass"), La Pérouse's flagship. Among them were naturalist and chaplain Abbé Mongès, versed in ornithology and entomology; the senior astronomer, Lepaute Dagelet, who also had experience as a plant collector; geographer Bernizet; Gaspard de Vancy, a draftsman of landscapes who may have been recommended by Marie Antoinette; artist Jean Robert Prévost the younger (nephew of the older Prévost); and Robert-Paul de Lamanon, whose broad interests included not only meteorology but "geology, quadrupeds, fishes, shells, and other aquatic animals." Also on the flagship was botanist and gardener Jean-Nicolas Collignon.

Before leaving on the expedition, La Pérouse chatted with the great naturalist Buffon at the Jardin des plantes in Paris about what kind of geographical observations should be made and what should be collected. Buffon gave La Pérouse a complete set of his *Histoire Naturelle* and colored bird paintings to aid in identification, and he referred La Pérouse to André Thouin, the chief gardener, with respect to botany. Thouin strongly urged that a gardener who knew how to collect, care for, and preserve plants should be taken along, and he enthusiastically recommended his bright teenage assistant Collignon, the son of another gardener. Throughout much of the long voyage Collignon would conscientiously keep in touch with his mentor by letter.

The naturalists enjoyed a special farewell dinner at Versailles in mid-June. On June 28, 1785, La Pérouse received his final blessing for the success of the expedition from Louis XVI, and two days later at Le Trianon he bade Marie Antoinette farewell. Behind the scenes there was a startling but untrue rumor that accompanying this expedition would be sixty hardened French criminals who would be put ashore on newly discovered New Zealand to claim that island group in the name of France. It had been the plan for this grandiose French expedition's two ships to depart on the extended around-the-world voyage from Brest on July 11, but unfavorable winds delayed their sailing until August 1.

The ships proceeded around South America and past Easter Island, spending the late spring of 1786 in the Sandwich Islands (Hawaiian Islands) and most of the summer months in Alaskan waters. In September, they finally sailed south along the west coast of North America through almost constant fog, except for a momentary glimpse of Northern California's Cape Blanco. They arrived opposite Monterey Bay on September 11, 1786, but did not reach safe anchorage until the fourteenth, and the scientists did not get ashore until the next morning. The French were warmly received by the Spanish inhabitants: supplies in abundance were dispatched to the two vessels without a demand for payment, horses provided for travel, garrison soldiers made available for any necessity, the astronomical and surveying instruments set up ashore. Pedro de Fagés, the governor of Alta California, received the Spanish minister's official letter from La Pérouse and extended the hospitality of his own home to the visitors. For Alta California, that coastal region north of Mexico, the French La Pérouse Expedition would represent the first non-Spanish European visit.

The naturalists' eagerness to explore the California scene was undoubtedly frustrated by a reception and dinner for the expedition members with the Franciscan missionaries of Mission San Carlos Borromeo de Carmelo, involving a horseback ride over the hill to Carmel Bay. The Frenchmen were welcomed "like lords of the parish making their entrance into their lands." The missionaries and Indians with great cere-

Figure I. Reception of La Pérouse at Carmel mission in 1786. Original by Gaspard de Vancy, copied by Tomás de Suría. (From Claire Perry, *Pacific Arcadia: Images of California 1600–1915* [New York: Oxford Univ. Press, 1999].)

mony massed in review for the French visitors, a moment memorialized by a de Vancy sketch (figure I). The missionaries, carrying the cross and banner, then led La Pérouse and his party to the church for a singing of the *Te Deum* in thanksgiving for the explorers' voyage. When the Frenchmen emerged from the church, the Indians were still standing in formation.

One can appreciate the naturalists' impatience to be out in the field, despite the lateness of the season. Already the bounty of the Monterey area had become evident. That initial entry into Monterey Bay had been greeted by great flights of Brown Pelicans *(alkatros)* above, with Sea Otters, seals, and numerous spouting whales on either hand. The enthusiasm of all La Pérouse's naturalists here was superbly expressed by "natural philosopher" de Lamanon: "I work more than twelve hours a day, and yet I am never beforehand in my work: fish to anatomize, quadrupeds to describe; insects to catch, shells to class . . . experiments to make . . . and nature to contemplate—I would that for all this I could multiply my existence twenty times over."

The gardener, Collignon—charged with noting the new flora from the standpoint of its usefulness for food, medicine, or ornament—did well in recognizing five different members of the genus *Artemisia* (sage-brush), as well as yarrow, mint, nightshade, and a few other plants. He collected what was undoubtedly a Monterey Pine cone, which made its way back to France in 1787; and eventually from twelve of its planted seeds young pines grew up. Most interesting historically, Collignon gathered seeds from a sprawling beach plant that had finished blooming. In the mission garden he was good enough to plant potatoes from Chile for the missionaries, as well as seeds of other exotic plants. Meanwhile, Captain de Langle of *L'Astrolabe*, noting how hard it was for the Indian women at the mission to grind grain using a roller on a flat stone, donated a grist mill from his ship. La Pérouse's account noted that "our lithologists [geologists] were equally zealous as our botanists," but met with fewer rewards. On the beach a few pieces of water-smoothed jasper were picked up. But they never realized that the granite they collected had been carried from Southern California by the Pacific Tectonic Plate, slowly edging northwestward.

The botanist Martinière, finding few flowers in bloom, became enchanted by the diversity of marine invertebrates in a bucket of seawater from one of the many tidepools. And he took time to pen "A Memoir Concerning Certain Insects," which was eventually published. The peninsula abounded with cottontail rabbits, Black-tailed Jackrabbits, and Black-tailed Deer. In addition, there was a variety of new and interesting birds that were collected and stuffed, but never described by their collectors: the Oak Titmouse, Chestnut-backed Chickadee, California Thrasher, Bullock's Oriole, Western Scrub-Jay and Steller's Jay, White-crowned Sparrow, Nuttall's Woodpecker, and California Quail, the latter encountered in coveys of three hundred to four hundred. Many more birds were recorded, including the Bald Eagle, Turkey Vulture, American Kestrel, Great Blue Heron, and a diversity of water birds. La Pérouse recognized the magpie but had no idea it was a new species (the Yellow-billed Magpie, *Pica nuttalli*) different from the one named by Linnaeus (Black-billed Magpie, *Pica pica*), which occurs in Eurasia. The younger Prévost, though obsessed with painting flowers, made the first painting of the California Thrasher, as well as that of a now more famous species, the California Quail, which would be named the state bird of California in 1931 (figure 2).

La Pérouse was a worthy natural scientist himself, who had become a full member of the Académie royale de marine in 1775. Following his instructions to make observations on the native peoples encountered during the voyage, La Pérouse in his journal described the Indians, their

Figure 2. "Male and Female Quail of California," by Jean Robert Prévost with the La Pérouse Expedition, 1797. (Courtesy California Historical Society, FN-30508.)

habits and habitations, and astutely assessed the nature of their local dialects, their relationship with the missionaries, the crops they were cultivating, even the nature of the "inexpressibly fertile" soil. He noted the presence of what would later be recognized as distinctive species of cypress and sycamore, thought local "sea ears" (abalones) were superb but other mollusks unexciting, and carefully wrote up the naturalists' observations in his own journal, while he and his engineers executed maps of the local coastline, streams, and mountains that remain recognizable to this day. La Pérouse would write of this small bit of California, "There is not any country in the world which more abounds in fish and game of every description." A century later, in *Overland Monthly,* an author averred that La Pérouse's "account of the natural resources of the country and its characteristics was never surpassed in fidelity by his successors" (Stillman 1869, 257).

On September 24, 1786, with a breeze out of the west, the two ships of La Pérouse's expedition sailed out of Monterey Bay and steered southwest, bound eventually for the southern coast of China. There, at the Portuguese trading center of Macao, one of the naturalists—the lucky Dufresne—was entrusted with some of the expedition journals and collections and given permission to return to France. By September 1787 the ships had reached Kamchatka, and more of the expedition's effects made their way overland to France with de Lessep, the Russian vice-consul from *La Boussole*.

In January 1788, after suffering losses (including the death of the talented naturalist de Lamanon) in an encounter with natives on Samoa, *La Boussole* and *L'Astrolabe* sailed into Australia's Botany Bay, so named after its unique flora by Captain Cook's First Expedition in 1770. There the La Pérouse Expedition was astounded at encountering the so-called Australian First Fleet under Governor Arthur Phillip, carrying the initial cargo of convicts from England. The First Fleet was equally astounded at the appearance of two French frigates! Some of the British ships, with the governor aboard, had already moved north to the great harbor where Sydney is located today. La Pérouse's vessels anchored at the northeastern edge of Botany Bay, which became known as Frenchmen's Cove. Here Father Receveur, one of the naturalists, died from injuries sustained in Samoa and was buried in a parcel of land later deeded to France and honored with a memorial.

During the six-week stay, two damaged longboats were replaced, a vegetable garden was planted, and short field trips were taken. La Pérouse had an opportunity to talk with some of the British officers, saying in praise of Captain James Cook that Cook had "left nothing to those who might follow in his track to describe, or fill up." Meanwhile, Abbé Mongès visited marine officer Ralph Clark on the First Fleet's *Friendship*, where he enjoyed looking at Clark's collection of insects and wished Clark would give him one of the grasshoppers.

Some of the French sailors tried (unsuccessfully) to stow away on the British vessels, hoping for a quicker return to Europe. At the same time, some British convicts transported north to Sydney Bay escaped, came cross-country to Botany Bay, and tried to board the French frigates, hoping to ship as sailors and escape from an extended incarceration "down under." The convicts were quickly sent back to Sydney, little realizing for the moment how lucky they actually were.

Fortunately, additional expedition records made their way back to Europe from Australia through the good offices of the English—because after the two French frigates departed from Botany Bay on March 10, 1788, they were never heard from again. This great mystery was not to be completely solved for nearly two centuries, despite continuing investigations by many nations over the years, although some shipwreck accounts and apparent artifacts were reported from the Santa Cruz archipelago by 1826. It was finally affirmed that the ships had indeed been wrecked near Vanikoro, some five hundred miles southeast of Guadalcanal.

Among the expedition's journals and collections that the naturalist Dufresne brought back to France from Macao in 1787 was a wooden case from Collignon. Among other things, it contained three packets of seeds, including some the gardener had collected on the beach near Monterey in mid-September 1786. From these seeds, planted at the Jardin des plantes in Paris, developed a number of "beautiful herbaceous plants" with rose-colored flowers, of a new species in the Four O'clock Family. These were eventually described and named *Abronia umbellata* by the great British botanist Aylmer Lambert. Thus seeds collected in mid-September 1786 from an unknown plant, which was already through blooming along the shoreline of the Monterey Peninsula, were planted in France, grew up there, and blossomed, becoming the first flower (Coast Sand-verbena) to be scientifically delineated from California, indeed from western America!

From the unfinished journal of La Pérouse that survived to be published in France in 1797 came the French explorer's tribute to the benefits of natural history: "The sciences, by softening the manners of men, have, perhaps, contributed more than wise and beneficial laws to the happiness of society."

The Spanish Malaspina Expedition with Naturalist Haenke, 1791

Very convenient for the exhaustless studies of the
naturalist . . .

—Alessandro Malaspina

Two Spanish corvettes, *Descubierta* and its consort *Atrevida,* comprising Spain's primary eighteenth-century scientific oceanic expedition,
sailed south in mid-September 1791 along the Pacific Coast, passing
Cape Mendocino and the Farallon Islands, and finally swung into Monterey Bay. When their cannonading request to enter the harbor at Monterey was not immediately acknowledged by cannons from the presidio,
they cruised on south past Point Piños (now generally called Point Pinos)
and anchored off present-day Pebble Beach. Early the next morning (September 13, 1791), guided back north through dense fog by one of their
small boats, they finally moored at eight o'clock, a fifth of a league off
the town of Monterey.

The commander of this Spanish expedition was Alessandro Malaspina,
a nobleman and brilliant young captain; like Columbus he was actually
Italian, but his birthplace in Parma was now held by Spain. Malaspina's
new corvettes, unlike the older vessels used by La Pérouse, had been specially built for scientific and polar exploration. According to an account
of the journey, they were well equipped with "the finest instruments," including thermometers, hygrometers, hydrostatic balances, two air guns,
two eudiometers (for analyzing gases), and even two "scaphanders"
(cork swimming suits) (Galbraith 1924, 221). The staff of naturalists had
been invited to recommend books for the expedition's library—botanical,
zoological, and mineralogical—a lengthy list including, of course, those
by Linnaeus and Buffon, and any by Cook expedition naturalists. The re-

quests by Spain for equipment and especially books went across Europe and in a sense brought Iberia into mainstream scientific endeavor. Archibald Menzies, George Vancouver's naturalist, was so impressed by the Spanish expedition that he wrote that they were "fitted out in the most ample manner for Discoveries with Astronomers Naturalists Draughts-men &c."

Spain's around-the-world Malaspina Expedition, after a number of months visiting especially Spanish dominions in South and Central American waters, received orders in early 1791 to search for the North-west Passage, not to mention a bit of snooping on competitor nations. After an unsuccessful summer's quest for a transarctic passage in the far north, the two vessels proceeded southward. Now at Monterey in September 1791, Malaspina became impressed by the diversity of plant and animal life, writing that it was "very convenient for the exhaustless stud-ies of the naturalist," and it would be "difficult to find another place bet-ter adapted"; "variegated hills with assorted greenery and flowers were found," where "peacefully whirled a thousand types of birds, while the rabbit, hare, and squirrel jumped lightly and timidly."

By this time Malaspina's expedition had only one naturalist left to pursue the natural history bounty, a Bohemian named Thaddeus "Tadeo" Haenke. When the two ships had earlier headed north for Alaska, Antonio Pineda, head of the natural history staff, and the very able elder botanist Luis Née, a Frenchman who had been a gardener at the Real jardín botánico in Madrid before joining Malaspina, had been left behind in Acapulco, Mexico. Haenke, though only in his late twen-ties, was both well educated and experienced in the natural sciences. Originally planning a career in music, he was an excellent bassoonist and trumpeter, and both sang and taught singing. But a debilitating bout with pneumonia turned him to science. At twenty-one he had received his doctorate from the University of Prague (in the present-day Czech Re-public), with Professor Joseph Mikan as his mentor. He pursued further training in medicine, physics, and botany, participated in two European explorations, wrote a treatise on the flora of the mountains along the Pol-ish border, and was awarded a silver medal from the Royal Czech Sci-entific Society. Haenke was conversant in five languages, was familiar with chemistry and mineralogy, and could draw, and his talent in music would be put to good use when he performed on shipboard and recorded the tunes of the Northwest Coast Indians.

Because Austria was chary to lose this talented young man, it had to be at the request of Holy Roman Emperor and King of Germany Joseph

II that Dr. Haenke was permitted to join the Malaspina Expedition, thus becoming the first Ph.D. to visit California. But the assignment turned out to be even more easily requested than accomplished. Haenke arrived in Cadiz, Spain, two hours after the expedition's ships had departed thence, on July 30, 1789. Trying to catch up, he sailed across the Atlantic to Montevideo (Uruguay) on South America's southeastern coast, where he was shipwrecked within sight of land. He swam ashore, saving only his collecting equipment and his copy of Linnaeus's *Genera Plantarum,* which he had tucked into his nightcap. Again he missed Malaspina, who was already sailing around Cape Horn for Chile. Undaunted, Haenke decided to walk westward across South America, through the Argentine pampa and over the Chilean Andes. Not only did he survive this arduous trek of about eight hundred miles, but he reportedly made a collection of 1,100 plants along the way, many of them new species. Then, quite by accident, at Santiago (Chile) he ran into Malaspina, who was on a side trip inland from Valparaiso, where the two ships had been anchored since February 1790. Haenke was finally able to join the expedition, about to set sail for the Arctic.

Haenke, assigned to Malaspina's flagship, the *Descubierta,* proved to be a valuable addition to the expedition. On shipboard he worked as an assistant to the naval surgeon, helped the Italian physiologist Felix Fontana study the declining purity of the air in the two vessels, entertained the assembled officers of the expedition with concerts on a "harpsichord," professed particular interest in antiscurvy plants, and expounded on his theory that the Earth was like a human body, the rugged, exposed peaks of the north representing the bones of its skeleton.

Ashore now at Monterey, Alta California, Malaspina's sailors quickly established an observatory on the beach, and observations commenced, including of the daily weather. Among their activities, a continuing endeavor involved measuring gravity's pull on a pendulum to determine if gravitational attraction varied around the globe. Soundings of the bay were made, as was a navigational chart, and latitude and longitude were determined. Malaspina had hoped to compare some of his expedition's scientific findings with those made by La Pérouse's, but La Pérouse's findings proved unobtainable. The accuracy of Malaspina's calculations was later commended by geographer Alexander von Humboldt, and John Charles Frémont verified the accuracy of Malaspina's Monterey Bay map.

Through the good offices of Father Fermín Francisco de Lasuén, father president of the Franciscan missions in Alta California, who had an interest in natural history, Indians from Mission San Carlos in Carmel were made available to lead hunting and fishing forays for Malaspina and Haenke. Even soldiers from the presidio and the expedition's seamen joined the Indians in collecting botanical specimens and valuable artifacts. In appreciation, Malaspina wrote Lasuén that he had "contributed considerably to . . . our collections for Natural History with various contributions of great value" bound for the Royal Museum of Natural History in Madrid.

Except for the reception at the Carmel mission, naturalist Haenke wasted no time during this brief sojourn in Central California. Even though autumn was approaching, he was surprised to experience a "second springtime," with many plants in blossom—a phenomenon, he thought, perhaps occasioned by the prevalent fog. The Bohemian was impressed by the great variety of vegetation, especially in the lower Carmel Valley. He took note of the oaks and pines forming woodlands on the hillsides, the sycamores in the creek bottoms, the "Garden Poplar" (Black Cottonwood) "nearby in all the arroyos," the "Wild Cypress" (Monterey Cypress) growing "very close to the beaches," the "Dogwood-scarce," and "el laurel" (probably the California Bay) bearing both flower buds and ripe seeds. He even collected several species of seaweed, labeling them with the notation "Regismontanae" for Monterey (figure 3). On the higher summits he observed "the red pine, a tree much taller than the rest": the Redwood, which he called "Red Cypress." Indeed, Haenke was the first scientist to obtain specimens of this magnificent conifer, and he recounted that the wood was valuable as lumber. A Redwood growing at the Alhambra in Spain, 125 feet tall in 1926, was authenticated as having been from his collection. The Alhambra Redwood has died, but contemporaries, undoubtedly Haenke's, have recently been discovered at another royal garden.

Haenke appraised useful as well as possibly poisonous plants, applying a mixture of recognizable and more obscure scientific names such as *millefolium* (Yarrow), *rhusradicans* (Western Poison Oak), *la cicutasiides* (Water Hemlock, poisonous), *sambucus-racemosa* (Red Elderberry), *la arthemesia-dracunculus* (Tarragon, a sage), and *solidago-cinerea* (a goldenrod). Altogether this energetic naturalist reportedly catalogued over 250 species. Into the expedition records went comments about the nature of the soil, description of the rocks on either side of the Carmel River and the crops the Indians were sowing, and tribute to the La Pérouse Expedition for having given the missionaries grain and fruit seeds, from which new plants were now beginning to flourish at Mission San Car-

Figure 3. *Cystoseira australis* (now *C. osmundacea*), a brown seaweed collected by Haenke with the Malaspina Expedition at "Regismontanae" (Monterey) in 1791. (Photo by Paul Silva, courtesy University and Jepson Herbaria, University of California, Berkeley.)

los. He collected crabs, dogfish, salmon, mackerel, skates, and certainly "an innumerable bunch of savory sardines," the latter an intimation that Monterey would later become the "sardine capital of the world."

The Malaspina Expedition report for the California visit also recorded some of the region's mammals, large ones such as Mountain Lions and bears and small ones such as rats and rabbits. However, the "tiger" was probably a Bobcat. Numerous birds were seen during the short stay, including hummingbirds, pelicans, mockingbirds, quail, and seagulls, some supposedly stuffed by sailors and eventually reaching Spain. The ships delivered a selection of plants from the Monterey area to Acapulco for botanist Luis Née to examine, including branches from two coastal trees, the Coast Live Oak and the Valley Oak, collected by a pair of *Atrevida* officers. These were scientifically named and described by Née and became the first western tree specimens brought to Europe.

When Malaspina had earlier discovered that a youthful Spanish cabin boy, José "Little Pepe" Cardero, possessed artistic ability, he was assigned to the *Atrevida* as one of the expedition's artists. It was Cardero who made sketches of the Farallons, the Monterey presidio, Mission San Carlos, and some of the local Indians. He also painted birds, including the California Quail, California Thrasher, Red-winged Blackbird, and Red-shafted Flicker. This flicker had first been recorded and painted in 1778 by Cook's Third Expedition at Nootka Sound, Vancouver Island, where Malaspina had sojourned during August 1791. Cardero would also serve as artist for the two Spanish survey ships that replaced Malaspina on the West Coast in 1792.

Another of Malaspina's artists, Tomás de Suría, had actually received training at the Royal Art Academy of San Fernando in Madrid before going to Mexico at age seventeen, where he became associated with the Academy of San Carlos and served as a designer for the mint. He was one of four candidates to be considered for the expedition; and when he was picked as artist-draftsman, his wife gave her thirty-year-old husband permission to join. In Monterey he sketched a number of the mission Indians, while Malaspina contributed his own valuable observations about these natives and their dress and habits. Also, Suría kept a personal journal, affording an invaluable fresh and unofficial contrast with the naval accounts, to which he had no access. Suría traveled aboard the *Descubierta* and never once mentioned his fellow artist, Cardero. His published journal unfortunately ends before the expedition's California visit.

The stay in Monterey Bay was only too short. On the morning of September 26, 1791, to the roar of presidio cannons at nine o'clock, anchors were hauled up and the Malaspina Expedition was on its way. Haenke carried with him the first collection of dried plants and seeds to reach Europe from Alta California; it included, besides many grasses, sedges, and rushes, species such as California Fuchsia, Durango-root, and Alkaliheath. Left behind in Monterey was John Green from Boston, who had been picked up at Cadiz by the expedition and became not only the first United States citizen to arrive in Spanish California but also the first to die there, on the afternoon of his arrival; he was buried the next day at the presidio chapel.

Malaspina paused at Acapulco to pick up the other naturalists, Née and Pineda, who had been busy collecting in Mexico. Then the two ships proceeded across the Pacific, stopping at oceanic islands and finally reaching the Philippines, where Pineda died of fever. Finally the expedition returned to South America in the summer of 1793.

War with the French Republic eventually put an end to Malaspina's worldwide expedition. In July 1793 Malaspina received special orders at Callao (the port of Lima, Peru) to return to Spain by way of Montevideo, where the ships were to be readied for battle. Encouraged by Malaspina, once again Haenke chose to cross South America on foot, joined by his Filipino assistant, Gerónimo de Archángel, a gunner from the ship, collecting more plants. He planned to join the expedition on the Atlantic side, but Haenke had not arrived by the time the ships departed from Montevideo on June 21, 1794.

Haenke remained in South America, eventually starting a botanic garden, cultivating silkworms, and ministering to the health of the inhabitants in the high-mountain town of Cochabamba (Bolivia), where he resided. He continued his field pursuit of natural history in South America. His memorable floral find, in 1801 along the Amazon River, was the Giant Water Lily, with six-foot-wide floating lily pads, eventually named *Victoria regia* in honor of Queen Victoria (now *Victoria amazonica*). At Kew gardens in 1852 the Waterlily House had to be constructed specially to accommodate Thaddeus Haenke's gigantic discovery. Although he always hoped to return to Europe, his dream was never realized. He died in 1816 or 1817 at a house on his own estate by poisoning when a maid accidentally gave him a toxic potion in place of his medicine. Many years after the Malaspina Expedition, Haenke's botanical collections were described in Karel Presl's two-volume *Reliquiae Haenkeanae,* based on specimens at the National Museum of Bohemia in Prague. Some of his plants eventually returned to America, and are now in herbaria from Harvard University to the Missouri Botanical Garden and Pomona College in California.

Arriving back in Spain, Malaspina came into conflict with the ruling minister, Manuel de Godoy, over management of the Spanish colonies. In an opera-type plot, Godoy's lover, María Luisa, the wife of Spain's King Carlos IV and currently miffed at Godoy, prevailed on Malaspina to write a secret letter about his ideas for reform, including criticism of Godoy's incompetence. When the queen frivolously made the letter public, Malaspina was arrested for treason and for nearly a decade was confined to a dank military dungeon. Finally, through Napoléon's intercession, he was released but exiled from Spain. His official report was not published for nearly a century, and many of the expedition's papers and collections remained untouched until recently, during the sesquicentennial. Malaspina died in Milan, Italy, in 1810.

The British Vancouver Expedition with Menzies, 1792–1795

Within me a particular desire of traversing unknown regions
in quest of my favorite pursuit . . .
                    —Archibald Menzies to Joseph Banks (1789)

Of all the early naturalists who visited California, Archibald Menzies surely had the greatest influence over the longest period of time. It helped that he lived almost to ninety, from his birth at Aberfeldy, Scotland, in 1754 to his death in 1842 at London. Yet he certainly did not obey the modern dictum "Publish or perish," since he published barely a thing during his long and busy career, which included a lengthy stint as a peripatetic surgeon with the British navy. Nor did he personally name and describe many botanical specimens. But his extensive plant collections were garnered from around the world, including the Americas' eastern and western coasts, and became an invaluable repository that was shared with others, generating excitement among botanists as well as among horticulturists and ordinary gardeners. Menzies's scientific pursuits and voyaging experiences at the end of the eighteenth century, and his advice and example for other naturalists during four decades of the nineteenth century, definitely made him, as the young Asa Gray proclaimed in 1839, "that Nestor of botanists" (in reference to the wise counselor King Nestor, who fought with the Greeks against Troy). It was Alexander Menzies who became the first naturalist to investigate the natural history of California's entire coast, from Trinidad Head to San Diego.

As a young man, Menzies was a student gardener at the Royal Botanic Gardens in Edinburgh. Professor John Hope of Edinburgh University made arrangements for him to pursue a medical degree there. He never actually received an M.D. degree from Edinburgh, but in 1799 he received an honorary M.D. from Scotland's Aberdeen University. Joining the navy as an assistant surgeon in 1782, Menzies served during the American Revolutionary War, his ship being one of those involved in the famous Battle of the Saints off Dominica in the Lesser Antilles, where the French fleet was defeated in the final sea battle of the war (April 12, 1782). In 1784 Menzies transferred to HMS *Assistance*, devoting, as had become his custom, his "vacant hours to Natural History" along the East Coast and in the Caribbean.

Meanwhile, Professor Hope suggested that his former student send some plant specimens to Sir Joseph Banks, president of the Royal Society, including some from the new United States. Menzies's first letter to Banks, dated May 30, 1784, contained seeds from New York City and was the first among many epistles and New World specimens transmitted to Sir Joseph, who was impressed by the young botanist. In late summer 1786, Banks made special arrangements to place Menzies as naturalist on a private fur-trading vessel commanded by James Colnett, who had served as midshipman on Cook's Second Expedition. Originally Menzies, together with other members of Colnett's crew, was not permitted "to trade or barter for any curiosities," but Banks intervened so that Menzies could make natural history collections. Colnett's *Prince of Wales* visited North America's northwestern coast in 1787 and 1788, but never journeyed as far south as California.

Menzies returned from this voyage in July 1789 with an admirable collection of dried plants that were divided between Banks and the Royal Botanic Gardens, Edinburgh. Banks next recommended Menzies as naturalist for a voyage to the Pacific on a new vessel under the command of Henry Roberts. The "sloup-of-war" HMS *Discovery* was named after one of Cook's ships on which Roberts had served. Conflicts with Spain delayed the voyage, however. After a tentative agreement between England and Spain at the end of 1789, the *Discovery* was assigned a new captain, George Vancouver, who had been a teenage midshipman on Cook's Second and Third expeditions. Banks's continued efforts on behalf of Menzies assured the position as naturalist with Captain Vancouver, several years his junior, "on a voyage round the world," per the British Admiralty's orders, departing from Falmouth on April 1, 1791. Banks, remembering his own unpleasant dealings with Vancouver, had words of warning in a letter to Menzies about this "tyrannical" commander: "I trust he will have too much good sense to obstruct [your]" duties as naturalist. Indeed, tension soon developed when Vancouver curtly turned down Menzies's request to be named surgeon as well as naturalist. Vancouver was further annoyed when Menzies, assigned to mess with the officers in the gunroom, complained about his cost for victuals.

Banks's instructions for Vancouver's naturalist were detailed: "He was to investigate the whole of the natural history of the countries visited. . . . All trees, shrubs, plants, grasses, ferns, and mosses were to be enumerated by their scientific" as well as local names; dried specimens were to be prepared of plants "worthy of being brought home"; and seeds and in some cases living plants were to be collected. Menzies was to examine the countryside for ores, minerals, and soils; take note of birds, mammals, and fishes of commercial value, especially Sea Otters, whales, seals, and wild sheep; and make detailed observations of natives encountered during the voyage, recording everything in a detailed journal. To abet Menzies's assignment, Banks arranged to have a wood-and-glass "plant-hutch," eight feet by twelve feet, assembled on the *Discovery*'s quarter-deck, a navy ship's hallowed site. The plant frame was for live plants and germinating seeds.

Several men with artistic ability were with this expedition, the most productive being John Sykes, a young *Discovery* officer who made 177 drawings, ten of which were published in Vancouver's *Voyage,* including all that portrayed California. Later in his career Sykes achieved the rank of admiral. Thomas Heddington was artist on the *Chatham, Discovery*'s consort, at sixteen the youngest of the expedition's officers. These two artists did only landscapes and headland profiles, mostly wash drawings in various shades of gray, with a few watercolors. The natural history illustrations were done by Menzies, with considerable skill and talent.

By the time the *Discovery* reached the Pacific Northwest in the spring of 1792, Menzies had already botanized with the expedition on two continents (Africa and Australia), and at New Zealand, Tahiti, and the Sandwich Islands (Hawaiian Islands). On April 17, 1792, the coastline of New Albion—North America—south of the Mendocino headlands was first sighted. Without stopping, the ship swung to port and sailed north. While Vancouver commenced his hydrographic survey, in the further quest for the Northwest Passage, Menzies described the Northern California shoreline as possessing "most beautiful prospects of hills and vallies varied with woods and pastures," regretting only that he had no opportunity "to examine it more particularly."

Vancouver spent the summer of 1792 in the Straits of Juan de Fuca and at Nootka Sound on Vancouver Island, where he accepted from Spain the relinquishment of its North Pacific claims of sovereignty dating back to 1775, claims and ensuing actions that had almost resulted in a war between England and Spain. Nootka Sound had been visited by Captain Cook in 1778. Ten years after Cook, the Spanish, concerned about foreign threats to a region they viewed as their territory, had sent ships to the area, seized developments ashore and foreign vessels (including one captained by Menzies's former captain, James Colnett), and established their own settlement. It was this settlement that was supposed to be turned over to Vancouver.

This dispute had been tentatively concluded in November 1790 by the Nootka Sound Convention, and Britain appointed Vancouver as its commissioner to meet with the Spanish commissioner, Bodega y Quadra, at Nootka Sound on what became known as Quadra and Vancouver Island. Quadra, a native of Peru and now commander of the royal navy for Mexico and California, had traveled along the West Coast during 1792, replacing the Malaspina Expedition. With Quadra were two young naturalists, botanist José Moziño and anatomist José Maldonado, and a youthful artist, Atanasio Echeverría, all trained at the Royal Botanic Gardens of Mexico and associated with the Royal Scientific Expedition to New Spain, headquartered in Mexico City. The trio was particularly employed in collecting plants for a proposed *Flora Mexicana*.

In October 1792, the *Discovery* and *Chatham* prepared to sail south along the coast to Alta California. Indisposed for months, Alexander Cranstoun, the expedition's surgeon, was officially replaced by Menzies, who then served as both naturalist and surgeon. Menzies had already performed many of Cranstoun's duties, including prescribing to Vancouver for a chronic thyroid deficiency (which resulted in the captain's death at forty). Menzies, benefiting from his new assignment, was given an additional cabin where he could house his increasing collection of specimens. At Nootka Menzies learned that Don Quadra had been suffering from severe headaches for two years, and now as ship's surgeon he provided some relief for the Spaniard, engendering a close friendship.

At Nootka two unusual passengers were picked up, Raheina and Tymarrow, attractive Hawaiian maidens who had been brought to Nootka for nefarious purposes. Gallantly, Captain Vancouver offered them passage home. On October 13 the ships headed south, with the *Chatham* taking a side trip to explore the Columbia River. The *Discovery* continued alone, battling "a series of perverse gales and stormy weather," finally arriving in pleasant weather at San Francisco Bay after dark on November 14 for a ten-day stay. Dawn's light showed this outpost of the

Spanish Armada descendants to be guarded by a single brass cannon tied to a log that overlooked the strait.

Members of Vancouver's crew now suffered from scurvy. Menzies had been ill, but with a cold, and did not leave the ship until midday. Once ashore, he took a short walk, noting that the countryside was not as barren as it had appeared from shipboard, but he was disappointed that there was "not a single plant in bloom." There were extensive thickets of medium-size evergreen oaks with holly-like leaves, an apt description of Coast Live Oak. And the hawthornlike small trees laden with brilliant red berries were undoubtedly Toyons (the local hawthorn has blue-black fruit). Menzies found Blue-blossom, a dwarf species of California Buckeye, Twinberry, probably California Bay, Yellow Bush Lupine, Arroyo Willow, Shining Willow, and many shrubby composites including Coyote-brush and Mock-heather. Also, in his tramping he flushed several birds that he did not recognize. On the sixteenth the *Discovery* moved farther into the bay to a better anchorage closer to the presidio. Menzies made several short junkets ashore, seeing a variety of plants new to him, but few in bloom. Again he regretted that "it was not my lot to visit this Country at a more favorable Season for Botanical researches." However, he did describe in detail an attractive quail he encountered, the California Quail.

Menzies did not feel up to joining the several-day trip with Captain Vancouver and others by horseback to visit Mission Santa Clara de Asís, so Vancouver provided the British view of the oak-woodland landscape down the peninsula: "We entered a country I little expected to find. For about twenty miles it could only be compared to a park which had been originally closely planted with the true old English oak." By Sunday, November 18, Menzies was feeling well enough to accompany Captain Vancouver four miles inland to Mission San Francisco de Asís (Mission Dolores) to enjoy an "elegant & plentiful dinner." The party rode horses thoughtfully provided by the Spanish, and coming along were the Hawaiian girls, who demonstrated great flair in mounting their steeds. They were an uninhibited pair, later at a presidio function sprawling out full length on the ground rather than squatting on their heels like the Spanish ladies.

The mission fathers showed Menzies their small garden of vegetables and fruit trees, none of the plants very vigorous, with much of the garden overgrown by weeds. Menzies wondered if the plants were simply unsuited to the climate and soil or just of poor quality. Near the mission were numerous beehive-shaped wigwams where lived the hundreds of Indians who cultivated the mission lands.

Three days after the arrival of the *Chatham* from its Columbia River exploration, the two ships left San Francisco Bay on November 25, 1792, bestowing farewell gifts of rum and wine in hogsheads on the presidio and mission, for a six-week stay at Monterey. Arriving in Monterey Bay late the next day, they found their supply ship *Daedalus* at anchor, and at the presidio Bodega y Quadra, who with his staff including naturalists had been awaiting Vancouver's arrival for more than a month, "to give us," as Menzies wrote, "fresh tokens of their friendship and hospitality." The expedition's astronomical instruments were quickly installed in the observatory tent erected on a "delightful spot" overlooking Monterey Bay, with a large marquee field tent nearby. But Menzies's "late indisposition" rendered him so weak he was unable to go ashore for about a week. When he did go, he was initially disappointed once more with the countryside, "so exceedingly dry & parched," and nothing in bloom.

The British were cordially received by the Spanish, as they had been at San Francisco Bay. On December 2 a large cavalcade including Menzies made a cross-country trek through an attractive hilly landscape interspersed with Monterey Pines, "coppices of stiff scrubb brush wood," and natural pasture teeming with cattle and horses to Mission San Carlos at Carmel for a welcoming reception and "sumptuous dinner." Hosting the party, held in an arbor at the mission garden, was Father Fermín Francisco de Lasuén, who, with his interest in natural science, had previously delighted hosting the expeditions of La Pérouse and Malaspina. After dinner, mission Indian men, dressed in hides and heads of Black-tailed Deer, pantomimed for the English visitors how they decoyed the deer, even imitating the browsing of bushes, until they could get close enough to kill their game with bow and arrow.

That evening at the presidio home of newly appointed governor José Arrillaga there was a dance performance, with a Spanish couple demonstrating a traditional fandango. But the highlight, at least for the men present, was singing and a sinuous native dance by Vancouver's two female Hawaiian passengers, Raheina and Tymarrow. The Spanish ladies, embarrassed, excused themselves from the performance. Vancouver missed this excitement, having retired early to the *Discovery*. Next evening the British threw their own party for the Spanish dignitaries and their wives aboard the *Discovery* at anchor, where the ladies became seasick from the undulations of the vessel. On another evening, Vancouver had to abandon a second attempt at a shipboard party when a storm came up, and the Spanish gentlemen and their ladies had to be quickly deposited back on shore. Not until the next morning did Vancouver's officers discover that the storm had forced the guests to spend the night huddled together in the *Discovery*'s shore tent.

By December 5 Menzies was well enough to accompany Lieutenant Robert Broughton, commander of the *Chatham,* and Lieutenant Peter Puget on a pleasant stroll west along the bayshore to Point Pinos and around to the ocean, through open stands of a pine new to the botanist, the Monterey Pine, mixed with aromatic coastal shrub in which he recognized "species of the Genus *Artimesia*" (California Sagebrush). On this trip Menzies collected on the ocean dunes south of the point the blossom from a yellow wallflower, later named *Erysimum menziesii* in his honor by British botanist William Jackson Hooker. Upon returning from the point through the pine woods on a different path, Menzies shot "a number of small Birdes, a new species of Hawk & several Quails." Two of the birds he supposedly collected on the Monterey Peninsula represented the first individual of their species ever to be both described and preserved (called the type specimen): the California Quail and the California Condor, the latter dining on a dead whale. These two, in the British Museum of Natural History, are the only remaining avian specimens brought back by Menzies from California. Menzies noted that the quail when eaten had a "delicacy of flavor" equal to that of the English partridge.

The following two days Menzies stayed aboard the *Discovery,* "examining drawing & describing my little collections" and those brought to him by members of the crew, and by the Spanish and Indians. On the morning of December 8, 1792, he was off once more, this time with a sporting party of Spaniards from the presidio. The party rode about five miles northward along the coast, passing several lagoons swarming "with Geese Ducks & other Aquatic Fowls," but these were difficult to shoot from horseback. The countryside beyond rose above the shoreline and supported thinly scattered clumps of dwarfed Coast Live Oak and a few grassy openings, but the predominant vegetation was "stiff low Shrubs," the coastal chaparral, some fragrant, some evergreen, and many species unknown to Menzies.

The British were often accompanied on their excursions by members of Quadra's entourage, including his naturalists. Quadra would bring along his "Plate & Cooking Equipage &c" so that all had "the luxury of dining in those retreats off Silver, on the best of everything he could afford." On December 9 the Spaniards feted the British at the presidio's inland garden, where a special arbor was erected. While dinner was being prepared, the governor set off with a party to amuse themselves on a bear hunt, but without success. Menzies never encountered any bears either on his various trips, although he did see cottontails, Black-tailed Jackrabbits, and Gray Foxes. But the most abundant mammal, especially around the presidio, was a "large variegated Squirrel which burrowed in dry sandy grounds," the California Ground Squirrel. Next evening it was Vancouver's turn to host a party, this time ashore, culminating in a fireworks display that impressed the Spanish. Not to be outdone, the day after Christmas the Spanish staged a bullfight for the British near the presidio.

Menzies's most productive botanical foray, on December 18, was into the shoreline sand dunes northeastward from the presidio, where he found two trailing plants in bloom: the Yellow Sand-verbena along with the pink Coast Sand-verbena, whose seeds had been collected by the La Pérouse Expedition. Menzies's painting of the two was published as a color lithograph in William Hooker's *Exotic Flora*. He was pleasantly surprised to discover a number of other plants in flower: an evening-primrose (Watson Willowherb), Coast Goldenrod, a buckwheat (Beach Knotweed), Hairy Oxalis, Bush Monkeyflower, Western Verbena, a new form of the California Poppy, and in a nearby pasture California Buttercup.

On December 21, one of Menzies's last collecting trips was in company with Lieutenant Broughton and others on horseback to the summit of a ridge southwest of the presidio, where he found three beautiful shrubs in the Heath Family he thought he recognized from the Northwest: Salal, Evergreen Huckleberry, and an "Arbutus" (manzanita). It turned out, however, that the manzanita Menzies collected was a new species common in the Monterey area, now *Arctostaphylos tomentosa,* to be named by Frederick Pursh, describer of the Lewis and Clark Expedition plants, from a Menzies preserved specimen with flowers.

In early January 1793, Vancouver's two ships and Quadra's *Activa* were being readied for departure from Monterey Bay. But a series of desertions by sailors from both the *Discovery* and the *Chatham* created a major dilemma and delay. It seems the life of a sailor at sea did not compare with the pleasures of the easygoing Spanish life ashore. Five sailors tracked down had a more imaginative excuse. Three were Catholics and claimed they had been prevented from attending Mass; the other two professed that they wanted to become Catholics.

The *Discovery* on January 5 received a special cargo, which would have a long-term impact on its immediate destination, the Sandwich (Hawaiian) Islands: "four young cows two Bulls & ten breeding Sheep." Menzies discovered that another of his duties was overseeing the livestock. Additional cattle and sheep would be carried from Monterey to Hawaii when Vancouver visited the following winter. It was he who "initiated" the cattle industry in the islands by obtaining a pledge from Kamehameha that none of the livestock gifts would be eaten for at least a decade.

At last, on the afternoon of January 14, the *Activa, Discovery,* and *Chatham* sailed together out of Monterey Bay. Aboard the *Activa* was Lieutenant Broughton, bound eventually for England with Vancouver's report on the treaty, now at a stalemate, with the Spanish and a request for an additional ship. He also carried a box of seeds from Menzies for "His Majesty's Garden" (Kew) and a lengthy letter for Banks ending with an urgent request for two more reams of papers for drying plants. When Broughton finally arrived home, he discovered that England was at war with France.

After a week of sailing together, the three vessels parted company, the *Activa* for Mexico and the *Discovery* and *Chatham* on their way to the Sandwich Islands with the herd of livestock. The two Hawaiian maidens had a pleasant homecoming surprise when Vancouver obtained for them two estates on Kauai to ensure their safe future. During the expedition's sojourn in the islands, on February 16, 1793, Menzies became the first European to reach the summit of Mauna Loa, an active volcano nearly fourteen thousand feet high; he underestimated its altitude by a mere forty-six feet. Artist Heddington started out on the climb with Menzies, but he succumbed to mountain sickness.

On May 2 the *Discovery* returned to California, making landfall at Cape Mendocino and then beating north to anchor at eight o'clock Sunday morning for three days in Porto de la Trinidad, just off Trinidad Head (figure 4). The main purposes of this stop were for the crew to repair a leak near the bow and to procure wood and water. Meanwhile, Menzies made several trips ashore, botanizing at the edge of the temperate rain forest. Much of his collecting took place during an excursion to the summit of the headland itself, where supposedly he discovered three "new" species of *Ribes,* the specimens preserved at the British Museum of Natural History Herbarium: Canyon Gooseberry, named after

Figure 4. Trinidad Head, California, visited by Archibald Menzies with Captain Vancouver in May 1793. (Photo by R. G. Beidleman.)

Menzies by Pursh; Chaparral Currant; and the striking Fuchsia-flowered Gooseberry. Probably the last was actually collected by Menzies at Monterey, since its present coastal distribution is south of the Golden Gate. He also collected Coast Silk-tassel, California Mist-maiden (to be named by Edward Lee Greene), Fringecups (another flower named by Pursh), Smallflower Nemophila, and Pig-a-back Plant (a saxifrage named by Pursh). At the summit of the head he was surprised to find a decaying cross, erected sometime earlier by the Spanish when they laid claim to the port in the name of Spain. What Menzies did not encounter here were the Redwoods, which occur farther inland. On May 5, 1793, it was up anchor in Porto de la Trinidad, and the *Discovery* was bound north again for Nootka Sound, where Vancouver was annoyed to find a newly built Spanish fort greeting the *Discovery*'s arrival.

Five months later Menzies returned from the Northwest to California, this time aboard the *Chatham* with Peter Puget, with whom he had been assigned "in pursuit of botanical information." On October 20, 1793, at a lagoon on the west side of Tomales Bay north of San Francisco Bay, Menzies observed Brown Pelicans and the less common American White Pelicans, as well as a variety of gulls and plovers. Nearer at hand there were hawks, eagles, crows, ravens, and a questionable "red-breasted lark." He managed to shoot some plovers and other small birds. A recent grass and brushwood fire on the headland had destroyed much of the vegetation. Most of what remained Menzies had seen before around San Francisco Bay and on the Monterey Peninsula, "excepting a new species of *Sisyrinchium* with yellow flowers [Golden-eyed-grass] of which I brought on board live plants for the garden." In the spring of 1992 Eric Groves, a British authority on Archibald Menzies, visited Tomales Bay, and down a defile on its west flank found a Golden-eyed-grass "in a damp flush—probably at the same spot where Menzies might have collected it. A great thrill!"

The *Chatham* finally reached San Francisco Bay, rejoining and anchoring next to the *Discovery* in the old berth. It was shocking to discover that during the Vancouver Expedition's absence from California, the Spanish government, and in particular the acting governor at Monterey, José Arrillaga, had become intensely antagonistic toward foreigners, imposing severe restrictions on travel ashore, even making it difficult to get as far inland as the mission. This particularly upset Menzies. The last time he had been here he had been ill, and now he was thwarted in his field collecting by regulations. Captain Vancouver did get permission for work parties to disembark, and for officers to go ashore for pleasure. But Vancouver arbitrarily refused to grant "shore leave" to the naturalist, or even to let him accompany the work parties.

On the afternoon of October 23, the *Discovery* and *Chatham*, attempting to sail out of the bay, collided with each other. They finally got under way the next morning, with the *Chatham* in the lead. Encountering bad weather—perverse winds, showers, calms, thick fog—resulting in the loss of some of Menzies's live plants, the *Discovery* at last cruised into Monterey Bay early in the morning of November 1, to encounter more regulations and receptions as unfriendly as those at San Francisco Bay, although the authorities did allow work parties ashore for water and wood. After only a week the expedition continued south along the coast, reaching Santa Barbara on November 10.

Here relations between British and Spanish, and between Vancouver and Menzies, proved more cordial. In a few days, with Vancouver's blessing, Menzies was allowed to travel into the hills east of the Santa Barbara presidio "for the purpose of collecting Plants & examining the natural produce of the Country." The naturalist encountered many "new & rare objects" but few plants in either flower or seed. But he enjoyed hiking through "beautiful groves of the Ever green Oak" (Coast Live Oak), much larger than those "crabbed" ones around Monterey. And "the thickets swarmd with squirrels & quails & a variety of other birds which afforded some amusement in shooting them as I went along." Along the shore Indians were observed catching fish, of "plentiful supply and diversity," including types that are favorites to this day, the Pacific Bonito and Yellowtail.

A few days later Menzies, with a party of officers, visited an extensive salt marshland (Goleta Slough) north of Santa Barbara, dominated by what Menzies called "Samphire," the English name for pickleweed. Around the edge of the marshland were three Indian villages, most of the inhabitants away in the hills collecting acorns. The next day, November 16, Menzies made "a solitary botanical excursion" west of the presidio, encountering along a creek in an oak woodland both plane trees and poplars. These were two new trees: the Western Sycamore, eventually to be collected here and named by Thomas Nuttall almost fifty years later; and the Black Cottonwood, named by John Torrey and Asa Gray as a subspecies of the boreal Balsam Cottonwood Menzies had earlier encountered at Birch Bay across from Vancouver Island. He also collected plants and seeds of a beautiful monkeyflower (Bush Monkeyflower), new to him "as I had not before seen it any where."

After their friendly reception, the Vancouver Expedition reluctantly left Santa Barbara on November 18, sailing south past the island of Santa Catalina and at last around the tip of Point Loma, where Menzies noted vast beds of seaweed (Giant Kelp) in the water. The ships moved into the protected harbor at San Diego, anchoring opposite the presidio on November 27. Vancouver at first limited Menzies to exploring no farther than the presidio, but work parties going ashore brought him "Plants & branches of the produce of the Shore that greatly excited my curiosity."

On one of Menzies's few longer trips ashore, he visited Point Loma on November 29 with a beach party headed by James Johnston, master of the *Chatham*. While Johnston surveyed atop the point, Menzies hiked up the brushy hillside and discovered "a vast variety of Plants that were entirely new to me," including half a dozen species of cacti, an evening-primrose, milk vetch, the ill-smelling shrubby Bladderpod, and an ice plant, probably Sea-fig, *Carpobrotus chilensis*. Although ice plants are considered native to South Africa, this early record for Alta California by Menzies lends weight to the current idea that this species might be a California native. Menzies was generally disgusted because "to my no small mortification I met with only two plants in flower & very few in seed." However, there were numerous dormant plants on the cliffs, which he imagined would produce quite a floral display in springtime. A few days later Menzies joined a hunting party traversing a brushland near the coast. The men were beating the bushes to scare up game, including cottontail rabbits, Black-tailed Jackrabbits, and California Quail. The hunting proved unexciting, but Menzies was delighted because the beating and trampling of the aromatic chaparral "impregnated the air with a refreshing fragrance."

Menzies was elated when the padre at the San Diego mission sent aboard for him a branch of a Mexican Palo Verde in bloom, and live plants and fruit of Jojoba, a plant that would be named, though incorrectly, by Thomas Nuttall. The padre was prophetic when he ascribed "many virtues" to the latter desert shrub, which would gain short-lived economic prominence more than a century and a half later in the Southwest for the oil from its seeds. Menzies was most pleased with the live plants, and he immediately placed them in the quarterdeck plant frame. The Jojobas survived the voyage back to England and were planted in the Royal Botanic Gardens, Kew. Menzies and two sailors spent two days ashore near the *Discovery* digging up other species and adding them to the plant frame "till all the vacant space was filld up with such plants as were likely to be a valuable acquisition to the same royal collection."

With everything accomplished ashore, on December 8, 1793, the expedition vessels prepared to sail but were delayed by strong southerly winds until the next morning. At last the ships proceeded south along the Baja coast to Rosario, surveying, and then southwest to their winter sojourn in the Sandwich Islands. During this visit, in late February 1794, Vancouver convinced Kamehameha and his islanders to become "subjects of Great Britain," although England failed to consummate Vancouver's remarkable agreement. The really sad news, learned months later, was that Bodega y Quadra, a good and respected friend of the Vancouver Expedition, had died on March 26, 1794, in Mexico City.

After winter in Hawaii and months in the North Pacific, the *Discovery* and *Chatham* proceeded south again to California, with Menzies's greenhouse now crowded with boreal plants. Continuous storms with heavy rain and violent winds dashed salty waves across the quarterdeck and destroyed most of Menzies's live plants. The two ships became separated. The *Discovery* was off Cape Mendocino on November 3, intending to attempt an exploration of Bodega Bay to the south. But another storm interfered, forcing the ship to sail on to Monterey Bay. It arrived there on the afternoon of November 6, 1794, to find the *Chatham* already at anchor.

During this visit the Spanish proved amiable and the weather "delightfully pleasant." There were numerous opportunities for trips ashore. Menzies noted that wild fowl were much more abundant than in past visits, and the expedition hunters were very successful in procuring ducks, geese, plovers, curlews, quail, and hares for the table. Because of scant winter rains, most vegetation was "dry and scorched." But on the hills north of Mission San Carlos Menzies found good specimens of evergreen shrubs, some of which he had collected in the Northwest but lost to the storms. So he gathered live plants of Evergreen Huckleberry and probably Salal, as well as the Shaggy-barked Manzanita, *Arctostaphylos tomentosa*, he had mistaken earlier for a northwest coast manzanita, and replenished his quarterdeck frame with these new collections.

For a week starting November 19, Menzies accompanied Vancouver and some officers on a horseback ride northeast toward the Salinas Valley. Early in the trip, "the sight of the most extraordinary mountain" was glimpsed off to the south; Vancouver said it resembled a "sumptuous edifice fallen into decay," with cream-colored sedimentary columns. Artist John Sykes, who accompanied the party, made a sketch of "Corral de Tierra" (the future setting for John Steinbeck's *Pastures of Heaven*, published in 1932). Sykes's picture, titled "A Remarkable Mountain near the River of Monterey," would be included as one of the plates in Vancouver's *Voyage* (figure 5). The "richness and fertility" of the Salinas Valley's soil was noted, an astute observation anticipating what would become a famous agricultural region. On the mountains that enclosed the broad valley Menzies noticed a new pine (Foothill Pine) and oaks (including the Valley Oak). Later interpretations of this excursion suggest the final destination was modern-day Pinnacles National Monument, some fifty miles southeast, but it seems unlikely that the party reached that interesting volcanic area.

On November 26, while repairs were being made to the *Discovery*, Vancouver sent sailors under Lieutenant Spelman Swaine and undoubtedly with Menzies in three boats across the bay to Mission Santa Cruz, among other things to obtain fresh vegetables. The party brought back a real botanical prize, a specimen of the Redwood, which became the type specimen for A. B. Lambert's description of this species in 1823. University of California botanist Willis Jepson actually saw this type specimen in England about a century later, and written on the back of the sheet was "Santa Cruz, Menzies." Jepson also tracked down the unnamed Redwood specimen collected earlier by Thaddeus Haenke with the Malaspina Expedition.

On December 2, 1794, with a northeasterly breeze and pleasant

Figure 5. "A Remarkable Mountain near the River of Monterey" (Corral de Tierra). Sketch by John Sykes, 1794. A field party with Captain Vancouver passed here en route to the Salinas Valley in November 1794. (Courtesy California Historical Society, FN-30521.)

weather, the Vancouver Expedition sailed out of Monterey Bay and south along the California coast. Vancouver had received word that the Nootka affair with Spain was finally settled. The ships stopped briefly several times in Baja California, then just past Cape Corientes on the Mexican mainland the helmsman swung the *Discovery* to a southwesterly course. In early February 1795 there was a three-day stop at Albemarle (now Isabela), largest of the Galápagos Islands. Forty years later, in September 1835, Charles Darwin, naturalist with HMS *Beagle,* would set foot on the same island. Unfortunately, Menzies's journal observations when ashore at Albemarle are among the missing. Vancouver then proceeded along the Peruvian and Chilean coastlines, with a stay at Valparaiso, Chile, where some of the northern plants in the greenhouse that had succumbed to the sultry climate were replaced with South American ones.

While in Valparaiso, Vancouver and five of the ship's officers, including Menzies, received an invitation to visit the captain-general (governor) of Chile at his home in the inland capital city of Santiago. At dinner the dessert was garnished with strange seeds that caught Menzies's eye. He pocketed some, later planting them in pots on his return to the *Discov-*

ery. Five of them grew to become trees at Kew gardens. Thus did Menzies introduce the famous Monkey-puzzle Tree from Chile into Britain. King William used to point out the group of bizarre trees to visitors at Kew. When the last tree died in 1892, a half-century after its collector, Menzies, had died, it was made into a table that became the cherished possession of another botanist. The expedition left Valparaiso and, prevented from stopping at Cape Horn because of the typical bad weather, continued on toward England. Along the way Menzies managed some collecting at St. Helena (off South Africa), to which in two decades Napoléon would be exiled.

The *Discovery* arrived at Deptford (London) on October 20, 1795. On the homeward voyage Menzies had two distressing experiences. During the expedition he had assigned to him a "servant" (seaman) "to aid and assist me in the different pursuits I was instructed to fulfil," including taking care of the plants in the glazed frame on the quarterdeck. After the brief stop at St. Helena, Vancouver, being short-handed, appropriated Menzies's servant to stand watch as an ordinary sailor. Throughout the years at sea Menzies had not objected to his assistant doing deck duty now and then, as long as he "executes my order respecting my garden." But on this particular occasion, when the *Discovery* encountered a nasty storm, the plant frame where Menzies's seedlings grew was left covered by his "servant," who was busy elsewhere, and some of the plants died. Menzies reported the dereliction of botanical duty to Vancouver, who declined to discipline Menzies's assistant. Irritated, Menzies complained "of being unjustly used in this proceeding," whereupon Vancouver "flew into a rage," used abusive language, and finally placed the naturalist under cabin arrest on July 28, 1795, for "insolence and contempt." After a few days Menzies apologized.

The other unpleasant experience involved Menzies's divided loyalty during the expedition—loyalty to Sir Joseph Banks, his mentor, as the assigned naturalist, and to Captain George Vancouver while serving as the expedition's surgeon. Menzies had instructions from Banks to turn over all of his journals, including a voluminous personal one, and other effects to the secretary of state. Vancouver, on the other hand, demanded that all of Menzies's journals be turned over to him. Menzies gave Vancouver only his surgeon's logbook, and the threat of a court-martial ensued. Menzies wrote Banks while Vancouver wrote the Admiralty. The court-martial was dismissed.

Back in England, Menzies began to go over his extensive collection of plants, animals, and a quantity of anthropological specimens, especially

artifacts from the Pacific Northwest. In fact, chided Vancouver, the naturalist's ethnology collection had almost threatened the carrying capacity of the expedition's vessels. Initially, Menzies's many collections went
to Sir Joseph Banks's home in Soho, although the living plants and seeds
were quickly sent on to Kew. It was remarkable that any live plants had
survived, what with wild waves, salt spray, arctic to torrid temperatures,
melted tar dripping from rigging, and foraging by the *Discovery*'s livestock. During the voyage Menzies had sent seeds to Banks on two occasions; some of them successfully germinated and thrived at Kew. Menzies began transcribing his personal journal, eventually filling five
volumes of manuscript, and organizing his many fine drawings. Surprisingly, he devoted more lines in his journal to nautical observations and
weather than he did to plants and animals.

Menzies's vast plant collection included the first large assortment of
species ever brought back from California. It was Menzies who introduced Toyon into England, where it is known as California Maybush but
has never done well. Menzies's seeds of the California Poppy, a flower
he mistook for a European species, germinated at Kew gardens but produced weakly plants that soon died. His small private herbarium, consisting mainly of grasses, sedges, and cryptogams, was eventually be-
queathed to the Royal Botanic Gardens, Edinburgh, while most of his
preserved specimens of plants became widely distributed among
herbaria, especially the British Museum's, and among individuals
throughout the world. The results of the Vancouver Expedition, *A Voyage of Discovery to the North Pacific Ocean, and round the World,* were
eventually published in three volumes in 1798. Two centuries after Menzies's adventures with the Vancouver Expedition, James Reveal wrote in
Gentle Conquest (1992) "that only now, in the 1990s, is the Menzies
collection being reassembled and studied."

Menzies continued an active correspondence with fellow botanists.
Increasingly, he became a mentor to a new generation of natural scientists who were invited to examine his plant collections, including David
Douglas, Thomas Coulter, and John Scouler. When Frederick Pursh was
working on his *Flora Americae Septentrionalis,* largely based on the
Lewis and Clark plants, Menzies was glad to share his North American
plant collections with him, and Pursh's 1814 description of Menzies's
specimen of *Arbutus menziesii* (Pacific Madrone, the "Strawberry Tree"
collected in Washington) was among the first scientific delineations of a
tree from the West Coast. Thomas Nuttall, when he was back in Britain
during the War of 1812, studied Menzies's plants from America in Sir

Joseph Banks's herbarium. Much later, when Menzies's interest had turned to ferns, the chief gardener at Kew received a note from him about a fern collected by the Beechey Expedition. In 1839 Asa Gray, the rising American botanist, met Menzies on several occasions in London, including for dinner at his home, viewing him as a "pleasant and kindhearted old man."

When Archibald Menzies died in 1842, his lifetime had encompassed every major oceanic exploring expedition visiting New Albion, from Great Britain's Captain Cook in 1778 to the United States Exploring Expedition in 1841. Always exhilarated by the floral quest, he exclaimed, when seventy-five, that "were another expedition going, I would immediately set off again." Menzies, "that Nestor of botanists," is buried at Kensal Green Cemetery near his Notting Hill home in the outskirts of London, in a grave now obscured by time and vegetation. On the original tombstone was engraved a factual inscription, yet so much more fitting would have been botanist Willis Jepson's tribute: "Few things escaped his naturalist's eye." Or, even better, one Sandwich Islander's cryptic description, mentioned by David Douglas in a letter, of surgeon-naturalist Archibald Menzies: "The Red faced man who cut off the limbs of men and gathered grass."

The Russian Rezanov
Expedition with
Langsdorff, 1806

A blind zeal for Natural History . . .
    —Georg Heinrich von Langsdorff

It had been sixty-five years since Georg Steller, a German naturalist with
Bering's second exploration of the North Pacific on behalf of the Russians, had collected a Steller's Jay specimen and jotted down (in Latin)
the West Coast's first botanical observations at Kayak Island off Alaska.
Now, coasting south past California's "Pinto de los Reys" on April 8,
1806, came the *Juno*, newly purchased in New Arkhangel (Sitka, Alaska)
and associated with Russia's first around-the-world venture, the Krusenstern Expedition, sent out by Emperor Alexander I. Aboard as captain
was Count Nikolai Petrovich Rezanov, and Dr. Georg Heinrich von
Langsdorff, a well-respected Prussian naturalist, served as Rezanov's personal surgeon.

Langsdorff was born in 1774 at the village of Wöllstein, southwest of
Frankfurt am Main. He received his medical degree at the age of twenty-
three from Göttingen's Georg-August-Universität in 1797, then for five
years traveled widely, spending profitable time—but little on his first
love, natural history—in Portugal, England, France, and Prussia, meeting numerous important scientists and becoming conversant with many
foreign languages, but not Russian. Encouraged by being named a corresponding member of the Russian Imperial Academy of Sciences, however, he applied and was accepted as a naturalist with the Krusenstern
Expedition.

Langsdorff joined Count Rezanov in serving with Adam Johann von
Krusenstern, an officer in the Russian navy, who in 1802 had finally re-

ceived permission from the Russian minister of marine to obtain ships for an around-the-world voyage. The special emphasis of this expedition was to strengthen Russia's role in the North Pacific, in support of the new Russian American Company. Two American ships were purchased, renamed *Nadezchda* and *Neva,* and well staffed, including some men of science and a naval cadet, Otto von Kotzebue, who would later lead two expeditions with naturalists to the West Coast.

The vessels departed from Kronstadt in the Gulf of Finland off St. Petersburg on August 7, 1803, picking up Langsdorff at Copenhagen. They sailed across the Atlantic to Brazil, around Cape Horn, and on to the Sandwich (Hawaiian) Islands. Thence the *Neva* proceeded to Sitka, while the *Nadezchda,* with Rezanov and Langsdorff aboard, cruised to Kamchatka. There Rezanov took command, had his ship repaired and provisioned, and in September 1804 proceeded to Nagasaki, Japan, to present himself as the Russian ambassador to Japan. The Japanese detained him, restricted his shore visitation, and subjected him to "revolting" ceremonials. The Japanese rejected the Russian gifts and Rezanov's assignment as ambassador, and finally informed him that no more Russian vessels would be welcome in Japan.

The *Nadezchda* then proceeded to the Siberian coastline. There Rezanov, his valet, Langsdorff, and two naval officers departed from the main Krusenstern Expedition, which would continue its trip around the world, among other goals to further the search for the "lamented La Pérouse." Back at Kamchatka, Rezanov received orders appointing him imperial inspector and plenipotentiary of the Russian American Company he had earlier helped organize.

Rezanov's party took ship on the Russian American Company's brig *Maria* bound for Sitka. In Alaska Langsdorff was able to exercise his talent as both cartographer and artist, drawing a detailed map of New Arkhangel (Sitka); a construction plan of a Russian *lodka* boat carried aboard the *Maria;* sketches of the natives, including dancing Tinklit women; and a beautiful portrayal of fur seals along a Pribilof beach. In Sitka, Langsdorff helped arrange the purchase of John D'Wolf's New England trading vessel, the *Juno,* and made a portrait of the ship. In the spring of 1806, the *Juno*—fully provisioned, loaded with trade goods, and also belowdecks a cargo of valuable sea otter pelts—sailed south to Alta California.

Naturalist Langsdorff was delighted to go to California, commenting that "it seemed so much a debt due to science to undertake a journey to parts so little known, and which had received so little scientific examination." Even at sea he managed to pursue natural history interests, his first California bird being a Red-shafted Flicker that accidentally flew onto the *Juno*. He mistakenly called it a Golden-winged Woodpecker, having undoubtedly seen the painting of the eastern Yellow-shafted Flicker (labeled "Golden-winged Woodpecker") by Mark Catesby in a 1750s German edition of the first book on American natural history. This woodpecker impressed Langsdorff as so rare and beautiful that he preserved it as a museum specimen.

In that early spring eventide of 1806 along the California coast, while the *Juno* was approaching the entrance to San Francisco Bay, dense fog forced the dropping of the anchor until morning. Upon their arrival in the harbor next day, April 9, the Russians were greeted with some old international news: England had declared war on Spain. The Spanish at San Francisco Bay initially feared that the *Juno* might indeed belong to their new enemy. Relief was followed by a cordial welcome, notwithstanding rumors of Spanish-Russian dissension, and the presidio even provided a dragoon as a guard of honor for the ship's officers.

The Russians remained in the Bay Area for six weeks, the *Juno* being badly in need of repairs as a result of the many wild storms at sea. But Langsdorff quickly became frustrated with his restricted opportunities for "scientific examination." He did make a few shore visits—to the presidio, the mission, and on one occasion as far away as the mission at San Jose. For the shorter trips the Spanish were pleased to provide a horse for the naturalist. Langsdorff's first excursion to distant San Jose with three of the *Juno*'s boats proved unsuccessful. Five days later Langsdorff embarked in a baidarka (a boat of hides on a wood frame), reaching Mission San Jose at sunset. His return on April 23 took two days, including an unpleasant overnight, but he spent some time ashore on foot both going and returning. The sole extent of his botanical observations on the return trip was noting a *salsola* (probably California Sea-blite) covering a shoreline bog near the mouth of Alameda Creek. During this excursion there had been very rough water in the bay, with waves splashing into the boat, and of the birds, mammals, and plants the naturalist collected, only three Sea Otters survived.

In San Francisco Bay much of Langsdorff's time was spent aboard the anchored ship, where the constant unloading of goods from Sitka and loading of foodstuffs from Alta California for Alaska proved disruptive to his naturalist activities. Indeed, the naturalist wrote, "My researches in Natural History met with more obstacles in California than in any other part of our expedition. To detail all the petty circumstances which crossed me in this way would appear prolix [excessive]. . . . I shall only . . . mention some of the principal." A number of mammal skins he laid on deck to dry were thrown overboard, and his plant drying papers were inadvertently but irretrievably buried beneath loaded freight. There was a general disregard for scientific endeavors by the ship's company. Live birds that Langsdorff brought aboard were "suffered to fly away," and heads were cut off a number of collected ducks and other water birds he had intended to prepare as museum specimens. When he asked a sailor to take him on a boat trip, he was told by Rezanov that the crew had better things to do, and "our expedition was not undertaken for the promotion of Natural History." He finally "relinquished all ideas of attempting farther labours in the science, and resigned myself to the wishes of the Chamberlain [Rezanov]." The reason Langsdorff was still able to make shore trips was that he had been assigned as interpreter in business dealings with the Spanish. Speaking Spanish was certainly easier for him than dealing with the shipboard Russians, since he could speak Spanish while the Slavic languages were not his forte.

There was one unusual botanical experience early during the stay. When Langsdorff paid a visit to Mission Dolores, a short distance inland from the bay, Father Joseph Uria, impressed by the visiting naturalist, led him by the hand to a large painting of an agave (century plant). Instead of a flowering stalk, there rose from the basal leaves the figure of the Holy Virgin, which the father assured the botanist was the source of "many extraordinary miracles." With tongue in cheek, Langsdorff complimented the lucky painter "who had seen so great a natural curiosity with his own eyes."

During the *Juno*'s six-week visit, another interesting development unfolded. In his official report, Rezanov digressed from his formal epistle, begging forgiveness at interjecting a romantic interlude into the otherwise serious document. It seems that when Rezanov and Langsdorff first visited the presidio to meet Commandant Don José Argüello (who happened to be away at Monterey) and his family, they were both smitten by one of the commandant's daughters, fifteen-year-old Maria de la Concepción Marcela Argüello. Langsdorff the surgeon could hardly contain himself, extolling "her vivacity and cheerfulness, her love-inspiring and brilliant eyes, exceedingly beautiful teeth, expressive and pleasing features, shapeliness of figure, and for a thousand other charms, besides an artless natural demeanor." Rezanov himself became so enamored that he visited the presidio daily and claimed that he had fallen madly in love;

he did not mention the possible international value of a union between a Russian chamberlain and a well-placed Spanish belle.

Within a fortnight Rezanov had asked the commandant for his daughter's hand, hoping that "fate decree the completion of my romance—not begun in hot passion, which is not becoming at my age." Concepción, thrilled with her handsome, important suitor and eager to escape California, was ecstatic over his offer of marriage. (Rezanov was, by the way, only in his early forties.) In Russia he had been married to the beautiful daughter of Grigor Shelekhov, who had initiated Russian settlement in North America. Heartbroken when she suddenly died and in poor health himself, Rezanov had decided to seek respite through adventuring, and thus joined the Krusenstern Expedition.

Before the marriage could take place, Rezanov and Langsdorff returned to Sitka after their sojourn in San Francisco Bay. There Langsdorff decided, with Rezanov's approval, to remain in Alaska. Langsdorff joined John D'Wolf, the Yankee from whom the *Juno* had been bought, and spent some time traveling with this American friend. Rezanov, meanwhile, caught a ship back to Siberia and, though ill and without his personal doctor, proceeded in early autumn 1806 by horseback westward toward St. Petersburg. He carried the treaty with Spain he had executed, and he hoped to obtain permission from the czar for his proposed marriage as a Greek Orthodox Catholic to a Roman Catholic. Near Krasnoyarsk in the depths of Siberia, five hundred miles northwest of Irkutsk, Rezanov, feverish and exhausted, was thrown from his horse and died on March 13, 1807. Meanwhile his betrothed, Concepción, faithfully awaited his return to Alta California, hoping she could escape this country with her Russian lover. Several decades later, despairing, she became a nun. Not until 1852—in Monterey according to one account—did she learn from the visiting governor of the Hudson's Bay Company, Sir George Simpson, that Rezanov had died dozens of years earlier.

Concepción died on December 23, 1857, and was buried at St. Dominica's Cemetery in Benicia. As a girl of fifteen, she had said of her native land, Alta California, that it was "a beautiful country, warm climate, an abundance of grain and cattle,—and nothing else." The daughter mimicked her father's feelings; he had told Rezanov that even if Spain gave California to the Russians, "This California is a cursed land, that causes nothing but trouble and expense." When the *Juno* departed from Puerto de San Francisco for Alaska in mid-May 1806, Langsdorff, like Concepción and her father, also expressed disappointment with California. But as the naturalists with the next Russian expedition were to find, "there was much room for new discoveries in this country, so seldom visited by learned men."

This first "official" Russian visit to California, however, was not without its eventual accomplishments. Abetted by the missionaries, Rezanov returned to Alaska with a shipload of needed supplies for the isolated Russian colony at Sitka: dried meat, beans, peas, grain, and tallow, and especially from Mission Santa Clara fresh fruits and vegetables. And he had arranged a treaty with the Spanish that would continue this arrangement of exchanging foodstuffs from Spain's Alta California for Alaskan furs and Russian merchandise. As Rezanov's official correspondence revealed, he hoped Russia could eventually claim and colonize portions of the northwestern Pacific Coast.

Naturalist Langsdorff began his return to Russia from Sitka, commencing in early summer 1806. After passing the winter in Kamtchatka, he sailed for the Siberian port of Okhotsk in May. En route his small ship was suddenly lifted up a meter out of the ocean on the back of a large whale. It was at Okhotsk that Langsdorff learned of the death of Count Rezanov. And on his way home through Siberia, Langsdorff passed by Rezanov's grave, then marked only by a large uninscribed stone shaped like an altar.

In 1812, the year that the Russian colony of Fort Ross in Alta California was established, Langsdorff was appointed Russian consul general at Rio de Janeiro, Brazil. He visited Russia briefly in 1821, to receive honors as a state councillor and knight of the Order of St. Vladimir. But he was soon back in Brazil to continue his botanical collecting for the Russian Imperial Academy of Sciences, to which he had been elected a full member. As consul general he headed the Imperial Russian Scientific Expedition, one of the first expeditions into the interior of Brazil. Innumerable bouts of malaria, however, began to take their toll, and he finally became "totally deranged in his mind." He was taken back to Europe by a friend in 1829, "weak in body and soul." Langsdorff died on June 29, 1852, in Freiburg, Germany, possibly having committed suicide.

Russia Visits Again

This unknown but beautiful country, richly endowed by
Nature . . .

—Otto von Kotzebue

THE *RURIK* VISIT WITH KOTZEBUE, CHAMISSO, AND ESCHSCHOLTZ, 1816

The Russians in 1815 embarked for geographic and scientific discovery
with a newly built ship, the *Rurik,* a two-masted, nine-gun brigantine.
The specific purposes of this expedition were to search once more for the
Northwest Passage (although optimism had declined), to cruise South
Sea islands, and incidentally to impress the Spanish in Alta California,
who had begun grousing about Russian developments at Bodega and
Fort Ross, up the coast from San Francisco Bay. The expedition was
under the command of Otto von Kotzebue, son of a famous German
writer and dramatist, an able amateur naturalist who had served with
both the Krusenstern Expedition and earlier with Krusenstern in the Arctic. Krusenstern himself supervised the construction of the *Rurik,* into
which were incorporated two innovations for the welfare of the expedition: an invention for preserving perishable foods (it proved ineffective),
and an ice machine. There was also a distinctive "safety boat," but it was
so heavy that it had to be abandoned before saving anyone.

Important from the scientific standpoint was the choice of the natural science staff. The draftsman selected was twenty-year-old Louis Choris, a Russian who had already served as artist for an expedition in the
Caucasus Mountains. His artistic efforts over a three-year period with
the *Rurik* would more than justify the choice, especially his realistic de-

pictions of California Indians (figure 6). The original pair of naturalists chosen were Dr. Karl von Ledebour, professor of natural history at the University of Dorpat (then a Russian town, now Tartu in Estonia), and his assistant and future brother-in-law, Dr. Johann Friedrich Eschscholtz. When Ledebour's health failed, the replacement was Adelbert von Chamisso, who in his mid-thirties became the oldest member of the expedition. He would serve as botanist and interpreter.

Chamisso was born in 1781 of French nobility. The French Revolution forced the family initially to Berlin, where young Chamisso grew up, traveling widely with his parents. He once served as page to the Prussian queen Louise, fought with the Prussian army, and studied philosophy and literature. His first unrequited love affair, with a governess, inspired him to write poetry, some of which was published in a journal he edited. Quickly recovering, he had another love affair, this time with a married writer, Helmina von Chezy. In France in 1810 the handsome aristocrat was drawn into the social circle of the famous French writer Madame Germaine de Staël at the romantic Château de Chaumont on the Loire River, where he enjoyed the evenings of conversation, music, and "games of love." Entranced with de Staël's "German seriousness, meridional fire, and French manners," he inevitably fell in love with her, to Mme. de Staël's delight. Other ladies cared less for him, and he was frequently sent to a remote chamber in the castle to smoke his large, smelly German pipe. Then together with one of de Staël's sons he became interested in botany. They both made plant collections, a pastime that continued when the de Staël party moved to Coppet, Switzerland, just north of Geneva. Here de Staël's new lover proved too much for Chamisso, who wrote a poem starting out, "Farewell, I leave forever," packed up his bags in disgust, and departed.

In 1813, studying the natural sciences at the University of Berlin, Chamisso became acquainted and went on field trips with German botanist Diedrich von Schlechtendal, who would eventually edit some of Chamisso's botanical work for the periodical *Linnaea*. Chamisso throughout his life had a flair for unorthodox field apparel, his most memorable being an old South Sea chief's garb, and on occasion he was seen coming back from field trips with his plant collections wrapped in a handkerchief. When Napoléon returned from Elba to France early in 1815 and regained power, Chamisso was ready to abandon Europe.

Dr. Eschscholtz, now in his early twenties, had been born at Dorpat in 1793. He studied zoology and medicine at the University of Dorpat

Figure 6. "Bateau du Port de Sn. Francisco." Ohlone Indians paddling a balsa in San Francisco Bay, 1816. Watercolor by Louis Choris with the *Rurik* Expedition. (Courtesy California Historical Society, FN-30512.)

and received a medical degree. Aboard the *Rurik* he served as ship's surgeon and as expedition zoologist, his particular interest being insects. On the voyage, Chamisso found the younger Eschscholtz both an excellent scientist and a kindred spirit—a zoologist, true, but one willing to help with plant collecting. As Chamisso wrote, "From the initial handshake on, he and I formed an intimate friendship whose sky will never be clouded. We shared all our studies, troubles and pleasures." The two naturalists' relationships with Captain Kotzebue proved at times to be less congenial, however, a situation common between many naturalists and their ship captains. Kotzebue would object to scientific collections being stored on board, and—shades of Rezanov—on occasion he even ordered them thrown overboard.

The *Rurik* departed from Kronstadt, a Russian fortress island in the Gulf of Finland, on July 27, 1815, bound for Alaska, Chamisso joining the ship at Copenhagen two weeks afterward. More than a year later, after the Arctic tour, the *Rurik* sailed south along the Pacific Coast. At midnight on October 1, 1816, Alta California's Point Reyes was sighted by moonlight, and the following afternoon the *Rurik* dropped anchor in San Francisco Bay off the presidio. Incidentally, Eschscholtz had almost missed the departure for California from Unalaska in the Aleutians. Engrossed in his plant collecting there while tramping in the rugged mountainous terrain, he had been overtaken by nightfall and was too apprehensive to make his way back to the ship in the dark.

With only a month for fieldwork in California, Chamisso and Eschscholtz found themselves busy because, as Kotzebue observed, "there was much room for new discoveries in this country." Indeed, much of volume 3 of the expedition's report includes contributions by the naturalists, although, surprisingly, the botanist's comments on plants were rather limited. Chamisso observed that in general the flora of the country was poor, compared with warmer climes, although "it offers much novelty to the botanist." Certainly the autumn season was not ideal for plant collecting, because little was in bloom. Chamisso complained that the countryside "now offered the botanist nothing but a dry, dead field." He did note that the "prickly-leaved oak [Coast Live Oak] is the most common and largest tree." And he paid tribute to the previous botanical efforts in the area by Menzies and Langsdorff.

Most of Chamisso's plants were collected in the hills and vales around the presidio and mission and included at the time two new genera and thirty-three new species. One flower that Governor Pablo Solá added to the botanist's collection especially delighted Chamisso because it looked just like one he knew from Berlin, the "five-finger cinquefoil." Indeed it was Silverweed, which has distribution across the Northern Hemisphere. Eschscholtz added about ten more plants, including what is now a subspecies of the honeysuckle Twinberry, which he named after his mentor at the University of Dorpat, Professor Ledebour. Chamisso and Eschscholtz generally proved to be compatible plant collectors (figure 7). But on one occasion when the zoologist had spied a green plant out in a marsh, he asked his botanist companion what it was, knowing that Chamisso particularly liked wetland plants and would certainly try to collect it, whereas Eschscholtz did not want to get his feet wet. In his journal Chamisso sarcastically jotted, "That's the kind of thing you can expect from your closest friend." The major excitement ashore during the stay was the special fight between a bull and a bear, staged for the benefit of the expedition.

The paper by Eschscholtz on the San Francisco Bay region's plants collected by the two naturalists during the 1816 expedition, "Descriptiones Plantarum Novae Californiae," finally appeared in 1826 in *Mémoires de l'Académie impériale des sciences de St. Pétersbourg*. It was not only the first scientific paper on California regional flora but also the first to use the word *Californiae* in a scientific paper title. Because Eschscholtz's ar-

Figure 7. Chamisso (left) and Eschscholtz collecting the California Poppy at San Francisco Bay in 1816. (Courtesy University and Jepson Herbaria, University of California, Berkeley.)

ticle appeared after his second visit to California in 1824, many later botanists assumed that it included plants from both the 1816 and the 1824 expeditions. In truth, Eschscholtz had already presented the paper at a meeting of the Imperial Academy in 1823.

Chamisso also published articles on his collections in *Linnaea* and elsewhere and gave many specimens to European specialists, who published reports on them. Among the two naturalists' 1816 collections were many that are familiar today somewhere around San Francisco Bay, including California Rose, Chamisso Bush Lupine, Blue Nightshade, California Blackberry, Yellow Sand-verbena, California Coffeeberry, and Blue-blossom.

During the short 1816 visit, despite Eschscholtz's abiding obsession with insects, he collected only a few, among them two common butterflies, the familiar Monarch Butterfly and a new species, the Common Checkerspot. The zoologist also made a collection of rocks and minerals that were later deposited at the University of Dorpat's Cabinet of Natural Curiosities, where they were described by geology professor Maritz von Engelhardt.

The famous flower from the 1816 expedition is the California Poppy, which in 1903 the California legislature designated as the state flower, symbolizing California's golden mineral wealth and sunshine. This was the blossom that, when in its grandest floral displays across the coastal landscape of Alta California, inspired Spanish sailors at sea to exclaim, "¡Tierra del fuego!" ("Land of fire!"). One of the common Spanish names for the poppy is *copa de oro,* the "cup of gold." The California Indians were more interested in its medicinal qualities, among other things using the juice as a mild narcotic. In October, during the *Rurik's* stay, few of the poppies were in bloom, not at all like the impressive poppy spectacles during springtime. In 1820, when Chamisso first described the plant in print, together with a life-size color painting, he named the genus in honor of his colleague (figure 8). Under botanical rules the genus *Eschscholzia* remains misspelled to this day.

Chamisso, in discovering other new California plants, honored the artist Choris by naming the Choris Popcornflower after him; the composite genus *Lessingia* for a German family of botanists named Lessing, who assisted Chamisso in describing many of his collections; *Frankenia salina* (Alkali-heath) after Johann Franke, an early professor of botany; and the genus *Horkelia* after the University of Berlin's prominent botanist Johannes Horkel. In turn, Chamisso over the years would himself be honored by numerous species names, perhaps the most appropriate being *Micromeria chamissonis,* whose common name is Yerbabuena, fittingly the first name for the tiny shoreline settlement that later became San Francisco. Its scientific name is now *Satureja douglasii,* honoring the British botanist David Douglas.

On November 1, 1816, All Saints' Day, the *Rurik* sailed out of San Francisco Bay into the Pacific Ocean. Chamisso noticed that the water of both the bay and the ocean was highly phosphorescent. In the samples he collected he could see, with his primitive microscope, multitudes of tiny organisms, "infusoria" he called them. In the future they would be recognized as free-living, luminous marine bacteria. The *Rurik* headed for the Sandwich (Hawaiian) Islands, then the Arctic, and finally back to Europe around the Cape of Good Hope. Almost two years later it came to anchor on the Neva River in Russia close by the palace of the expedition's patron, Count Nikolai von Romanzoff. Chamisso named a genus of plants in the Waterleaf Family *Romanzoffia,* whose species occur from Alaska south to San Francisco.

Although botanists may well claim Chamisso as one of their own, he is really more famous in other areas of endeavor. Shortly before he joined the *Rurik* he published his most famous story, *Peter Schlemihl's wunder-*

Figure 8. Friedrich Guimpel's painting from Chamisso's original specimen of the California Poppy. (From *Horae Physicae Berolinensis,* edited by C. G. Nees von Esenbeck [1820].)

same Geschichte (Peter Schlemihl's Wonderful History), about a young man who sold his shadow to the devil and wandered around hither and yon in seven-league boots. During Chamisso's later years, he obtained an honorary doctorate and, thanks to the recommendation of Alexander von Humboldt, became a full member of the Russian Imperial Academy of Sciences. Even while serving as superintendent of the Royal Botanic Gardens and Herbarium in Berlin, he gained fame for his innumerable patriotic and romantic ballads, as well as for political and social poems. Two years after his death in 1838, composer Robert Schumann set to music Chamisso's lyrical love-poem cycle *Frauenliebe und Frauenleben.* Chamisso also became an authority on Australasian languages, penned among the first poems in German on American Indians, and as a result of his brief exposure to the Sandwich Islands, where he wished he could have spent a year, compiled a book on Hawaiian grammar. But his zoological colleague Eschscholtz would probably have been most impressed by the botanist's recognition, from his oceanic experience, of the alternation of generations (metagenesis) in marine organisms called tunicates.

THE *PREDPRIATIE* VISIT WITH
KOTZEBUE AND ESCHSCHOLTZ, 1824

In the autumn of 1824 Captain Kotzebue returned to Alta California in command of the Russian ship *Predpriatie.* Although the vessel's main charge was to protect the Russian American Company of Alaska against smuggling by foreign traders, it was well supplied with astronomical and other scientific instruments. Accompanying Kotzebue again as naturalist was Eschscholtz, and also aboard were a mineralogist, Hoffman, and an astronomer, Preus. Remaining behind in Alaska was Heinrich Emil von Lenz, a German physicist who during the sojourn in the North Pacific was the first to recognize that the cold waters of deep ocean basins originated from polar regions.

The new status of Alta California, separated from Spanish control in the spring of 1822, became evident when the *Predpriatie* sailed into San Francisco Bay on September 27, 1824, and received no salute from the shore battery, now flying the Mexican flag. The first official who came out to the Russian ship asked to borrow gunpowder for the presidio's cannons so that a salute could be fired. But this welcome in no way diminished Kotzebue's continued appreciation for "this unknown but beautiful country, richly endowed by Nature," whose governor in Monterey was now Don Luís Argüello.

The expedition made its headquarters in San Francisco Bay from late September until late November 1824. The day after arrival, the captain and Eschscholtz went by boat south to Santa Clara. Wild ducks and Bald Eagles were observed, the latter seemingly pursuing the numerous quail and rabbits along the shoreline. But in the bay the only active "fishermen" appeared to be American White Pelicans, no Mexican boaters being seen. After lunch on an island, the party proceeded to the Santa Clara mission on horseback, passing flocks of ducks and geese, including Snow Geese feeding in corn stubble fields. Some of the birds were "so

tame that we might have killed great numbers with our sticks." Kotzebue perceptively conjectured that they were "all birds of passage, spending the winter here, and the summer farther north." They spent three
days at Santa Clara before proceeding to the mission at San Jose.

The next excursion was north to the Russian settlement at Fort Ross
(figure 9). The group proceeded by boat up San Rafael Bay to a landing
at the small, new mission San Rafael Arcángel. From there off to the
northeast in the distance could be seen the newer mission San Francisco
Solano, northernmost of Alta California's mission chain. Accompanying
the party, which included Kotzebue, Eschscholtz, and Hoffman, were a
number of horses and four well-armed Mexican dragoons to guide and
for protection against "los Indianos bravos." In the lead was Don Es-
tudillo, commandant at San Francisco, laden with a dagger, heavy
sword, pistols in his belt, and staff. The commandant, Kotzebue observed, was "a good personification of an adventurer of the olden
times." The cross-country route toward Fort Ross was "of a very romantic though wild character," with Black-tailed Deer abounding.
Kotzebue's stay at the fort lasted only two days, but Eschscholtz decided
to remain there, "intending to prosecute some botanical researches" before returning to San Francisco Bay by one of the many Aleutian
baidarka boats out in the ocean after Sea Otters. When an autumn storm
roared in out of the southwest, uprooting trees and blowing off roofs
around San Francisco Bay, the fleet of twenty baidarkas carrying Eschscholtz was forced to seek refuge at Point Reyes. Those aboard the *Predpriatie* in San Francisco Bay were concerned over the zoologist's failure
to appear, but Eschscholtz was certainly not upset over the unscheduled
stop at a new spot. Ashore he collected a jet-black beetle with impressive
mandibles, a new species he would later describe and name (California
Black Tiger Beetle). Finally, after about ten days Eschscholtz arrived at
the new anchorage of the *Predpriatie* in San Francisco Bay off the small
village of Yerba Buena. The naturalist was "not in the slightest degree
disheartened by the difficulties he had undergone" and eagerly looked
forward to another adventure.

Speaking of Fort Ross and beetles, the fort during its existence would
become a lodestone for many Russian naturalists, especially beetle collectors, including Russian America's governor general Wrangel, who
thereabouts caught the Gigantic Eleodes beetle (almost one and a half
inches long). Ilia Voznesenskii, from the Russian Imperial Academy of
Sciences, visited in 1840 and "began to collect a rich harvest of acquisitions in all branches of natural science and ethnography," including

Figure 9. Russia's Fort Ross, circa 1827. Sketch by Captain Auguste Duhaut-Cilly of the French Duhaut-Cilly Expedition. (Courtesy California Department of Parks and Recreation.)

birds, fishes, and 350 plant specimens, along with pencil sketches and paintings. George Tschernikh, who managed the fort's agricultural pursuits, collected the explosive Bombardier Beetle, which was named after him. Voznesenskii and Tschernikh supposedly hiked to the top of an unnamed summit with Fort Ross Colony manager Count Alexander Rotchev's beautiful wife, Princess Helena, who christened the peak St. Helena after Russia's early patron saint.

On November 12, Kotzebue embarked on a river trip that was a first for California natural history. In fine weather two boats, a *barcasse* and a shallop, "provided with every necessary," with Marco as pilot and a soldier from the presidio as guard, set off across the bay, transporting Kotzebue and Eschscholtz on the longest "scientific" trip into the Alta California interior to date. By way of the Sacramento River the party explored to within about thirty miles of California's modern capital.

On the inland trip there was much to be seen and enjoyed. At one point, just beyond Suisun Bay, Kotzebue and Eschscholtz climbed up a "slate/quartz" hillside just "to admire the beauty of the prospect." Black-tailed Deer, geese, ducks, and Sandhill Cranes were plentiful, and large fish were jumping. Kotzebue remarked that there was "such a su-

perfluity of game, that even those among us who had never been sports-
men before, when once they took the gun in their hands, became as eager
as the rest." Because of the abundance of fish in the river, pelicans were
common. When an Aleut crewman killed one of them with his javelin,
the other pelicans vigorously attacked the sailor until beaten off. Vines
of California Wild Grape, with small but very sweet fruit, grew ram-
pantly and festooned the riparian corridor of trees, leading Kotzebue to
predict for Alta California that "the vine might be cultivated here to
great advantage." He and Eschscholtz consumed great quantities of
grapes without sustaining what the two called any "inconvenience."

At night the boaters bivouacked onshore, building a large fire so that
their errant hunters would have a beacon to find their way back. More
than lost hunters appeared: namely, bears chasing deer right into the
campsite. The second morning, a small Northern Pacific Rattlesnake was
encountered. Kotzebue remarked that it could have been a "dangerous
neighbour" except that Eschscholtz quickly captured it. At the end of the
second day on the Sacramento River—as it swung to the north, flowing
among numerous small islands and with a powerful current resulting
from heavy autumn rainstorms to the east—only ten more miles were accomplished. On the advice of the pilot, the exploration party concluded
its voyage and turned back, leaving, as Kotzebue penned, "the farther
prosecution of these inquiries to some future traveller."

At the last upriver camping site, beyond the present-day town of Rio
Vista, there was a recently deserted Indian village, with burning embers
in several of the fire pits and two abandoned canoes made of reeds. In
the distance on some of the river islands smoke could be seen rising from
other Indian fires, signs of either the Chupcan or the Hulpun tribes, according to the pilot. From this spot the explorers could see off to the east
"the majestic chain" of the snow-capped Sierra Nevada, while the low
land in the foreground, covered by water, "spread out like a verdant
sea." What remained of the day was again "passed in sport," with numerous deer shot, although many a venison steak was appropriated during the night by scavenging coyotes. The next morning, while preparing
for departure, Kotzebue reflected on the "lovely and fertile plains,"
where someday "many thousand families might live in plenty and com-
fort, but which now, from their utter loneliness, leave a mournful impression on the mind."

The boat trip back was an easy one, and soundings were diligently
made of the Sacramento's depth, ranging from up to sixty feet in midstream on the east to about fifteen feet at the mouth. On November 23

the exploring party arrived at the *Predpriatie,* anchored off Yerba Buena, with a welcome cargo of venison for the whole crew. After Kotzebue prevailed on a frigate to carry letters back to Europe by way of Alaska and Siberia, the expedition made immediate preparations to leave. The camp onshore was broken up and instruments brought aboard. The last night proved memorable because a skunk that wandered into someone's tent was killed, with "abominable" results. On the morning of November 25, 1824, at the ebbing of the tide and into a fair-weather northwesterly wind, the *Predpriatie* proceeded out of San Francisco Bay, barely escaping a crash into the rocks at the headlands.

For Eschscholtz, his second California stay had been highly productive. Not having to assist Chamisso, he had been able to focus on animals. Among his insects—the first major insect collection to be made in Alta California—were nearly two hundred beetles (mostly from Russia's Fort Ross up the coast and from the Santa Clara trip), all but one of them new species. Included were a large snail-eating ground beetle found at San Francisco; a Ciliated Sand Beetle from the coastal sand dunes; the Western Twelve-spotted Cucumber Beetle, which can exhibit devastating population explosions; another species that would become a West Coast economic pest, the Potato Wirebeetle, whose yellow larvae infest potatoes; and for variety a very different kind of insect, the native Linear Earwig. All of today's familiar earwigs here are European in origin. Eschscholtz was also an enthusiastic beachcomber, collecting and naming California seashells, among others a number of limpets including the common white Dunce-cap Limpet.

By the end of the California visit the naturalist would garner the first significant collection of reptiles and amphibians made in California, including a Western Yellow-bellied Racer, Valley Garter Snake, the giant Marbled Salamander from Redwood country, several slim earthworm-shaped salamanders (California Slender Salamander), California Newt, and what was probably a Western Spadefoot Toad. Two of the salamanders turned out to be new species and were originally given their scientific names by Eschscholtz. The zoologist noted, incidentally, that the most successful collecting technique was to overturn boulders. Eschscholtz also wrote about numerous mammals he encountered, including the Broad-handed Western Mole and Brush Rabbit. He caught a shrew, shot a weasel along the Sacramento River, and collected a Coyote. Both near the Sacramento River and in the vicinity of Fort Ross he observed through a telescope "herds of animals of the shape of goats . . . with short straight horns"; these he judged to be the "often spoken of"

Mountain Goat, but in reality they were European domestics. When he traveled by boat from Fort Ross with Aleut Sea Otter hunters, he was amazed at the abundance of the valuable otters. He recorded forty different birds, including several kinds of blackbirds, the Red-shafted Flicker, the American Crow, the California Quail, and a Sandhill Crane. Eschscholtz thought the crane was "probably from the north," but some are in fact residents in the Central Valley of California today.

Upon Eschscholtz's return to Europe, he became a professor of zoology and professor extraordinary of medicine, as well as director of the zoological museum, at the University of Dorpat, his alma mater. Johann Eschscholtz passed away in May 1831, much too soon for such a prodigious naturalist.

HMS *Blossom*'s Visit with Beechey, Collie, and Lay, 1826–1827

You will pay every attention in your power to the preservation of the various specimens of natural history.
—British Admiralty orders to Captain Beechey

Captain Frederick William Beechey's trip south to California in 1826 with HMS *Blossom* was in a sense for rest and relaxation, but especially for supplies. During the summer, Captain Beechey had been cruising eastward above the Arctic Circle in the hopes of encountering Captain John Franklin's Expedition, which was sailing from the east, and thus of completing at last the Northwest Passage, albeit at a very high latitude. The two expeditions never met, even though Franklin, despite being thwarted by bad weather, had gotten within 146 miles of Beechey's farthest point east (Point Barrow).

The *Blossom*, a British sloop, had been specifically designed for worldwide exploration, particularly in polar waters, and was well supplied with scientific equipment. Despite the fact that it carried sixteen guns, it was ordered not to commit any hostile act, even in the event of war, since the *Blossom* was, by order of the Admiralty, "sent out only for the purpose of discovery and science, and it being the practice of all civilized nations to consider vessels so employed as excluded from the operations of war." Such was the case with Captain Cook's Third Expedition when it was guaranteed freedom of the sea by the Yankees during the American Revolutionary War.

Beechey's expedition of some one hundred men included a noteworthy array of "gentlemen," all interested in natural history. The captain himself, only thirty years old, son of the famous portrait painter William Beechey, had already served on polar expeditions with Sir John Franklin and William Edward Parry and had surveyed along the North African coast. Many years later Beechey became president of the Royal Geographical Society and was remembered as the first ethnographer of the western Eskimos. Lieutenant Edward Belcher, the *Blossom*'s assistant surgeon, later commanded his own expeditions and surveys, and in 1852 he would direct a search for the lost Franklin Expedition. Throughout his life, from the time of his scientific studies as a teenager, Belcher made contributions in natural history, especially mineralogy. A species of gull (now Heermann's Gull) was at one time named in honor of Belcher, "to whose science and enterprize zoology is indebted for many important acquisitions in various departments" (Richardson et al. 1839, 39). This expedition also had an artist "of elegant pencil" aboard, William Smyth, mate of the *Blossom,* who produced beautiful landscapes and some sketches of the animals encountered. His illustrations appeared in the published narrative of the expedition.

Captain Beechey collected both animals and plants, as well as executing maps and writing observations on the natural history of the voyage. There were also two "professional" naturalists on board, George Tradescant Lay and Alexander Collie, both assisted in their work by Belcher. Lay, a botanist and the official naturalist of the voyage, bore a middle name recalling the great botanical father-son Tradescants of early seventeenth-century London. About a decade after the *Blossom* voyage, Lay, who by then had become an agent of the British and Foreign Bible Society in China, embarked from Macao on another sailing expedition on the private ship *Himmaleh,* financed to aid "missionaries in circulating religious books on the coasts of China and neighboring countries" (Bretschneider 1898, 290). He became recognized as an authority on things Chinese, and after serving on a special British mission to China (1840–42), he was appointed British consul at Canton, then first British consul in Fu Chow; he was finally posted to Amoy, where he died from pernicious fever (malaria) in 1845.

Alexander Collie, the ship's surgeon and zoologist, was a Scot who like so many others received his medical training at Edinburgh. He then spent several years in Europe studying botany, mineralogy, and chemistry. The same year that he was assigned to the *Blossom* he became a fellow of the Linnean Society. Collie was a diligent journal writer, so much so that what he wrote, especially on intricate dissection of various animals, greatly exceeded what could be published in the expedition's final report. In naming a spectacular Mexican jay (Black-throated Magpie-Jay) after Collie, the report's dedication rightly recognized that Collie's "exertions contributed much to our knowledge in every branch of science" and his notes "form the most valuable part of the present observations." The last of this "ingenious" zoologist's life would be spent in southwestern Australia, where he continued his preoccupation with natural science. Collie died at the settlement of Albany on King George's Sound in 1835, where he had been serving as the first government resident.

The Admiralty's instructions for the Beechey Expedition emphasized the importance of "collecting rare and curious specimens in the several departments" of natural history, with "at least" two specimens of each article. Their extended stay in San Francisco Bay—almost two months— seemed to afford this opportunity, but unfortunately botanist Lay was still in "Owyhee" (Hawaii), where he had been left to recuperate from severe dysentery when the expedition had sailed for Alaskan waters. Thus the initial natural history work in California fell mainly upon Collie's shoulders.

On November 7, 1826, a beautiful morning, the *Blossom* sailed through the narrow entrance into San Francisco Bay and anchored "in the very spot where Vancouver had moored his ship thirty-three years before." Two days later a party of three from the ship—George Marsh the purser, Collie, and Mr. Evans, who was versed in the Spanish language— plus a guide, two vaqueros, about ten loose mission horses, and a dragoon armed with a musket and a bull's-hide shield decorated with the royal arms of Spain (despite Mexican independence), left on horseback for Monterey to procure medicines, provisions, and other stores. This nine-day round trip represented the first extended foray on land by any oceanic expedition party in Alta California to date. The group proceeded past Mission Dolores and scaled Mount Bruno for a magnificent vista of the bay region. They spent the second night at the Santa Clara mission, where they received new horses, new vaqueros, and a dragoon with a huge broadsword and were plied with drams of "aquadente" by the hospitable padres. Mission San Juan Bautista proved even more hospitable, adding an additional day to the festivities because of a holiday. A highlight at Bautista was Padre Arroyo's demonstration of his unique invention, a "water clock" that could set off an alarm bell at his bedside. From here the party traveled over the low range to the southwest and onto the sparsely vegetated Llano del Rey to Salinas, crossing creeks and ditches alive with ducks. The group continued to Monterey across a countryside that Beechey's narrative described as "covered with fragrant southernwood, broken here and there by dwarf oaks, and shrubs," with many pasturelands filled with cattle. At five in the evening of the fourth day the men arrived in the square at Monterey and met William Hartnell, an English merchant associated with Begg & Company of Peru, to whom Captain Beechey had written from San Francisco Bay on November 13 about obtaining supplies.

Collie—perhaps because of the season, the pace of the trip, and his stronger interest in zoology—collected few plants on this excursion. But he did make general observations concerning the vegetation, about plane trees (sycamores) and Redwoods and "a low tree with a smooth reddish-brown bark, bearing red berries" that had extremely hard wood (Pacific Madrone). He also mentioned the dwarf oak woodland near Monterey Bay, with its interspersed chaparral, and "yedra," "the most remarkable shrub in this country . . . which produces tumors and violent inflammation," an apt observation about Western Poison Oak. The wetlands of San Francisco Bay were alive with geese, ducks, shorebirds, wading birds, and myriads of Red-winged Blackbirds. Near Santa Clara in the brushland were numerous coveys of "handsome" quail. Collie actually captured a live "California partridge" (California Quail) that eventually was joined by others for the voyage back to England. The females died, but at least one male survived to "enjoy good health" in the London Zoological Society's gardens, a California first for Great Britain.

At Monterey the three *Blossom* men were disappointed to discover that no extra medicines and stores were available. They procured what they could and arranged for the preparation of salted meat, all to be picked up when the *Blossom* later put in at Monterey. The party returned the same way, not on a different route as they had hoped. There was another day's delay at San Juan Bautista because no horses seemed to be available, and the padre was much less cordial than before. On November 17 the field party was back on the *Blossom* in San Francisco Bay.

As the remaining days went by, Beechey reflected, "There was no society to enliven the hours, no incidents to vary one day from the other." The "picturesque scenery" had lost its appeal, hunting "had lost its fascination," the treks to the mission and presidio had "grown tedious and insipid." Also, there was the sad incident of the marine who fell into the water while duck hunting and drowned, not knowing how to swim. With surveying temporarily completed, the *Blossom* was relieved to sail out of the harbor on December 28, anchoring off Monterey on New Year's Day 1827. A shore party set to work cutting a supply of spars, and, through Monterey's local mercantile firm owned by Hugh McCulloch and William Hartnell, such provisions and naval supplies as were available were taken aboard.

Beechey and others among the "gentlemen" were entertained at the Hartnell home. Dr. Collie volunteered medical assistance to the community's residents, especially to Hartnell's seventeen-year-old wife, Teresa, who was pregnant again while still suffering from her recent childbirth. A party from the *Blossom* traveled to Mission San Carlos through pastureland and open woodlands of tall Monterey Pines and Coast Live Oaks. While at the mission, Captain Beechey saw the sketch by de Vancy depicting La Pérouse's visit and tried unsuccessfully to wheedle it from Monterey alcalde José Ramón, even offering to pay a good price for it.

On January 5, the *Blossom* departed for the Sandwich (Hawaiian) Islands. The day prior to leaving, Captain Beechey penned a letter to the British consul general in Mexico City emphasizing the importance of having a British agent in Alta California and recommending Hartnell for the position of vice-consul. His suggestion was never acknowledged. Following its winter stay in Hawaii, a spring and summer voyage took the *Blossom* to China and eventually back to the polar sea, where there was still no evidence of the Franklin Expedition. The *Blossom* lost three sailors in Arctic waters, and altercations with coastal Indians severely wounded six marines. When the expedition returned to Monterey on October 29, 1827, escorted by a line of American White Pelicans, a number of the crew were suffering from scurvy. They were greeted by the news that the Duke of York had died, paving the way for his younger brother to become heir apparent (and later, in 1830, King William IV).

It was good to be at Monterey once more. This time Captain Beechey was actually able to reside in Hartnell's home, the Hartnells being away visiting in Santa Barbara with their young son and new baby. Monterey made up for what Yerba Buena lacked socially, offering a round of house parties, musicals, dancing, picnicking, and the "pleasure of the chase." Beechey's account notes that botanist George Lay, finally back with the expedition and in fairly good health, was "ready to resume his occupations," and helpful sailors enjoyed "finding a variety to enchant the naturalist." Together Collie and Lay pursued their field collecting, while artist Smyth made a number of sketches.

On November 17, after a most enjoyable stay, the *Blossom* left Monterey for San Francisco Bay, where the crew was able to obtain a larger and better supply of water than at Monterey. The Monterey water was not only scarce but brackish, bathed in by the Indians and mingled with soapsuds used by the washerwomen. On December 3, 1827, with fresh snow frosting the mountaintops around the bay, the Beechey Expedition sailed onto the open sea on its way toward Cape Horn and eventually England.

This expedition of potential scientific promise had proved somewhat disappointing with respect to California's bountiful flora. As Beechey himself wrote, "Our knowledge of the natural history of this country cannot be expected to be very extensive." A major publication was completed in 1841, titled *The Botany of Captain Beechey's Voyage,* written by William Jackson Hooker and George Walker-Arnott and dedicated to Captain F. W. Beechey, "by whose zealous encouragement, the plants described in the present volume were chiefly collected." Twenty-one pages of the publication were supposedly devoted to plants collected at San Francisco Bay and Monterey during the expedition, and there was a lengthy supplement at the end of the publication about California flora. But the latter, written by the same authors in 1840–41, primarily summarized West Coast collections by Menzies, Douglas, Coulter, Scouler, Nuttall, Chamisso, and even Tolmie, who had never even visited California.

The twenty-one-page section is mostly concerned with Monterey Peninsula plants, including some new species. But often only one specimen had been obtained, contrary to the Admiralty's specific instructions to collect at least two articles. When *The Botany of Captain Beechey's Voyage* was being compiled, the "closet-naturalist" authors could not refrain from adding derogatory comments: "a very indifferent and solitary specimen"; "only one specimen, and that far from a good one"; "of this only a fragment exists in the Collection"; and, for a Cow Parsnip, "a wretched specimen of this obliges us to put a mark of doubt to the name."

The specimen of California Fuchsia justified a special comment: "this very remarkable genus, with flowers in shape and colour resembling a Fuchsia but with the fruit of an Epilobium." Plant collections had been made both in "sandy plains in the Bay of Monterey" and in "the valleys and tops of the mountains" on the peninsula. Collie saw the "Red Cedar" (Redwood) at the latter sites. He was informed that between Santa Clara and Santa Cruz some of these trees stood 150 feet high and 25 feet in circumference, and the wood of young specimens was so full of sap that it sank in salt water. Collie noted the Monterey Pines as "abundant on the granite hills close to the sea on the south side of the Bay," and he said they "gave the name Pine Point to that area" (Point Pinos). He observed that these "pitch pines" could reach eighty feet in height, with a large diameter, and thought they should be "very fit for masts," but they decayed when buried in the ground. Today the Monterey Pine, introduced around the world, has become one of the world's most abundant and important lumber trees, but as Malaspina previously noted, it was not suitable for masts.

By 1827 there were already many plants on the Monterey Peninsula introduced from Europe. Among those collected by the *Blossom* naturalists were sweet clover, bur clover, mint, probably some cultivated gourds, and Sea-fig, *Carpobrotus chilensis*. This species, also collected earlier by Menzies in San Diego, was recently featured on a U.S. postage stamp celebrating California's sesquicentennial. Whether it was introduced into California or is a native remains controversial.

If the botanizing had left a bit to be desired, the zoological pursuits proved more productive, especially around Monterey. Among the many birds encountered, Collie collected three new species here: the Pygmy Nuthatch, undoubtedly from a Monterey Pine woodland, and in the Monterey coastal chaparral and woodland understory the California Scrub-Jay and California Towhee. His new subspecies included a Band-tailed Pigeon, Red-shafted Flicker, Black Phoebe, and Northern Mockingbird. At San Francisco he got a new subspecies of Bewick's Wren. Collie is presumed to have obtained the first specimen of the Black Turnstone, a new species. It does occur along the Monterey rocky shoreline, but Collie's original specimen was probably collected in the Northwest. The Douglas Quail, another new species supposedly collected by Collie in Monterey, is actually a Mexican species.

Collie was thorough with his avian endeavors. In the San Francisco area he killed a Great Horned Owl in the daytime, after having been attracted to it by a number of harassing birds. His detailed dissection of the owl turned up the makings of an owl pellet in its "stomach . . . a compact mass of feathers, bones, and portions of the skins of birds, and the whole leg and foot of a duck." He noted numerous large flocks of Red-winged Blackbirds, and he not only described plumage differences between juveniles and adults but cut open their stomachs to discover larval insects, "farinaceous roots," and gravel. He had good opportunities to view shorebirds in San Francisco Bay, including a wintering American Avocet of which he made an elegant painting. Because this species appears so different in winter, for some time Collie's specimen was thought to be a second North American species and was named the White Avocet. A San Francisco mission padre shot an Anna's Hummingbird for Collie and informed him that hummingbirds remained during the winter around San Francisco Bay, which is indeed true for the Anna's Hummingbird.

There were other collecting opportunities. Several species of mammals were observed and a few procured, particularly around Monterey, including an unlikely specimen of the Virginia Opossum Collie described right down to the prickles on its tongue. But "no specimen was brought home" by the expedition, and today California mammal books aver that the Virginia Opossum is an exotic that was certainly not introduced into the state before 1880.

Collie noted several kinds of fish, commented on the many sardines in Monterey Bay, and collected a new species of chimaera (Spotted Ratfish) in Monterey Bay that was named after him by Lay, *Hydrolagus colliei*. The naturalists received reports of reptiles, but none was actually seen. Beechey collected some marine shells, including limpets, a chiton, and two species of abalones on the rocks at Monterey. Lieutenant Belcher gathered numerous marine crustaceans, while Lay observed great numbers of jellyfish floating in the ocean off San Francisco Bay and expressed regret that he had spent too much time with the flora and not enough on the habitats and inhabitants "that people the ocean."

The expedition report included a good description of the local geology, with a colored map showing geological formations of the San Francisco Bay region, prepared by Belcher with Collie's assistance (figure 10). Belcher not only surveyed the main part of the bay but proceeded east into San Pablo and Suisun bays. The geological mapping was rather good by the standards of that time, even indicating main rock outcrops in different colors, including areas of "jasper" (probably chert) and serpentine, as well as topographic features. This historic California geologic map was not only the first for the Golden State but probably only the second for North America, the first having been a map of eastern United States by William Maclure. There is a full-page reproduction of Belcher's map in *Geology of California* (Norris and Webb 1990). Meanwhile, Collie compiled the information that "earthquakes are rather common, and one in 1806 so shook the buildings of Mission Santa Clara, that a new one was obliged to be erected." Some of Collie's earthquake reports were current: a shake at the San Francisco presidio in April 1827, and a smaller quake at ten o'clock on December 26 in the same year at San Jose. All in all, the *Blossom* savants touched on many aspects of the natural history of this portion of California in the time at their disposal.

Nicely enough for botanist George Tradescant Lay, when the California specimen of a member of the Sunflower Family was critically examined, William Hooker and George Walker-Arnott wrote, "If this were to form a new genus, we would propose the name of *Layia*, after one of

Figure 10. Geologic map of the San Francisco area by
Edward Belcher of HMS *Blossom,* assisted by natural-
ist Alex Collie, 1826. (From Robert M. Norris and
Robert W. Webb, *Geology of California,* 2nd ed. [New
York: John Wiley & Sons, 1990].)

the discoverers." The type specimen became *Layia gaillardioides* (Woodland Layia). The sole salamander collected at Monterey was thought for
a time to be a new species and was named in honor of Beechey, *Sala-
mandra beecheyi.* The specimen eventually made its way to the museum
of the College of Surgeons in London, where it was reidentified as a California Newt, the type specimen of that species having been collected in
Alta California by Eschscholtz in 1824.

For Captain Beechey's visit to Alta California, there are three note-

worthy lasting tributes. During his stays in San Francisco Bay, he produced detailed and accurate charts of the harbor not to be improved
upon for a third of a century (except that he moved the 1775 name Al-
catrazes—"Island of Pelicans"—from its original island, now called
Yerba Buena, to the present infamous penitentiary rock). His fine collection of marine shells would eventually be identified by John Gray and
George Sowerby and described in an 1839 publication. Meanwhile, all
along the California coast as well as inland today is found the familiar
California Ground Squirrel, *Citellus beecheyi,* first collected by Collie at
Monterey Bay.

HMS *Sulphur*'s Visits with Belcher, Hinds, and Barclay, 1837 and 1839

Sulphur was the school in which I more particularly studied geographical botany.

—Richard Brinsley Hinds

The British naval vessel HMS *Sulphur* was new, commissioned in 1835, but its captain was an old salt, Frederick William Beechey, who had brought HMS *Blossom* to California a decade earlier, with Collie and Lay as his naturalists. In 1836 Beechey embarked on another British surveying expedition in the Pacific "for the advancement of science and the safety of navigation" on the HMS *Sulphur,* with her consort HMS *Starling* commanded by Lieutenant H. Kellett, sailing for the Hydrographical Office of the Admiralty.

Aboard were two assigned naturalists, Richard Brinsley Hinds and George Barclay. Hinds, with an amateur's interest in botany, was the expedition's official "surgeon-naturalist." Barclay—the official collector sent out by the Royal Botanic Gardens, Kew—whose assignment was to gather seeds for propagation at Kew and secondarily to obtain herbarium specimens, was promised that he could get ashore for his collecting as long as it did not interfere with the naval survey. As a civilian, Barclay was assigned to mess with the warrant officers rather than the ship's officers and was listed on the *Sulphur*'s book "for victuals only." Hinds, as a naval officer and the *Sulphur*'s surgeon, outranked and "outperked" Barclay, much to the civilian's dismay, while Hinds was irked that Barclay was getting paid a comfortable salary by Kew solely to collect plants. Grumbled Hinds: "I deny any claim that he had as a man of science, and I never discovered that he had any sentiments which belong to such." Adding to the atmosphere of competition on the expedition was another surgeon and plant collector, Dr. Sinclair of the *Starling*.

Captain Beechey's ships departed from England in late December 1835 and started the survey in South American waters. But by February 1836, when Valparaiso, Chile, was reached, Beechey realized he was too ill to continue in command and was invalided back to England. Lieutenant Kellett temporarily assumed Beechey's responsibilities, but a year passed before a new captain, Edward Belcher, would be assigned to the expedition. Belcher had served as assistant surgeon under Beechey on *Blossom* in 1826 and 1827.

Captain Belcher assumed command in Panama in February 1837, with Lieutenant Kellett moving back to the *Starling*. The two ships carried out surveys along the Central American and Mexican coasts, then sailed to the Sandwich (Hawaiian) Islands and Alaska. Even though a professed purpose of the expedition was "for the advancement of science," a closer reading of the Admiralty orders revealed that "large collections of natural history cannot be expected, nor any connected account of the structure or geological arrangement of the great continent which you are to coast; nor indeed would minute inquiries on these subjects be at all consistent with the true objects of the survey." Yet perhaps because Belcher himself had an interest in natural history, especially conchology (shells), commendable collections were amassed despite the admonition of the Admiralty.

On October 9, 1837, the *Sulphur* and *Starling* headed south from Nootka Sound (Vancouver Island) along the North American coast. During their last day at sea, off Point Reyes, they were welcomed by what was probably a Gray Whale. It played around them all afternoon until Belcher ordered a cannonball to be fired at it, whereupon the whale dived and was seen no more. The *Sulphur* sailed between the headlands and shortly after midnight on October 20 came to anchor in "Yerba Buena Bay," the informal name of the cove at the settlement of Yerba Buena (San Francisco). Less fortunate was the *Starling*, which was caught by a sudden gale and "experienced very unpleasant weather" before finally getting into the harbor. On terra firma early the next morning, Belcher, notorious for his acerbic outlook, thought that everything ashore was, as he wrote, "going to decay" and seemed much worse than on his previous visit with HMS *Blossom*.

For the expedition naturalists, as with previous visiting botanists, autumn was not the ideal time to collect plants. They spent several days on foot near the anchorage. Then on the twenty-fifth, during a visit to San

Francisco's Mission Dolores, Barclay noted the "coppice-Wood," consisting of Coast Live Oak and several shrubs, including "Yerba Oso or bears herb" (Oso Berry), Skunkbrush, and probably Toyon, growing on the area's only wooded ridge between the mission and the presidio. There were literally hundreds of cattle grazing in the meadows. Around the mission the ground was strewn with the carcasses of bullocks and horses, killed solely for their hides and tallow and now attracting a host of scavenging crows, ravens, and vultures. Also very numerous was the Coyote, which Hinds described as "quite a pest to the residents, for its daring predatory habits."

The next day Belcher, with his two surgeon-naturalists and Barclay, left by boat for the Santa Clara mission with a letter of introduction, to make arrangements for obtaining "table supplies." The pilot of the boat got the small party lost, and they were forced to stay overnight in a marsh. To make matters worse, the father president of the Alta California missions, whom Belcher needed to see, turned out to be in San Jose. During the Santa Clara trip Dr. Sinclair did manage to collect two late-blooming plants, a white Hayfield Tarweed and the Western Goldenrod. Another specimen of the latter in blossom was collected at Yerba Buena by Barclay.

Kew's botanist Barclay was slighted when Belcher decided to take a party, including surgeon Hinds, up the Sacramento River. Belcher's plan was to survey beyond where he and a crew from the *Blossom* had ventured in 1826, beyond where the Sacramento and San Joaquin rivers entered Suisun Bay. Hinds would now have the opportunity to collect both plants and animals during an excursion traversing a region where no naturalists had traveled before. The survey contingent boarded the *Starling*, with its flotilla of five boats and five from the *Sulphur* including cutters, pinnaces, and gigs, and sailed across San Pablo Bay, through Carquinez Strait ("Estrecho Karquines") into Suisun Bay until the pilot felt that further travel by the big vessel would be difficult. The ship anchored, with a fine view of Mount Diablo to the south, and surveying by the small boats continued for a few days in the vicinity. The sailors out hunting encountered a variety of large ducks and the smaller teals, snipe, and assorted shorebirds lumped together as "curlews." The geese liked to spend the day in the meadows, returning at twilight in such immense clacking flocks as to darken the sky to their roosts in the marshes.

On October 31, 1837, the small boats left the *Starling* and proceeded upstream to the Sacramento River, missing the mouth of the San Joaquin just to the south. The guides were two mission Indians who had been

trained to "hunt for Christians," namely, to collect vagrant natives—
"Christians nolens volens," as the guides said, or in other words, un-
willing recruits—for work at the missions. On the second day of the river
trip the Sacramento had widened to a third of a mile. The tall banks were
beautiful, "belted" with willow, cottonwood, oak, Oregon Ash, Brown
Dogwood (which Hinds collected), sycamores of immense size (a sample obtained by Hinds), and the Northern California Black Walnut (the
northern variety, named much later by botanist Willis L. Jepson to honor
Hinds, *Juglans californica* var. *hindsii*), whose nuts on occasion provided
dessert for the travelers.

Autumn was on the land, and the leaves of the deciduous trees, "producing beautiful varieties of tint," were beginning to fall. Masses of California Wild Grape vines festooned the trees, and Hinds collected some.
Everyone was eating the grapes, despite their small size and relatively
large seeds. Beyond the river and the adjoining marshes, the countryside
appeared parklike, a meadowland with a scattering of the impressive deciduous oaks that in this area replaced the evergreen oaks of the coast.
One of the largest oaks measured twenty-seven feet in circumference; another was sixty feet tall, "truly a noble sight." The men noticed that the
Indians had a tendency to build fires at the base of the finest oaks, and
"thus prematurely and wantonly destroyed" them. Hinds's specimen of
this oak would be named *Quercus hindsii* in his honor by George Bentham, but it later proved to be the Valley Oak already described. The survey members agreed that the acorns, together with the grapes, "formed
a very considerable part of our people's daily food," as was equally true
for the local Indians. Also noted was some type of amole or soap plant,
whose root was not only roasted and eaten by the Indians, "sweet as
chestnuts," but was reportedly used in washing.

Wildlife was abundant all along the river. Large fish, which Belcher
presumed to be sturgeons, were incessantly jumping out of the water.
Today both White and Green sturgeon occur in the Sacramento–San
Joaquin delta but ordinarily do not move this far upriver to spawn until
late winter. During shoreline stops at 9:00 A.M., noon, and 3:00 P.M. to
set up astronomical stations, the survey party had opportunities to collect animals. Abounding in the riparian corridor were "very beautiful
ducks, owls, hawks, and other birds," including kingfishers. On one occasion Dr. Sinclair killed forty-eight geese and eight ducks. Captain
Belcher, when in a capricious mood, would shoot ducks flying over; but
without his express permission, no one else was permitted to shoot at
them. No wonder it was said of Belcher that "no officer of equal ability

has ever succeeded in inspiring so much personal dislike" (Pierce and Winslow 1969, xii).

Specimens of Coyote, Gray Fox, Black-tailed Deer, Raccoon, River Otter, Long-tailed Weasel, and Western Gray Squirrel were collected on the Sacramento River voyage, and some of them made their way back to the zoological collections of the British Museum. There were plentiful signs of Beaver, and during the day ashore the Coyotes "showed a bold front," while producing "the most terrific howlings" at night. The expedition's final report has plates portraying the Coyote, the Raccoon, and a Raccoon skull, probably the first portrayal of these mammals from California (figure 11). Belcher fired twice at what he called a small "tigercat" (probably a Bobcat), but his small shot seemed to make no impression on the thick fur. There were immense herds of elk, some with racks of antlers six feet across. The most singular hunting event involved killing an elk with blows from an oar. Instead of saving this elk as a specimen, the men ate it. This was doubly unfortunate, because not only was the meat tough, but the mammal they consumed was actually a new kind of elk, the Tule Elk, eventually named by C. Hart Merriam more than half a century later (now recognized as a subspecies of the American Wapiti).

Belcher and his boats likely reached the junction of the Sacramento and American rivers before turning back from what they called Elk Station (present-day Sacramento), "where the river forked." Here Indians had been camped, but with the arrival of the flotilla they vanished, abandoning their fires, acorn caches, and other belongings. Belcher left presents for the Indians and watched them through the telescope as they cautiously returned.

Late afternoon one day toward the end of November, the exploring party turned back down the river, encountering several parties of Indians as well as seeing abandoned villages and many Indian skeletons scattered about. The guides reported that there used to be many more villages of what they called the Wallock tribe. The villages seen, locally called rancherias, were situated on piles of dirt about a hundred yards across, built up above what was obviously a floodplain, with moats around them to carry floodwaters away. The mud lines ten feet up the riverbank trees testified to periodic flooding. The excursion party returned to the *Starling*, and the ship proceeded downstream, arriving at the *Sulphur* in San Francisco Bay on November 24, "having been absent on this interesting but harassing service thirty-one days" (Pierce and Winslow 1969, 51).

Figure 11. Sketch of a stuffed Raccoon collected by Hinds along the Sacramento River during Captain Belcher's HMS *Sulphur* Expedition in 1837. (XF86t s593 p. 11, courtesy Bancroft Library, University of California, Berkeley.)

The sailors who accompanied the river trip vociferously complained about the "constant and harassing duties they had to perform and the caprice and malice of the officer commanding," whereas those who remained behind with the *Sulphur* felt that with the captain away it was "the most comfortable time ever spent on board." One evening's entertainment involved a vivid aurora borealis display extending in broad rays, unlike the merely shimmering ones seen earlier. Barclay had spent most of his time botanizing around Yerba Buena, except for a three-day excursion north to Sausalito, then on to Angel Island and east across the bay to Pinole. Also between November 7 and 15, he did some collecting upstream at the *Starling* anchorage. The weather had been beautiful, which the locals said was unusual during late autumn, the few rain showers coming in on southwesterly breezes. Captain Belcher opined that the San Francisco Bay climate was "one of the finest in the world."

On the final day of November the wind was up strongly from the northwest, the temperature had dropped to forty-two degrees, and fresh snow topped the highest slopes around San Francisco Bay. It was a good time to set sail for Monterey Bay. Eleven deserters remained behind, including a sergeant and a corporal of the marines. There were rumors that a renegade rifle brigade of "desperate ruffians and wild American hunters from across the Rocky Mountains," as well as ship deserters, was being assembled, with the possible intent of taking over Alta California from the Mexicans. There had already been an abortive revolution against Mexico by Alta Californians the previous year. It did seem suspicious that the *Sulphur*'s deserting sergeant had previously been a shipmate of the rifle brigade leader. According to Belcher's report, the old-time Spanish Yerba Buena residents surprisingly emphasized that "to Great Britain their hopes are directed."

The *Sulphur* and *Starling* arrived in Monterey on December 2, 1837, to discover that the French surveying frigate *Vénus*, which they had encountered at Oahu the previous July, had been anchored at Monterey from October 18 to November 14 and had depleted some of the supplies Belcher had hoped to obtain. The captain, Abel du Petit-Thouars, had planned to go on to San Francisco Bay earlier for water but decided to remain in Monterey, giving his crew a rest during their four-year surveying voyage. Fortunately, Belcher was able to scrounge what he called "table supplies" locally, and much of the rest he was able to obtain, at a price, from his friend the United States consul to the Sandwich Islands, John Coffin Jones Jr., who happened to be at Monterey "on a matrimonial expedition." While there Jones had bought a recently wrecked whaling ship and was shrewdly selling its contents. The most difficult necessity to obtain was fresh water. Midshipman Francis G. Simpkinson noted in his journal that the only thing nature had not supplied the Monterey Peninsula with was water, and "the country cannot be irrigated and sufficient water cannot be spared to supply a ship."

Barclay and Dr. Sinclair went ashore the first day in Monterey and proceeded over the hills to visit the mission, but with the lateness of the season they did not bother collecting any plants. Captain Belcher found the establishments at Monterey a great improvement over those at Yerba Buena and indeed better than when he had been here with the *Blossom*. All were delighted with Monterey, at the edge of a beautiful bay ringed by sand dunes and oak-and-pine-wooded hills beyond. There were herds of cattle grazing in the meadows, California Quail everywhere; and as Midshipman Simpkinson noted, the countryside presented "a picture of the prodigality of Nature never to be equalled by the hand of man." Here news reached the ships that King William IV had died in June, with the young queen Victoria succeeding.

Early on December 6 the *Sulphur* and *Starling* sailed south for San

Blas on the Mexican coast, the *Starling* in the lead. At San Blas they hoped to obtain more supplies and find letters from home. Neither wish was fulfilled, but at least they beat the *Vénus* into San Blas by one day.

Nearly two years later, the *Sulphur* and *Starling* returned to California after having surveyed around the Pacific. On the morning of September 20, 1839, the tops of trees came into view through thick fog, and then, within gunshot distance, the Russian settlement of Fort Ross appeared. Picking up two pilots, the British ships continued south about twenty miles to a safe anchorage in Bodega Bay, where there was a storehouse and residence belonging to the Russian colony. Leaving the *Starling* there to survey, Belcher proceeded to San Francisco Bay and his old anchorage off Yerba Buena. There he encountered the governor of Fort Ross, Mr. Alexandre, an old acquaintance from Sitka, who informed him of the strained relations between Russia and Britain. Within forty-eight hours the *Sulphur* returned to Bodega Bay on a quick survey. Though not venturing on to Fort Ross, Belcher had previously spied on it with his telescope and obtained further information on the Russian colony from a friend who lived at the fort.

On September 25 the expedition ships sailing into San Francisco Bay were delayed again by dense fog but finally reached their old anchorage. Previous surveying was quickly reviewed, and a supply of bullocks and vegetables was purchased. Returning south, the expedition stopped briefly in Monterey on October 5, completing observations by the next morning and sailing on to an Alta California port that was new for them, Santa Barbara. North of Santa Barbara near Point Arguello, Belcher noticed a scent in the land breeze reminiscent of a ship on fire. He conjectured that it might arise from naphtha (asphaltum) floating on the surface of the ocean, since he remembered that Vancouver half a century earlier had reported a similar smell here and had observed floating tar. It was at this spot that the major offshore oil field of Point Arguello was discovered in 1981. By evening of October 9 the *Sulphur* reached Santa Barbara, having made its way safely through the line of Giant Kelp. After a brief landing, the expedition continued on to San Pedro, arriving on the eleventh.

At San Pedro, Barclay and Hinds had time to scramble up steep cliffs rising above the little harbor. Some shell fossils, chert, and chalk were noted, as well as deep fissures in the earth presumed to be from frequent earthquake shocks. The two made the first botanical collections from this portion of California, including three species of buckwheat and several composites. But when the *Starling* set out to examine Santa Catalina

Island, neither Hinds nor Barclay was invited. The expedition paused down coast at San Juan Capistrano to survey the poor anchorage, below a rocky promontory noteworthy in the future as Dana Point. A party visited the mission, about five miles inland, reputed "to abound in garden luxuries, country wines, and very pretty damsels" fondly called Juanitas. On October 17 the ships reached the port of San Diego, again after sailing through extensive kelp beds, "three miles in length by a quarter broad," lying off Point Loma, which Belcher learned were sometimes driven into San Diego harbor by strong southwesterly gales.

The two naturalists were the second, after Paolo Botta in 1827, to have arrived in San Diego as part of an oceanic exploring expedition since George Vancouver came with Archibald Menzies in 1794, although a few individual naturalists had visited the area. The pair collected a variety of plants, many of them different from the ones that had been encountered farther north and having a definite southwestern affinity. There were three species of saltbush, including Fourwing Saltbush shrubs up to nine feet tall, collected in the local hills, and several euphorbias. Although no cacti were collected, they were common, especially what Belcher called "the Turk's head variety" (probably the California Barrel Cactus), "abominably abundant, very much to the discomfiture of the surveying parties." The party visited Mission San Diego de Alcalá, in ruins like all the other California missions that had been removed from the control of the Catholic church in 1834. Belcher prophetically concluded that "it is not improbable that the whole country will ere long either fall back into the hands of the Indians, or find other rulers."

Early mornings before breakfast in San Diego found the hunters out after cottontails and Black-tailed Jackrabbits, ducks, and *cordoneces,* Spanish for the "quail of California." When Midshipman Simpkinson first encountered these beautiful birds around Yerba Buena, he had written, "The partridges are I think the most beautiful species I ever saw. . . . I should like much to have possessed a pair of them stuffed; they would have been a good present to a sportsman in England." The expedition's fishermen, meanwhile, obtained a bounty of excellent fish for the table by seining in San Diego harbor; and Hinds, having become an enthusiastic all-around naturalist, busied himself collecting numerous seashells, usually by dredging. He obtained one "in five fathoms, among mud," and later described the attractive Hinds's Scallop.

During the five-day stay, a "tolerable" survey was made of the harbor, and the chronometers were checked. On the afternoon of October 22 the expedition departed from California for the last time, traveling southwest across the Pacific. In Fiji, Belcher's two surveying ships encountered the squadron of the United States Exploring Expedition, with its company of scientists. The *Starling* had just broken two rudder pintles (support pins) at a reef, and Belcher sent the *Starling* to ask if the Americans had any extras. The American commander, Charles Wilkes, himself came over with three pintles, and the two expedition leaders had a guarded conversation over dinner, both reluctant to share much information. Wilkes's junior officers heard comments about what a tyrant Belcher was, but they responded that he could not be so bad by comparison with "Commodore" Wilkes.

Belcher continued to the Asian mainland, where his vessels became involved in the British war with China (the so-called Opium War), finally returning to Spithead, England, on July 19, 1842. This was, by coincidence, the same month that the last of the United States Exploring Expedition ships pulled into New York Harbor, Lieutenant James Alden, captain of the USS *Porpoise*, having been delayed because he had to pick up five boxes of the American scientists' specimens in Rio de Janeiro. Alden wrote in his journal what may on occasion have occurred to Captain Belcher with respect to the activities of his own naturalists: Could the delay have been "for punishment? Ans. No. Scientific purposes. Bah!"

Back in England, botanist Barclay sent his collections to William T. Aiton, superintendent of the Royal Botanic Gardens, Kew. Aiton, in turn, sent at least 405 specimens collected in 1837 on to Robert Brown at the British Museum. Pursuant to instructions, Barclay had kept a numbered and detailed field record of all the plants he collected. William Jackson Hooker of Kew also received a partial set of Barclay's collections but without detailed notes, as well as Sinclair's few specimens. Hinds, on the other hand, noted only locations for the plants he collected, his specimens going to George Bentham for Bentham's large, at that time private herbarium. Eventually all the expedition specimens went to Bentham, who with Hinds produced *The Botany of the Voyage of H.M.S. Sulphur* in 1844, the botanical descriptions done by Bentham, with Hinds as editor. The *Sulphur* plant collections are now at Kew, including four thousand collected by Barclay.

The zoological collections were detailed in *The Zoology of the Voyage of H.M.S. Sulphur* (1844), edited by Hinds. John E. Gray, keeper of the zoological collection in the British Museum, wrote the section on mammals, including pertinent comments on California's Coyote, Raccoon, weasel, ground squirrel, jackrabbit, and cottontail, with five plates. John Gould, the "Old World Audubon," wrote the bird section, but unfortunately no California species were included. The surprising part of this report was the lengthy section by Hinds on shells, including a number of California species, both from the ocean and from up the Sacramento River. There were numerous illustrations of these, and a few of the scientific names that Hinds gave them are still in use. It has been said of Hinds that his attention to the small shells in a few genera exceeded normal conchological diligence for the period.

The year previous, 1843, Belcher's *Narrative of a Voyage round the World, performed in Her Majesty's Ship Sulphur* had appeared. In volume 2 Hinds added an important phytogeographical appendix, "The Regions of Vegetation; Being an analysis of the Distribution of vegetable forms over the surface of the globe in connexion with climate and physical agents." This included a lengthy discussion of the California region, emphasizing that while the botany of Northern California "had already been tolerably examined," for Lower California (or New California, as he called it), "we trod in no footsteps, as none had preceded us." This was not quite true with respect to Santa Barbara and San Diego. Hinds rationalized that "the great end and aim of geographic botany . . . is the naturalization of plants."

For Southern California, Hinds noted that despite the late season, "after a day's wanderings we return[ed] with a dozen or more different species in flower." Here were found many plants of Mexican or tropical form. He recognized that in semiarid regions there was a general lack of trees but there were numerous bushy composites, euphorbias, and cacti. Many of the plants were fragrant, and a number had milky sap. The cacti did not seem to occur farther north than San Pedro (34 degrees north latitude), although they seemed to range as far south as 42 degrees in Chile. He was struck by the fact that many genera such as *Lupinus, Vaccinium, Ribes,* and *Rubus* had representatives along the entire length of the Pacific Coast; that some species that were deciduous in the north were evergreen in the south; and that different species developed at different times of the year. Hinds's observations about western American plant ecology were ahead of his time, although he lacked the attention to detail shown by Barclay when it came to collecting specimens.

Barclay returned to South America as a horticultural collector after the *Sulphur* voyage and is presumed to have died in Buenos Aires while still a young man. Belcher continued his naval career, surveying the European Channel Islands and serving in Southeast Asia, where he even experienced a spirited combat with pirates. His last command, in 1852, involved a fleet of ships searching for Sir John Franklin in the Arctic from both the Pacific and the Atlantic oceans. Two of the ships actually completed the long-sought-for Northwest Passage, but did so north of the Arctic Circle. Because Belcher abandoned his ship in polar waters in 1854, he was court-martialed, though eventually acquitted. He died in 1877.

Hinds, the "amateur" naturalist, rose to the challenge of dealing with both flora and fauna. He had penetrated farther inland in Northern California than any naturalist before him, and had become more familiar with the Southern California coast than any previous naturalist; and with publication of his own perceptive considerations, he laid the foundation for phytogeographical studies not only of California but of other areas in the Pacific region. He wrote that "Her Majesty's ship *Sulphur* was the school in which I more particularly studied geographical botany." Richard Brinsley Hinds, who died in 1847, remains today a relatively overlooked natural scientist.

The French Visit Again

THE VOYAGE OF THE *HÉROS*, 1827–1828

Members of the scientific corps were to seize every
opportunity for making observations on hydrographic
and other special matters.

> —Orders for the *Héros* expedition (1827)

In September 1786, the two ships of La Pérouse's grandiose expedition
had sailed into Monterey Bay. In the intervening forty-one years, momentous happenings had beset France: the French Revolution had come
and gone; Napoléon had come and gone; the Louisiana Purchase had
come and France's hold on the American hinterland had gone. In late August 1827, just over four decades later, the initial fate of the La Pérouse
Expedition was finally revealed at Vanikoro in the Santa Cruz island
group, nearly five hundred miles southeast of Guadalcanal, with artifacts
and word-of-mouth verification of the expedition's two shipwrecks.

In that same year a French naturalist once more set foot in Alta California, Dr. Paolo Emilio Botta, from the French trading vessel *Héros*.
Botta, whose parents had emigrated from Italy to France, was born in
1805 and received his medical training at Hôtel-Dieu in Rouen from the
father of author Gustave Flaubert. Aboard the *Héros*, under the command of Auguste Bernard Duhaut-Cilly, the pleasant, easygoing young
Botta served as the ship's surgeon, and as naturalist he made collections

for the Musée d'histoire naturelle of Paris, especially birds and reptiles, as well as any plants that appeared interesting. This was essentially not a scientific expedition: the around-the-world trading and exploring venture was encouraged by French businessmen interested in an assessment of Pacific mercantile markets and California's natural resources, with an opportunity to peddle some merchandise ashore.

When a weeklong dense fog finally dissipated, the *Héros* entered San Francisco Bay, having sailed up from Baja California, on January 26, 1827. By the time she and her crew departed from Alta California on July 27, 1828, they had experienced the coastline from Fort Ross and Bodega Bay south to Santa Cruz, Monterey, Santa Barbara, San Pedro, Los Angeles, San Gabriel, San Luis Rey, and San Diego. The expedition sailed on into Mexican waters and eventually overwintered from late autumn into late spring in Peru, where there was much butterfly collecting. No earlier expedition had ever visited as many coastal areas of California as did the *Héros*.

The educated and articulate Captain Duhaut-Cilly—described as "a gentleman of pleasing personality and many accomplishments," literary and artistic as well as nautical—and naturalist Paulo Botta, sharing kindred interests in science, would often go into the field together on collecting junkets, with guns over their shoulders. In January and February 1827 the pair encountered along San Francisco Bay's sandy and muddy beaches myriads of beautiful shorebirds, while in inland woodlands they saw "several fine species of hawks," with the Yellow-billed Magpie, blackbirds, sparrows, and "several fruit-eating birds" in the understory. A hunting trip south down the peninsula garnered geese and a variety of ducks, as well as Great Blue Herons, the latter considered a delicacy by residents.

In the coastal heath the two observed "a pretty species of hummingbird . . . a little ball of glowing iron throwing off rays of sparks. When several of them light on the same branch, the Arabian amateur of marvels might take it for a bough covered with precious stones, as in a dream from The Thousand and One Nights." Botta's description captures the essence of the Anna's Hummingbird, one of the eye-catching residents of the coastal chaparral the year around. San Francisco is the type locality where Botta collected the original specimen on which French ornithologist René Lesson based his description in 1829. Lesson named the hummer after Anna, wife of Napoléon's bird-collecting general Massena, Duke of Rivoli. America's dashing, French-blooded naturalist John James Audubon, after meeting Anna Massena in September 1828, described her as a "beautiful young woman . . . extremely graceful and polite." Audubon, who never visited California, first became familiar with this western hummingbird based on a female specimen and its nest loaned to him by Thomas Nuttall, who visited the region in 1836. Audubon's 1838 painting of Anna's Hummingbird for his famous *Birds of America* folio depicted, in addition to Nuttall's female and nest, four male specimens the artist discovered in, of all places, an Englishman's collection of stuffed birds (figure 12).

On March 7 the *Héros* set sail from San Francisco Bay for Monterey Bay, with a brief stop at Santa Cruz. Officially this stop was to pick up grain from the mission for transport across the bay to Monterey, but in reality it was to vend the ship's merchandise. Surrounding the Santa Cruz cove and its stream was a pleasant woodland. As Captain Duhaut-Cilly glimpsed a sky-blue Belted Kingfisher on its perch above the quiet stream, a white egret, and a Cinnamon Teal, he anticipated that if "Dr. Botta were to repeat often his work of collecting birds skins from California," no peaceful avian scene would remain. Over the two days in Santa Cruz, Botta indeed "threw a little confusion into the habits of these poor creatures," and the captain had to admit that "I took part in this cruel aggression." Departure from Santa Cruz was hastened by the approach of a winter gale. Dr. Botta, ashore tending to a sick villager, had to be recalled to the ship by cannon fire. Arriving across the bay in Monterey, the Mission Santa Cruz grain was discharged and more merchandise sold. The main excitement, financed by a few piastres from the *Héros,* was a fight between a captured bear and a bull, scheduled at the presidio on Sunday after Mass. The bull was victorious.

En route south to Santa Barbara, the captain noted that smell in the air of petroleum that many earlier expeditions had experienced, and yellow and bluish patches of "bitumen" were seen dotting the water. At San Pedro, Botta, Duhaut-Cilly, and some of the other officers went ashore for a hunt. They encountered a "forest of mustard, with stems higher than a rider's head, and forming what seemed like dense walls on both sides of the way"—a forest infested with rattlesnakes. The plant, which can reach a height of twelve feet, was Black Mustard, earlier introduced from Europe throughout Alta California. The abundance of rattlesnakes dampened the hunters' enthusiasm until open grassland was reached, where several rabbits were shot and observations made of "a kind of owl [Burrowing Owl] that has its nest in the ground and lives there in families."

On April 17 the *Héros* was off for San Diego, where Botta collected,

Figure 12. Anna's Hummingbirds, painted in 1838 by John James Audubon (original number 420). The female specimen and nest were collected by Thomas Nuttall in California in 1836. (Courtesy collection of the New-York Historical Society, 1863.17.425.)

near the ship's anchorage, the type specimen of one of the Southwest's most distinctive birds, the Greater Roadrunner. An early scientific name applied to this unique bird was *Saurothera bottae.* Botta wrote that the "charia" is a "running bird to which is attributed the ability to kill snakes for food. It has a long tail which it raises to an almost perpendicular position. It seldom flies but runs almost as fast as a horse." A word phonetically similar to *charia* was commonly used by the Mexicans for the roadrunner. Its use supposedly dated back to a Franciscan monk who in 1790 described the "Churca" as a kind of pheasant that runs with four toes (two in front and two behind, characteristic of the Cuckoo Family).

For the Frenchmen in San Diego the chief source of amusement was hunting at Point Loma. Hardly had the shoreline been reached than bevies of California Quail took flight, while "hares and rabbits moved in bands across the fragrant and flowering fields that carpeted the slope of the hill." The hunters never went more than a hundred feet without encountering a Black-tailed Jackrabbit. "The mere difficulty of choosing a victim can be troublesome," and sometimes a single shot procured two victims.

After a brief side trip along the Baja coast and on to Mazatlán, the *Héros* was back in Alta California waters, during late spring and into midsummer of 1827 at San Diego, Mission San Luis Rey de Francia, and Santa Barbara. The ship stopped again in San Francisco Bay at her old anchorage near Yerba Buena by the early evening of July 17. Deciding to visit the last of the five established San Francisco Bay missions, Duhaut-Cilly invited Botta to accompany him to Mission San Francisco Solano in Sonoma with William Richardson, the presidio commandant's son-in-law, as guide and nine of the ship's crew. The *Héros*'s launch left at 4:00 A.M. on August 4, crossing north past Alcatraz Island, where a gunshot sent up a great cloud of pelicans "with a noise like a hurricane." Numerous deer were seen on the mainland, against a backdrop of dry grass slopes and scattered clumps of oaks. There seemed "few happier sites" than that of the Sonoma mission, but by this date the newest of the main mission chain was of "small account" despite its setting of mountains, streams, and bay. Duhaut-Cilly purchased all of the mission's tallow, made from Black-tailed Deer shot by some of the mission's young hunters.

The last week in August, the *Héros* again sailed south from San Francisco Bay, making brief stops along the coast. At San Pedro there was another inland trip, this time going beyond the pueblo of Los Angeles to Mission San Gabriel Arcángel, with its splendid vineyards and immense herds of livestock certainly the richest mission in all of California. While having tea at the nearby home of Don José Carrillo at the end of September, the party experienced a strong earthquake shock, causing everyone to rush out into the courtyard. A few days later when the *Héros*, delayed by threatening weather, finally weighed anchor at San Pedro near the end of the Palos Verdes Peninsula, a sudden landward gust sent the ship toward the rocks, dragging its anchor and brushing its keel. Only quick work averted a shipwreck.

After the *Héros* had reached San Diego and almost completed its provisioning of water and wood, Captain Duhaut-Cilly, returning on the evening of October 12 to the ship, was thrown from his horse. When he was gotten aboard, Dr. Botta discovered that Duhaut-Cilly had broken his collarbone. The captain and his crew were probably relieved to set sail on October 20, bound for a summer season of recuperation in Peru.

During the *Héros*'s second spring in California, the ship arrived in Monterey on May 3, 1828. By month's end she was on her way to the Russian colony north of San Francisco Bay, hoping to sell some of her cargo there. While she was lying several miles off Bodega Head searching for a safe harbor, three Alaskan baidarka boats approached, one of them carrying the Russian colony commandant, Paul Shelekhov, who not only got the French ship into Bodega Bay but made arrangements for Captain Duhaut-Cilly and Dr. Botta, accompanied by several Russians including the pilot, to take a horseback trip north some miles to Fort Ross, the Russian outpost. The men crossed the Russian River on a small baidarka arranged for by Shelekhov, while the horses swam across. Duhaut-Cilly likened the shaky baidarka experience to a crossing of the River Styx! The little party finally reached the fort on June 4 at eleven in the morning.

Shelekhov provided "the most refined hospitality" for the French party while at Fort Ross and showed them the agricultural and timber productions of the Russian establishment. During the tour, Captain Duhaut-Cilly was impressed by the lumbering of the Redwood, the *bois rouge* or *palo colorado,* and by the trees' immense size. One measured 230 feet from stump to crown, and another was 20 feet in diameter at the base. The captain noted that "one can figure the enormous quantity of boards that could be produced from a tree of such girth." It was most likely here among the Redwoods that Botta collected his second remarkable California vertebrate, a reptile related to the giant tropical boa constrictors. Botta's find, the tan-colored Rubber Boa, *Charina bottae bottae,* is only about two feet long and rolls over to play dead when confronted by something larger. Yet it possesses the pelvic girdle and reduced hind appendages of the larger boas. While Botta was busy collecting, Duhaut-Cilly made the most famous sketch of the Russian outpost, which was later published in his *Voyage autour du monde.* Among the few plants naturalist Botta collected in California were the California Buckeye and the Punch Bowl Godetia, now *Clarkia bottae,* first collected at Monterey. Both were initially named and described by Edouard Spach of the Paris Musée d'histoire naturelle.

At the beginning of his scientific career, Paolo Botta would have thought it difficult to become more famous than his father, Carlo Botta, who had been trained in medicine and was an eminent historian who even wrote an imaginative history of the American Revolution. The elder Botta's arrest as a spy in Italy encouraged his move to France, which led to service as a doctor with Napoléon's conquests in Italy and a stint as vice president of the French legislative assembly. Eventually he became a detractor of Napoléon, and ended as a rector of the academy at Rouen. But Paolo, later in his own life, became equally famous in a field not directly related to biology, archaeology. In 1842, while serving as French consul at Mosul in Mesopotamia (Iraq), he began excavations at Nineveh, and then at the palace of Assyrian king Sargon II nearby, receiving the first official permit for archaeological exploration in the Near East. Botta's classic four-volume *Monuments de Nineve* appeared in 1846–50.

Paolo Emilio Botta is seldom mentioned as a naturalist in California, despite his choice discoveries. He did publish "Observations sur les habitans de la Californie" in 1831, which was translated into Italian by his father in 1841 and eventually into English in 1952. On the other hand, in the *Oxford Companion to Archaeology* thirty-five lines are devoted to his biography, but with no reference to his natural history accomplishments; and his most famous collections—archaeological—reside in the Louvre, not in a natural history museum. Botta died in 1870 in Lebanon.

THE VOYAGE OF THE *VÉNUS*, 1837

Collect natural history specimens from the countries visited.

—Orders for the *Vénus* expedition (1836)

Duhaut-Cilly's *Voyage autour du monde* was published in 1834–35. The next year, almost a decade after the visit of the *Héros,* the French government announced that it was again time for "the presence of a national vessel on the western coasts of America" to inspire new "respect for the French Flag." Plans were laid in 1836 for the French frigate *Vénus,* under Captain Abel du Petit-Thouars, to embark from Brest for a Pacific expedition, primarily to investigate the whaling industry and protect French interests, but also to "acquire all possible information respecting the actual condition of the various countries visited." The "scientific observers" with the *Vénus* were French naval officers "who would make hydrographic and astronomical observations and collect natural history specimens from the countries visited." The *Vénus* surgeon was Adolphe Simon Néboux, and it was he, the captain, and Lieutenant Chiron who carried out most of the natural history work. The voyage track ran to Brazil, around Cape Horn, to Callao in Peru, to Hawaii, Kamchatka, and Alaska, and then down the coast to California, arriving at Monterey by daybreak on October 18, 1837, with gun salutes from the *Vénus* that were answered by an equal number from the Mexican shore battery.

The *Vénus* arrived in Monterey Bay none too soon, since many of its crew of three hundred, reportedly outnumbering the residents of Monterey, were suffering from scurvy. In a much appreciated act of hospitality, David Spence, a Scots merchant whose adobe cottage and store were the first on what would become Monterey's main street (Alvarado), provided an empty house as a hospital, as well as a site for the ship's observatory. All the scurvy victims were nursed back to health on an antiscorbutic diet, with the exception of one sailor, Durand, who was also afflicted with tuberculosis. When Durand eventually died, he was buried with military honors at Mission San Carlos.

Governor Juan Alvarado received the officers of the expedition "with his usual hearty politeness," and during the stay a grand ball was held for all. The *Vénus* needing to renew its supply of water, and a drought prevailing on the peninsula, Captain Petit-Thouars arranged to send a party in Captain William Sturgis Hinckley's Hawaiian ship *Kamamalou*, anchored in the harbor, up to San Francisco Bay on October 20 to procure water and vegetables. This expedition gave Lieutenant Chiron, hydrographic engineer de Tessan, and Cadet Mesnard a chance to reconnoiter the San Francisco port. Meanwhile, at Monterey provisions were procured and bread baking proceeded night and day, with flour obtained from as far away as San Juan Bautista.

Meteorological observations were carried out, variations of the magnetic compass needle were recorded, and a topographic survey of the harbor was accomplished through soundings. One day the captain carried his instruments up the western end of Jacks Peak. When the fog bank over the ocean began rapidly moving inland, he hurriedly gathered up his equipment, only to see the fog flow over a gap in the ridge below him and down into Monterey, as it still does today. Probably the greatest excitement in the bay itself was the frequent invasion by schools of migrating Humpback Whales on their way south to wintering-breeding grounds in Mexico, the cetaceans cruising among the anchored vessels, filling the air with a pervasive and nauseating aroma from their blowhole spouts.

On October 30, 1837, with fine weather, Spence led a hunting party from the *Vénus* on foot across to Carmel Bay and the mission, giving the scientific observers a chance to load up the one horse with their collections as well as with meat for the table. Among the animals were Blacktailed Deer, somewhat reminiscent of the Roe Deer of Europe; Blacktailed Jackrabbits; and a great number of California Quail, "a very pretty sort indigenous to this country." The commonly encountered "zorillo," a "curious little animal," elicited the comment that it "even sometimes lets itself be taken and petted." But soon the black-and-white mammal would spray a liquid "whose odor is so stinky and so acrid that it is impossible to bear it without nausea, and even vomiting . . . cloths often have to be burned!" Each expedition apparently had to learn about the North American genus *Mephitis* (skunk) on its own, there being none in Europe. The Spanish name *zorillo* is appropriate, however, because in many unforested regions of Africa lives the Striped Polecat or Zorilla, a distant relative that resembles the Striped Skunk and delights in spraying enemies with vile fluid from its anal glands.

The mission itself was a disappointment, already falling into ruins. It was a sad Father José Maria del Real who showed the Frenchmen around, taking them to what remained of the once fertile mission garden, which used to extend down the slope to Rio Carmelo. Now there were hardly any signs of cultivation, and the river itself was just a trickle. Gone, too, were the hordes of mission Indians, who had been set at liberty after the secularization of the missions in the mid-1830s. But a few ethnological tidbits were picked up. The Indians reportedly used rock-strewn steam-bath *temescals* (huts) to improve their well-being, and arrows tipped with poison oak sap to discourage their enemies. The visitors were shown the portable organ given to the mission president by George Vancouver, and Father Real told them about the disappearance of the memorable La Pérouse painting. He went on to relate that Captain Beechey, during the visit of the *Blossom*, had offered a considerable sum for the painting but had been emphatically turned down.

On this cross-country junket the men recorded observations of the Monterey Pines, "whose tops were lost in the sky and great evergreen oaks, scattered without undergrowth, resembling a park." The gray-green sprays of vegetation festooning the pines were recognized as lichens. When the naturalists visited the rocky shoreline of the peninsula, they found large limpets, the most impressive being the Giant Keyhole Limpet nearly four inches in length. The giant abalone shells they described, one of them a foot long and nine inches across, were undoubtedly those of the Red Abalone. The bay waters themselves teemed with fish and a great variety of shellfish.

By mid-November, loaded with water and all the provisions they could procure, and with observations for the mapping of the bay completed, it was time for the *Vénus* to sail. Captain Petit-Thouars expressed satisfaction that the prolonged stay had been of value not only to the health of the crew but for scientific reasons as well, the trips on land having been especially "profitable for our own natural history collections." When a calm prevented the *Vénus* from getting under way, whaleboats from about the harbor congregated to tow the ship, "proof," as the captain reflected, "of favorable regard" shown by the inhabitants and whalers toward the French voyagers.

On November 14, as the *Vénus* passed the rocky point below the presidio they had christened Point Venus, the presidio battery fired eleven guns, which were answered by the ship's cannons. By ten that evening, on the open ocean en route to Mexico, the final farewell came with a spectacular aurora borealis in the northwestern sky. Up the coast at San Francisco Bay the aurora borealis was also observed by the crew of Captain Belcher's HMS *Sulphur*. Belcher's two exploring vessels would arrive in Monterey on December 2, 1837, only to discover that the supplies they had hoped to obtain were scarce because of the visit of the *Vénus*. Belcher had been in the Sandwich Islands at the same time as the *Vénus*, and mutual friendships had arisen. The HMS *Sulphur* and HMS *Starling* would again ride at anchor with the *Vénus* after leaving Alta California, this time in San Blas Bay along the Mexican coast. At San Blas it was the *Vénus* that was short of supplies, and once again a number of her crew suffered from scurvy. The French and English were cheered by their reunion, although the lack of spirits, especially wine (unavailable at San Blas), dampened the party atmosphere.

Eleven volumes on the voyage of the *Vénus* would be published by 1855. There was much material on the whaling industry, but also included was a review of the expedition's natural history findings: zoology, botany, hydrography, and geography. Although the coverage for California was limited, there were plates portraying the California Gray Squirrel, "Lynx Cat" (Bobcat), "zorillo" (Striped Skunk), and memorable portraits of a Monterey Indian man and woman and a Mexican man and woman in local costume. Four hundred species of shells were procured during the entire voyage of the *Vénus*, undoubtedly many from the western coast of the Americas, and many mollusk drawings without descriptions were included in the zoological atlas for the expedition. But few specific localities were mentioned, although some of the shells undoubtedly had been found at Monterey.

The collection at Monterey of a Northern Caracara, a large bird of prey, represented the first record for California. But the two most intriguing natural history finds by the *Vénus* involved a Swallow-tailed Gull reported from California and a tarweed (a composite) reported from the Galápagos Islands. The *Vénus* zoological atlas contains a

painting of the gull based on a specimen supposedly taken in the road-
stead at Monterey. For years this was the solitary record for California
of a species that specifically breeds on the Galápagos Islands and along
the Colombian coastline and winters at sea off northwestern South
America. Meanwhile, the *Vénus* specimen of Coast Tarweed, named by
Néboux in 1840 and reportedly collected in the Galápagos Islands, actually occurs only along the California coast, including Monterey, where
it can be encountered blooming in autumn even today. As a California
botanist conjectured in 1935, the records must have been confused, as
the tarweed was really from Monterey and the gull from Galápagos. Yet
in June 1985 a Swallow-tailed Gull was observed twice in Monterey Bay
and photographed, and still another was reported west of the Farallon
Islands on March 3, 1996.

The United States Exploring Expedition in California, 1841

To extend the bounds of science and to promote knowledge . . .

—U.S. Congressional Act (1836)

On the evening of September 28, 1841, an overland party from the United States Exploring Expedition made camp for the twenty-fourth time, in a broad valley on the mountainous 42nd parallel right on the border of Alta California. Several of the group's riding horses had given out by this time. Everyone in the expedition had experienced intermittent fever, but botanist William Rich and Midshipman George W. Colvocoresses were still extremely ill with high fever and chills, for which quinine proved of little help. "Colvo" indeed was so delirious that his gun was taken away from him. That night the temperature dropped well below freezing.

Next morning the men were greeted by a beautiful distant view to the south of snow-capped Mount Shasta, and they were concerned that a column of smoke might be the Rogue Indians "telegraphing" the Chasty (Shasta) Indians about the Americans' approach. The explorers had been warned that the Shasta Indians planned to kill all the intruders. Later in the morning Lieutenant George F. Emmons scaled an isolated flat-topped butte, rising "like a tower" atop a summit to the east. Taking the latitude, he verified that the United States Exploring Expedition was indeed in Alta California. Emmons's Peak of the report is today known as Horn Peak.

The Yankees, though a bit behind other countries, were not to be outdone when it came to an around-the-world exploration. In some measure it all started with the imaginative John Cleves Symmes Jr. and his provocative Circular Number 1, dated at St. Louis on April 10, 1818. In this epistle, given wide circulation, Symmes proclaimed that the earth was hollow, with holes at each pole and "habitable within . . . stocked with thrifty vegetables and animals, if not men." What was needed was an expedition made up of enthusiastic explorers that would set sail for, say, the North Pole, disappear into Symmes's Hole, and navigate to the civilized core.

By the mid-1820s there were nine "memorials" before Congress asking the government to underwrite such an expedition, all of which died a legislative death. Then a newspaperman, Jeremiah Reynolds—who doubted the earth was hollow but did think a polar expedition would be good for the nation—came to Symmes's support. By 1828 Reynolds succeeded in getting a resolution passed by the House of Representatives to dispatch a public ship for oceanic exploration purposes. The USS *Peacock,* a rebuilt sloop-of-war, was made available, and a young navy man, George F. Emmons, came aboard to begin service with the *Peacock* that would last for thirteen years. Three natural scientists were also initially assigned to the ship's complement, including as assistant naturalist Titian Peale, a son of the great painter and museum man Charles Willson Peale of Philadelphia. To Titian would eventually fall the task of organizing a scientific corps for what would become known as the United States Exploring Expedition.

Ten years later, on August 18, 1838, under the command of Lieutenant Charles Wilkes, the United States Exploring Expedition—with five navy ships headed by the *Vincennes* and the *Peacock,* plus a large and noteworthy scientific staff informally called the Scientifics—sailed out of New York harbor. The expedition cruised southward beside North Africa, recrossed the Atlantic, and traveled around South America, where one of the ships went off on its own; the other four sailed over the Pacific past Samoa and north to the Sandwich (Hawaiian) Islands. Then it went south past Fiji to Australia, circling into Antarctic waters (never finding Symmes's Hole), up to New Zealand and north for a stop at the Fiji Islands. There the Wilkes expedition encountered Captain Edward Belcher and his two British ships, which had been on their oceanic survey for five years, including time along the West Coast. Some of Belcher's officers prophetically warned the Americans about the dangerous waters at the entrance to the Columbia River. From Fiji, then, Wilkes returned northeastward to Hawaii.

Here the remaining expedition ships separated. The *Peacock,* with most of the Scientifics and their collections, and the *Flying Fish* continued a mid-Pacific survey, while the flagship *Vincennes,* with Wilkes in command, and the gun brig *Porpoise* proceeded to the mouth of the Columbia River, arriving on April 28, 1841. So ominous did the river entrance appear, especially in the current bad weather, that Wilkes with his two vessels continued north toward Puget Sound, anchoring on May 2 just within Juan de Fuca Strait, exactly forty-nine years to the day after Vancouver had anchored at the same spot. Surveying commenced in the strait and sound, and Wilkes sent several parties ashore for inland excursions.

With Wilkes now were only three of the Scientifics: naturalist Charles Pickering, botanist William D. Brackenridge, and artist Joseph Drayton. Pickering, a descendant of early-day American statesman Timothy Pickering, with an M.D. from Harvard, had been a librarian and later curator at the Philadelphia Academy of Natural Sciences. Pickering, wrote his academy colleague John Cassin, "justly ranks with the most eminent of American Zoologists." Brackenridge, for years a head gardener in Scotland and at the Royal Botanic Garden, Berlin, had come to the United States and worked as a nurseryman in Philadelphia. He was appointed to the expedition as assistant botanist and horticulturist on the recommendation of Secretary of War Joel Poinsett, who felt there should be someone aboard who could take care of live plants collected by the expedition. Philadelphian Drayton was an experienced artist and engraver.

"Commodore" Wilkes, as Lieutenant Wilkes now titled himself, had already begun formulating plans for a party with naturalists to proceed inland just to the boundary of Alta California. On May 11 Wilkes left the *Vincennes* at Fort Nisqually and headed south with a horseback party including his purser, R. R. Waldron, and artist Drayton, to Astoria near the mouth of the Columbia River, hoping to meet the *Peacock* and the *Flying Fish.* The ships still not having appeared, Wilkes proceeded upriver to visit the Hudson's Bay Company's Fort Vancouver and the American settlements in the Willamette Valley. In early July, still with no word of the missing vessels, Wilkes returned to Fort Nisqually, having left Waldron behind at Astoria with orders to organize the proposed inland expedition.

On July 18, 1841, the *Peacock*—commanded by the well-liked lieutenant William Hudson and accompanied by the smaller *Flying Fish*—had completed its oceanic surveys and arrived within five miles of the Columbia River mouth, where the river's current could already be felt in the ocean. In early afternoon the *Peacock,* working into the Columbia River, struck a sandbar and within three hours was sinking, despite continuous pumping and throwing nonessentials overboard. By midnight the order came to abandon ship. Six boats were put in the rough water, the sick in the first; then the Scientifics, charts, books, and ship's papers; then the marines; and finally the crew and officers. Lieutenant Emmons led the boats safely into the shelter of Baker Bay on the south shore. Unfortunately, many of the expedition's valuable worldwide collections were lost with the sinking *Peacock*.

Emmons on July 22 received Wilkes's orders for the proposed inland trip from Purser Waldron, with Emmons assigned to command. The next noon Emmons and his small party, including zoologist Peale, botanist Rich, and artist Alfred Agate, were on their way up the Columbia from Astoria in two large canoes, bound for Fort Vancouver; later they moved south up the Willamette River to a staging encampment near the American settlements.

On July 27, when Wilkes learned of the *Peacock*'s wreck, he abandoned further surveying in Puget Sound and sailed south with the *Vincennes* and the *Porpoise* to ascertain the situation, as well as to survey the Columbia and Willamette rivers himself. By now he had decided to enlarge Emmons's overland party and have it proceed cross-country all the way to San Francisco! Emmons received notice of this important change in plans on August 4, and Wilkes's two vessels arrived off the Columbia River three days later. Before Wilkes himself commenced surveying the Columbia with the *Porpoise*, he assigned Lieutenant Cadwallader Ringgold to command the *Vincennes*, with the *Peacock*'s captain, Lieutenant Hudson, as a supernumerary and accompanied by naturalist Pickering and artist Drayton; they were to sail south to San Francisco Bay. There Ringgold was to survey the bay and the Sacramento River, then meet Emmons's inland expedition coming down the Sacramento River. At Astoria, meanwhile, a replacement vessel for the *Peacock* was procured and rechristened the *Oregon*.

The *Vincennes* arrived at Yerba Buena on August 14, 1841, and shortly moved to a better anchorage north across San Francisco Bay at Whaler's Harbor (Sausalito Bay). An observatory was set up on shore, and Lieutenant Ringgold with some officers, sailors, and Pickering, bearing provisions for thirty days, set off in six boats with an Indian pilot for the survey. They planned to meet Emmons's overland party at Sutter's Fort (on the site of present-day Sacramento) or beyond. Lieutenant Hudson remained behind in command of the *Vincennes*.

As Ringgold's boats made their way eastward, the San Pablo Bay hillsides were yellow with autumnal wild oats, while upriver the luxuriant tule marshes extended far inland on either side. The expedition narrative relates that "the scenery was very much admired, and Mount Diavolo [Diablo] . . . added to its beauty." The boat trip to Sutter's Fort, which was reached on August 23, took three days, including time for surveying, but Emmons's group was nowhere to be found. However, Ringgold's party did encounter the obstreperous sergeant and marine guard who had deserted from Captain Belcher's HMS *Sulphur* in 1837. Ringgold's men continued up the Sacramento River, still hoping to find the overland travelers, but beyond Sutter Buttes the snags and sandbars discouraged further travel. Ringgold took the latitude, and the next day the party turned back.

During Ringgold's upstream journey, while the Americans were meeting with a group of Indians on shore, Pickering's bowie knife and pistol were stolen for the second time during the expedition. Unfortunately, the Indian who was held hostage after the robbery managed to escape. Numerous large mammals were seen, including Lynx, herds of Tule Elk, Black-tailed Deer, Gray Foxes, Wolves and bears, and an aggressive Badger unsuccessfully pursued by Pickering. Later, when the unarmed Pickering was strolling along the river's edge looking for plants, a huge Grizzly Bear rushed out of the bushes at him. He ran back toward camp, yelling, but before his companions could muster, the bear had disappeared. The officer who chased a skunk was not so lucky. After the unfortunate incident, he was forced to abandon all his clothing. No mammals were collected during this trip, and the major botanical collection was an epidemic of poison oak rashes.

Meanwhile, Emmons and his overland party were still up in northern Oregon, experiencing their own adventures. The men had set up a temporary camp across from Reverend Jason Lee's Methodist Mission Station (north of present-day Salem, Oregon). After five uncomfortable weeks of waiting, reorganizing, indecision, and illness, on September 7, 1841, the overland contingent of the United States Exploring Expedition finally headed south up the Willamette River toward Alta California. It was an interesting assortment of people who constituted this now large company. There were three naval officers. In charge was Lieutenant Emmons, born in Vermont but now a resident of New Haven, Connecticut, who had been associated with the exploring expedition since its inception. He would return to California during the Mexican War, serve in the Union navy during the Civil War, and eventually become an admiral (1872). Passed Midshipman Henry Eld Jr. had earlier been with Wilkes on coastal surveys and, by coincidence, was a New Haven neighbor of

Emmons. Another passed midshipman was George W. Colvocoresses, who had been born in Greece but grew up in Vermont.

Emmons's inland party, which now numbered thirty-nine people, included the majority of the expedition's scientific corps, those from the shipwrecked *Peacock* and assistant botanist Brackenridge from the *Vincennes,* who had just arrived a few days earlier. Missing were Pickering and Drayton, who were with the *Vincennes,* and two of the Scientifics who had earlier left the expedition, conchologist Joseph Couthouy and philologist Horatio Hale. Not since La Pérouse had there been such a large number of scientists attached to a West Coast exploring expedition, and in particular representing such an array of scientific disciplines.

The senior scientist accompanying Emmons was zoologist Titian Peale. As a teenager Peale had served as assistant naturalist and artist with America's first scientific expedition to the West, which in 1820 proceeded to the Colorado Rockies under the leadership of Stephen Harriman Long. Subsequently Peale accompanied expeditions to Florida and South America, worked as manager of the Philadelphia Museum, and in 1828 as assistant naturalist had begun planning for the proposed United States Exploring Expedition.

Asa Gray, the rising star of American botany, had originally been appointed botanist for the expedition but tired of waiting and resigned. The position was filled by his assistant, William Rich. Rich, who had once served in Congress, had been recommended by the secretary of the navy. In some circles he was viewed as a political appointee, and during the Wilkes expedition, John Torrey later claimed, he reputedly did "nothing but smoke cigars & tell Stories." In fairness to Rich, he felt himself an amateur botanist and never wanted to replace Gray. After the expedition he failed to prepare the botanical material for publication and shortly disappeared from the scene. However, six-footer William Brackenridge, the horticulturist, proved to be an able and hardy field botanist. To him fell the responsibility of transporting and caring for the live plants gathered during the expedition, and he would work on the final botanical report.

Alfred Agate, employed as the botanical draftsman, had originally specialized in miniatures, living in New York City after several years' painting experience in France and Italy. He made many fine natural history sketches and paintings during the expedition. Agate died of tuberculosis in 1844 at thirty-three, before readying many of his illustrations for publication. James Dwight Dana, trained as a geologist and chemist at Yale under Benjamin Silliman, had taught mathematics on

shipboard for the navy, and then returned to Yale as Silliman's assistant, writing in 1837 what remains famous as *Dana's Mineralogy*.
Dana was the first professional geologist to accompany any West Coast
exploring expedition, although some earlier naturalists included geology among their interests.

The planned route south (almost that of today's Interstate 5) would
be up Oregon's Willamette River, over a divide onto the Umpqua River
(where David Douglas had botanized a few years earlier), then south to
the Rogue River. When that river swung east, the expedition would continue southeast on the route of the Oregon-California Trail and through
the Siskiyou Mountains (south of present-day Ashland, Oregon) to the
border of Alta California. The three-week trip to Alta California was
beset with troubles: rugged travel, recalcitrant horses, illness, broken instruments, threatening Grizzly Bears and Indians. The naval officers
were irked by the naturalists' wandering off in search of specimens. In
return, the scientists ridiculed the navy men who on dry land continued
to ply their nautical vocabulary, steering east by south, and making two
knots an hour. Brackenridge summed up by remarking that a sailor on
shore was like a fish out of water.

During this first inland exploration of Northern California by natural scientists, zoologist Peale, botanist Brackenridge, and geologist Dana
kept detailed journals for the route south to Sutter's Fort and eventually
on to San Francisco. Peale's first notation after crossing the border into
Alta California was about the large numbers of migrating Western Bluebirds and his collecting seeds of two kinds of evening-primroses.

"Commodore" Wilkes had originally anticipated that the overland
party would be encountering the boats from the *Vincennes* somewhere
up the Sacramento River by the last day of September. In reality, on that
date Emmons and his men were still miles from even the headwaters of
the Sacramento. They were instead traveling south through the so-called
Boundary Mountains toward the Klamath River, the distant vista of
Mount Shasta becoming lost in the haze from Indian fires. The night's encampment was on the north shore of the Klamath River, giving the sick
a brief chance to recuperate. In the stream light-colored salmon were
abundant but, in Brackenridge's view, "not very delicate to the taste."
Peale, as was his habit, wandered away, spotting a Great Blue Heron and
several mallard ducks, as well as abundant California Quail; but ammunition was too precious to expend on any of these.

The geological terrain was exciting to Dana as the party moved south
between the Cascade volcanics to the east and the much older Klamath

Mountains to the west with their scattered conifers, oaks, and chaparral. The season was ebbing, but Brackenridge noted a few plants, including sunflowers, Skunkbrush, Wax Currant, Antelope Bush, and, near the evening campsite in the wet meadow along the river, a plant known as quamash (Common Camas) that Lewis and Clark had collected and Frederick Pursh had described, and whose bulbs the Indians relished. Although Brackenridge did not know what the quamash was, he did recognize that it resembled a European lily.

Next morning, crossing the Klamath, the men spent five hours traveling south across what became a broad, undulating prairie, with a stupendous view of Mount Shasta ahead. The day afterward, in six miles the trail crossed an east-west ridge of sandstone, then went onto dry plains—"miserable country"—with black volcanic blocks scattered over hillocks up to two hundred feet high. There were intermittent salt-encrusted depressions—"salt marshes"—where Brackenridge collected some alkali-loving plants, including a mustard, Short-podded Thelypodium. Zoologist Peale recorded a dozen Pronghorn on the prairie, mammals he had painted in Colorado's shortgrass prairie during the 1820 expedition to the Rockies when he was only twenty-one years old.

On October 3, 1841, the expedition left the Shasta Valley and passed through mixed coniferous forest on the west side of Mount Shasta, or "Shasty Peak" as Dana called it. Close by was the steeply sloping cone of Black Butte (conspicuous beside today's Interstate 5), presumed by Dana to be "evidently an extinct crater, made of ejected cinders," looking fresh enough to be recent. Black Butte is actually an extrusive volcanic plug dome 9,360 years old. Dana regretted that time did not permit a closer examination of Shasta. No smoke wisped from this fourteen-thousand-foot glaciated volcano, but he had been told of a hot springs on the east side where one could boil eggs. From the expedition's evening encampment there was a beautiful sunset, the rays of the setting sun, pink through the smoky atmosphere, illuminating the impressive mountain.

Artist Agate took time to sketch the imposing peak. Meanwhile, everyone wondered what had become of the threatened massacre by the Shasty (Shasta) Indians. Six had, in fact, shown up before dinner back at Little Shasta River, dressed in deerskins and armed with arrows tipped with volcanic glass (obsidian). They demonstrated their skill with bow and arrow, traded fish and their bows and arrows for buttons and knives, received a few presents, and drifted away into the night. Ruminated Brackenridge, "I would as soon . . . have a musket discharged at me as an arrow from one of these Indians Bows." Peale, trying to get the Indians settled down for a portrait, finally showed them a painting of his mother, thoroughly frightening them since they had never seen a painting before. Deciding that Peale was the medicine man of the party, they feared he had put some spell on them.

The party commenced down the "Destruction River" (a headwater of the Sacramento), southeast of present-day Mount Eddy. On October 5, the men were suddenly set upon by a group of Indians. On the run, Brackenridge grabbed what he later considered the trip's most exciting find, a handful of three-foot-long, weird-looking leaves and a long seed stalk from near a spring. Further examination reminded him of the eastern pitcher plant, with hooded, insect-eating leaves. A decade would pass before another specimen, this time in bloom, was collected hereabouts in May 1851 by army surgeon Dr. G. W. Hulse. In 1854 the new species California Pitcher Plant, known locally as calf's-head or cobra lily, was described by John Torrey and named *Darlingtonia californica,* after William Darlington from West Chester, Pennsylvania. He was "one of the worthiest of American Botanists" and an individual who had promoted the U.S. Exploring Expedition.

For the next week the expedition traversed old mountain masses, a trek Dana described as "now up a rocky steep several hundred or perhaps a thousand feet high, and then to the bottom of the craggy valley," the slopes covered by coniferous forest interspersed with stands of oak and brushland. Today, 29,500 acres of this complex canyon countryside have been inundated by the manmade Shasta Lake (1945).

Although the transit through this mountainous terrain was trying, with packhorses falling into creeks and soaking supplies of sugar and tea, not to mention gunpowder, there were things for naturalists to exclaim over. For dramatic geological scenery nothing could surpass, as Dana wrote, "the magnificent spectacle of needle peaks and lofty pinnacles of granite [granodiorite]," visible to the west from the encampment on the night of October 4 (now Castle Crags State Park). Also, the route provided a botanical and zoological harvest despite the late season. Botanist Rich himself was still too ill to be interested, but Brackenridge and Peale reckoned that there was "a greater variety of singular plants than we have yet seen." Brackenridge collected more than forty different species, including Western Redbud, Hollyleaf Redberry, California Coffeeberry in fruit, the lavender-blue-flowered *Penstemon azureus,* scarlet-flowered Fleshy Sage, and the striking California Fuchsia. Meanwhile, zoologist Peale observed what he thought were two new marmots (probably Fish-ers, a large relative of the Marten), some small birds unfamiliar to him, and, most impressive, a number of California Condors. The next day he was entranced by an American Dipper that he watched bobbing up and down on rocks at the stream's edge, then actually walking on the bottom where the transparent water was half a foot deep, and popping back to the surface without any apparent effort. He became so engrossed that he forgot to collect the bird.

Dana filled his field book with descriptions of the complicated geology of the Klamath Mountains' ridges and canyons. Some outcrops were almost vertical, others dipped sixty-five degrees to the northwest; flat-topped river terraces were twenty feet high. The rocks were slates, schists, and shales, granite and rhyolite and gneiss, hard sandstones made up of tiny quartz grains almost too small for the hand lens, conglomerates and limestone and puddingstone, river deposits and glacial deposits, copper-blue serpentine-colored exposures and igneous dikes and quartz veins, hornblende crystals three inches long . . . but nary a fossil. For Dana, traveling through perhaps the most complex geological venue of the entire Wilkes expedition, interpretation of what he saw was nigh impossible. A century and a half later, geologists would write of this region, in phrases foreign to the young Yale geologist, that "the presence of accreted terranes, oceanic crustal slivers and subduction complexes has made the Klamath region the center of many debates on California and global structural history" (Norris and Webb 1990, 128).

On October 10, several hours after leaving camp, the members of the expedition, with the mountainous country finally behind them, looked south across gravel slopes of oak and "scraggy" bushes into the broad Sacramento Valley, with volcanic Mount Lassen and the snow-crested Sierra to the east. They were about 160 miles from Sutter's Fort at the junction of the American and Sacramento rivers. There were numerous Indians in the countryside, entirely naked and very friendly rather than threatening, who were busy gathering and drying acorns, which, as they demonstrated, were then pounded between two stones into meal, mixed with Pacific Madrone berries, and made into tarts. At the first valley encampment along the now broad Sacramento River (figure 13), Brackenridge noticed what he thought was a new species of sycamore, unaware that the Western Sycamore had recently been described and named by Thomas Nuttall during his 1836 visit to coastal California. Peale saw his first Yellow-billed Magpie, also collected by Nuttall. Peale had painted the West's other magpie, the American Magpie, east of the Rockies while with Major Long's 1820 expedition.

Figure 13. "Encampment on the Sacramento." The inland party of the U.S. Exploring Expedition, October 1841, by Alfred Agate. (From *Narrative of the United States Exploring Expedition*, vol. 5.)

After the expedition's guide momentarily got the party lost, his Indian wife found the trail by recognizing a hollow tree where she had cached a stone on a previous trip. Continuing south, they encountered increasingly verdant countryside, with open groves of oak. Brackenridge made a quick survey of the two species of oak, the Valley Oak and the Interior Live Oak, calculating "20 good trees to an acre." Game abounded, with elk (wapiti), Pronghorn, Black-tailed Deer, Wolves (keeping up a perpetual howl at night), Grizzly Bears, and evidence of Beaver activity on the side creeks. There were numerous California Condors, Turkey Vultures, Common Ravens, flocks of Red-winged Blackbirds, Great Blue Herons, Great and Snowy egrets, Sandhill Cranes, Mallards and other ducks, Double-crested Cormorants, and curlews, but fewer plants. On the evening of October 12, after five Grizzlies were killed, Dana jocularly observed that the supper consisted of "bear stewed, bear roasted, and dough fried in bear grease."

On October 16, 1841, the "Sacramento Bute" or "Three Buttes" (now Sutter Buttes), off to the southeast of the encampment, broke the monotony of the flat river valley. Dana made a quick, accurate outline sketch, from North Butte to Goat Rocks, commenting that the "Bute was evidently a volcanic cone." Where the group crossed the prairie to the base of the buttes, there was evidence that during the rainy season the plains had been under ten to fifteen feet of water. The travelers entered the craggy buttes through a gap in the volcanics, traversed an inner valley, and finally bivouacked at the eastern base of the highest ridge. During the night hurricane winds collapsed several of the tents, making everyone uncomfortable but particularly artist Agate and Midshipman Colvocoresses, who had relapsed with the fever. Dana, intrigued by the local geology, wrote in his field book about crystals of feldspar, hornblende, and quartz in rocks he called "trachytic porphyry," an acceptable approximation of the existing andesite. But the complex origin of the Sutter Buttes would remain unsolved for almost a century and a half.

Two days later, at the edge of the Sierra Nevada, the party crossed the Feather River just above its junction with the Sacramento, almost losing the guide in quicksand. Human bones strewn in all directions marked an Indian village site where supposedly some fifteen hundred Indians had perished from an epidemic in the summer of 1833. On the nineteenth the American River was forded, and there the overland expedition was met by Captain John Sutter (originally Johann August Suter), who had settled on a large Mexican land grant he called Nueva Helvetia in 1839 and just recently had purchased the Russians' Fort Ross establishment north of San Francisco. Inviting the travelers to his home for refreshments, Sutter recorded in his diary that "a party of Comodore Wilkes' Exploring Squadron, arrived from Oregon by land, consisting of the Scientific Corps. . . . I received them as well as I could and then the Scientific Corps left by Land for San Jose."

It was the same day the overland expedition encountered Captain Sutter (October 19) that Wilkes arrived at San Francisco Bay from the Northwest in the *Porpoise,* with the newly acquired *Oregon* appearing a few days later. With Wilkes's arrival, all onshore and shipboard frivolities among the *Vincennes* crew ended, and no longer could Lieutenant Hudson maintain his "hotel afloat" for visitors to Whaler's Harbor. When Wilkes took exception to Ringgold's survey of San Pablo Bay, Ringgold's crew was sent out again for nine days in the launch to repeat the survey. Although Ringgold was piqued at Wilkes, he did profit from this repeated survey, because when the gold rush exploded, his survey chart became a best seller.

From Nueva Helvetia (that year the site of Sutter's Fort, and later Sacramento) Lieutenant Emmons, the assistant surgeon Dr. John Whittle with the sick, and Dana were to proceed downriver in Captain Sutter's boat to join the squadron in San Francisco Bay. But they were able to leave a day earlier (October 21) in the boat of an American who had some Indian trappers with him. Downstream they encountered Ringgold and the survey launch, which they boarded for the remainder of the trip. Geologist Dana was intrigued by the bluffs of alternating soft sandstone and clay at Carquinez Strait (where today Interstate 80 crosses the river). What particularly caught his eye were the sandstone layers, ranging from one inch to four feet thick, at one point dipping to the west seventy degrees and then within a few hundred yards changing to a dip of thirty-five degrees, while nearby the layers were vertical. Dana remarked that this was an area of numerous faults (indeed, the Calaveras Fault crosses here). On October 24 the launch with its party of tired, dirty, bearded explorers reached the *Vincennes* at Whaler's Harbor.

At Sutter's Fort the thirteen remaining members of Emmons's overland party—including Midshipman Eld, botanist Rich, who was now feeling better, and Titian Peale, with an intelligent young Spaniard, Romero, as guide—set out about noon on horseback the same day as Emmons's group, bound cross-country for San Jose. Brackenridge also decided to accompany the land party, it "suiting my pursuits better," since they would travel through a region never before visited by naturalists. The route was straight down the San Joaquin Valley, the weather warm and dry. The first night's camp was at a pond within a grove of oaks. A horse guard was set during the night against mischievous Indians, who loved to steal horses, but no Indians were encountered. The next day, in crossing a rapid stream, Peale spotted the largest grizzly track he had seen during the entire trip. Thirty-two miles were made this second day, and at the night's stop Peale added a Raccoon and a Coyote to his collection, as well as a few of the hundreds of Mallards in the many small ponds around the camp.

About eleven o'clock the next morning, after the party crossed the San Joaquin River, the guide called a halt for the day at a large lagoon filled with ducks and geese, and on the shore Sandhill Cranes and Pronghorn. He warned that on the west side of the San Joaquin Valley there would be a very long journey without water or grass. Midshipman Eld, "wishing," as Brackenridge surmised, "to bring into play the very little power with which he was invested," immediately took command of the group. Disregarding the guide's advice, he ordered them to push on. They finally ended up in cold foothills at the edge of an impenetrable swamp, after more than ten hours and about thirty miles without water, wood, or grass for the horses, and having seen only one miserable habitation. The landscape was cluttered far and wide with the carcasses of cattle and horses, slaughtered only for their hides and tallow. The expedition horses' hooves crunched on the scattered skeletons.

Though deprived of lunch and dinner, the men did enjoy their beds among the rushes, too tired to be kept awake by the raucous squawking of the marsh birds. All day Brackenridge had collected only four plants, and Peale had seen only a few herds of wild horses, Pronghorns, and elk. Peale had seen elk in Colorado in 1820, but he was unaware that these in California's Central Valley were Tule Elk, a different subspecies. When the party started to cross the Diablo Range on the way to San Jose through marine sedimentary outcrops northeast of Mount Hamilton, in some of the hillocks "organic remains" abounded, with fragments of enormous oyster fossils, *Inoceramus,* scattered about on the ground.

Next morning, the travelers had a gluttonous Sunday breakfast of Canada Geese and a slaughtered bullock before heading into more mountainous terrain. About noon the men came upon a large Indian village where both men and women were busy grinding acorns into meal, many of them having been granted permission by the mission at San Jose to return to their old village for the acorn harvest.

The reception at the San Jose mission proved less than hospitable, even though Eld had letters of introduction. Luckily J. Alexander Forbes, a Hudson's Bay Company agent, invited the party to his nearby home, and the naturalists slept in beds for the first time since leaving Fort Vancouver in midsummer. The next morning, continuing toward San Francisco Bay, the group visited the Santa Clara mission, where no one could believe the group had come all the way overland from the Columbia River. The request to see the mission garden was curtly dismissed by the superintendent, who apparently assumed this rough-looking lot were renegade trappers wanting to eat the garden's fruit. But botanist Rich, surprising his companions by cajoling in good Spanish, not only arranged a garden visit for the group, but everyone ended up eating fresh fruit and drinking celebratory toasts with freshly fermented pear spirits. Before dark the land party reached San Francisquito (near present-day San Mateo).

It was October 26, and all the men were eager to get back to the expedition ships. But the horses "being much broke down," in Brackenridge's words, and forage abundant in the area, it was decided to lay over for a day. This gave Peale and Brackenridge a chance for some leisurely fieldwork. The botanist busied himself collecting more plants and seeds. Many here near the coast were new to him, since so far he had traveled only inland. Among others, the oak he found with prickly leaves was the Coast Live Oak, the shrubby orange monkeyflower he thought might have been the one introduced into the East by Thomas Nuttall was actually Bush Monkeyflower, the *Rhamnus* was the common California Coffeeberry, and the forty-foot-tall tree with fruit like a damson plum was the aromatic California Bay. Peale, who had been shooting Canada Geese, took some of the fruit pits and packed them with dry clay in a goose skin, hoping to plant them when he got back to Philadelphia. He had also been pursuing the abundant bayside birds, adding to his list Green-winged Teal, Bufflehead, Common and Hooded mergansers, Snow and Greater White-fronted geese, a few American Kestrels, numerous Short-eared Owls and Northern Harriers, and large flocks of Red-winged and Brewer's blackbirds. Peale even caught some Sandhill Cranes with a lasso!

By dark it started raining heavily, and even in their tents the men were thoroughly soaked by water running beneath their floorless shelters. Toward morning it cleared, and the group hoped to make it to San Francisco Bay soon. But again the horses gave out, only twelve miles from Yerba Buena, so the party had to bed down in wet blankets with a supper of recently killed geese and ducks, since the last of the beef was gone. En route Eld ran into one of the sailors from the *Vincennes,* who was collecting wood, but so bearded and uncouth did Eld appear that the sailor spoke to him in broken Spanish, mistaking him for a Mexican.

Shortly after sunrise on Thursday, October 28, with the weather better, the excited company was off, pausing briefly to pay their respects to the padre at the dilapidated Mission Dolores. When they reached Yerba Buena, they chanced to look through a window of the only tavern and saw members of Wilkes's crew playing billiards. It was indeed a delight to find one of the expedition's boats awaiting them at the landing. Before nightfall they were welcomed aboard the *Vincennes,* anchored off Sausalito in company with the *Porpoise* and the *Oregon.* The overland party had taken a full month longer to reach San Francisco Bay than Wilkes had expected.

Titian Peale, William Brackenridge, and William Rich had traveled overland about eight hundred miles from Fort Vancouver to Yerba Buena, the most extensive inland excursion by naturalists on the West Coast to date. Their colleagues, James Dwight Dana and Alfred Agate, who had returned by boat from Sutter's Fort to Yerba Buena, had gone almost as far.

Finally all members of the United States Exploring Expedition were once again united. The good news was Wilkes's decision not to send a large party, including the Scientifics, overland from California to the East Coast! At three in the afternoon of October 31 the flagship *Vincennes* upped anchor and with its consorts headed on the ebb tide for sunset at the mouth of the bay. Wilkes's ships were still attempting to depart at 2:00 A.M., waves now as high as thirty feet sweeping the deck of the *Vincennes* and even carrying away boxes of California plants. At 4:00 A.M., James Allshouse, one of the marines, was caught by a "combing roller" and died later that morning. At last the Wilkes Expedition was off for the Sandwich Islands again, then to the Philippines, Java, Capetown, and home to New York City.

The scientists, except for Rich, were occupied for years preparing the many volumes of publications on the expedition. Unfortunately many copies were destroyed in an 1856 fire, along with many of the collections. However, a number of specimens did become, as Emmons had hoped, the nucleus for the United States' first national museum, the Smithsonian Institution. Brackenridge, on the return of the expedition, was charged with the erection of a greenhouse in Washington, D.C., and the care of the living plants brought back. This collection eventually evolved into the National Botanic Garden near the capital. During the worldwide expedition Brackenridge had collected some ten thousand plants, the preserved specimens becoming the nucleus of the U.S. National Herbarium. For many years he served as superintendent of public grounds in Washington, later establishing his own nursery in Baltimore. He continued to be active in horticultural affairs and writing, was editor of *The American Farmer,* and inspired the establishment of the Maryland Horticultural Society. William D. Brackenridge died in 1893. For some obscure reason Wilkes had always referred to him as J. D. Brackenridge.

Charles Pickering, who died in 1878, spent the last fifteen years of his life producing a voluminous (1,200-page) *Chronological History of Plants.* James Dwight Dana served as a distinguished geology professor at Yale from 1849 until his death in 1895, the same year as the fourth edition of his classic *Manual of Geology.* He was one of Wilkes's last Scientifics. Before James Marshall actually discovered gold in the American River in January 1848, Dana had conjectured that there might be gold flakes found in that river and others arising in the Sierra Nevada. In 1863 members of the California Geological Survey—Whitney, Brewer, and Hoffmann—honored Wilkes's geologist by naming an impressive peak near Yosemite Mount Dana.

Titian Peale, in a long and productive life spanning from 1799 to

1885, would continue his pursuit of natural history and painting, working at the Patent Office until 1873 and becoming one of the nation's early photography enthusiasts. Singularly, Peale had been the youngest naturalist with the United States' first inland western scientific expedition in 1820, and the oldest with its first around-the-world oceanic scientific expedition in 1838–42. It was Peale who provided the apropos postscript for the United States Exploring Expedition: "to elevate our country in the rank of Civilized nations . . . exploring . . . a west coast for our people who followed there, and now count by millions."

PART TWO

The Early Peripatetic Naturalists

STIMULATED BY EXCITING REPORTS FROM the oceanic exploring expeditions, individual naturalists, eager to share in the adventure, began appearing in Alta California, arriving by ship and even overland. Some received encouragement from governments, institutions, and/or patrons, while others came solely because of a personal fascination with the diversity of nature. Before arriving in California, Scotsman David Douglas thoroughly explored the Northwest, and Englishman Thomas Nuttall with zoologist John Kirk Townsend were the first naturalists to cross the continent from Pennsylvania to Oregon. In California, Irish botanist Thomas Coulter was the first naturalist to cross the Colorado Desert to the lower Colorado River, while Nuttall's youthful American protégé William Gambel first explored the Southwest, crossing the Mohave Desert to reach the coast. German Karl Hartweg was the first to penetrate the northern Sierra Nevada, energetically collecting plants. Two of these naturalists died during their travels, while Nuttall won a niche in California literature. Among the men's collections were Gambel's Quail, Coulter's Matilija Poppy, Bristlecone Fir, Wrentit, Coast Silk-tassel, Coulter Pine, Yellow-billed Magpie, Western Sycamore, and Scarlet Fritillary.

David Douglas

Don David El Botánico

Early as was my arrival on this Coast Spring had commenced.
—David Douglas to William Hooker (1830)

On December 22, 1830, the Hudson's Bay Company brig *Dryad,* from the Columbia River, came to anchor in Monterey Bay, Alta California. Undoubtedly the passenger most eager to get ashore was a short, sturdy Scotsman. Baggage that accompanied the chap—nine gallons each of Madeira wine and of brandy, two Barnagore silk handkerchiefs, a large moose skin, one Jew's harp, and 150 Spanish dollars—provided little clue to his vocation; but the package of one hundred dried plant specimens suggested a botanist. David Douglas indeed was a plant collector, underwritten by the London Horticultural Society and on this trip also by the British Colonial Office, the Zoological Society of London, and the Hudson's Bay Company.

By this time Douglas was no stranger to either plant collecting or North America. In 1823 the youthful gardener from the Glasgow Botanic Garden had been sent to the East Coast in search of new fruit trees and oaks, visiting New York City and Philadelphia and junketing inland through upstate New York into Canada, west to the Detroit River, and then back to New York. Many were the dignitaries whom he met: former New York governor DeWitt Clinton, who was a fellow of the London Horticultural Society; Dr. David Hosack of the New York Horticultural Society; and Professor John Torrey of Columbia College. Douglas received a gift of Oregon-grape seeds from a Lewis and Clark plant cultivated at the Linnaean Botanic Garden on Long Island, toured Peale's Philadelphia Museum, and had his coat, vasculum, cash, and botany book stolen by his assistant in Canada. But most memorable of all, he was introduced to the notable Thomas Nuttall at the Philadelphia Academy of Natural Sciences. The two visited the famous Bartram Gardens, and botanist Nuttall enthralled Douglas with accounts of his western frontier travels.

In April 1825, again working for the London Horticultural Society, Douglas arrived in the western wilderness aboard the *William and Ann* at Fort Vancouver on the Columbia River. The ship's surgeon-naturalist, Dr. John Scouler, was a friend and fellow botanist from Glasgow days. Once ashore the pair immediately collected their first plant, Salal, with its flowering stalks of delicate pink bells. Douglas, whom friendly Indians called the Grass Man, then embarked on an exploration of the Northwest that extended into 1827 and encompassed 6,037 miles, with his belongings usually in a pack on his back and in a handheld kerchief, his precious seeds protected in a tin box. Among his finest discoveries were the tallest pine in the world, the Sugar Pine, and the conifer that bears his name, the Douglas-fir (whose species name, *Pseudotsuga menziesii*, honors Douglas's friend Archibald Menzies). In mid-March 1827, rather than return to Britain by Hudson's Bay Company ship, Douglas, the better to collect, chose to traverse the wild country from western Canada across to Hudson Bay.

Back in Britain Douglas received an enthusiastic reception, both for himself as an intrepid wilderness explorer and for the botanical treasures he had brought back. He reportedly introduced more plants into England than any earlier plant collector, and it was said that sales of the Red-flowering Currant by the London Horticultural Society more than covered the entire expenses of his trip. The most memorable visits Douglas paid in England were to Archibald Menzies's home in the outskirts of London, where the two Scotsmen shared their equally adventurous northwestern experiences. The high moment was when Dr. John Scouler showed up late one afternoon—"The three North West Americans under the same roof," as Douglas described the scene to William Hooker—and the trio dined that evening as the guest of Joseph Sabine, honorable secretary of the London Horticultural Society, at his club.

By October 31, 1829, Douglas—with his faithful traveling companion, Billy the little Scottish terrier—was once more bound for the Pacific Northwest under the auspices of the London Horticultural Society. Arriving at the Columbia River on June 3, 1830, he was soon traveling and collecting inland, frequently with Hudson's Bay Company's George Barnston, an amateur botanist with whom he had enjoyed botanizing on his previous visit. But by late July his thoughts turned to Alta California.

As Douglas wrote to his mentor William Hooker at the Glasgow Botanic Garden, "If I can venture thither in safely by land I will do so; if not I shall go by sea to Monterey." Toward the end of November, having recovered from an intermittent fever that killed twenty-four of the Hudson's Bay Company personnel, Douglas finally departed in the company brig *Dryad* for California. He had already sent three chests of seeds in a vessel bound for England, adding a small bundle of special and exceedingly beautiful botanical treasures, including, as Douglas noted, "by far the finest," the Noble Fir, in whose forest he had sojourned for three weeks.

Now at Monterey just before Christmas, Douglas was eager to explore Alta California and collect some of its diversity of plants. His first find was choice, the Fuchsia-flowered Gooseberry. Douglas, seeing it in bloom among the understory of the oak woodlands, rightly praised it as "a flower not surpassed in beauty by the finest Fuchsia." But when it came to getting away from Monterey, Douglas immediately encountered problems. The first deterrent to his botanizing involved government instability. Mexico had achieved independence from Spain in 1821, but the interim years had not been tranquil. By the time Douglas arrived, revolts had already taken place, and more were in the offing. In the Mexican province of Alta California Governor José Echeandía, at odds with the government, had resigned but continued to be an agitator. When the new governor, Manuel Victoria, showed up a month after the arrival of the *Dryad,* it took weeks of interviews between the naturalist and the suspicious military politician before Douglas was able to obtain a travel permit. For all Governor Victoria knew, the naturalist might be one of those foreign renegades hoping to overthrow Mexican control.

Thanks to a letter of introduction from Captain Beechey, Douglas was comfortably housed at Monterey, as Beechey had been for a time, with the community's influential English merchant William Hartnell, now a naturalized Mexican citizen. The Hartnell home, newly refurbished and a showplace, stood upslope half a league south of the bay, at the edge of the Monterey Pine forest overlooking the little capital. Douglas stayed in one of several guest rooms in a new wing of the comfortable adobe–mud brick residence, La Casa Arnel, the unofficial British embassy.

A family member at the house with whom Douglas became good friends was Hartnell's nephew Juan de la Guerra y Noriega, who had been educated in England. Among the Guerra heirlooms today is the small dictionary Juan borrowed from Douglas's traveling library, *El Curso completo, un diccionario universal de agricultura.* Douglas had been studying this during his 1823 visit to the East Coast. Shortly before Juan's unexpected death in 1833, he gave a rare possession to his sister Angustias, a small painting of La Pérouse's reception at Mission San Carlos in Carmel, the very picture that Beechey in 1827 had sought, unsuccessfully, to purchase.

While Douglas was initially restricted from traveling beyond the Monterey Peninsula, he busied himself during January making astronomical observations and investigating the local area's flora. Always interested in conifers, he became familiar with the abundant Monterey Pines. He recognized the Monterey Cypress as a cypress and placed that name on what would become the famous ocean promontory, Cypress Point at Pebble Beach. On the same day that Douglas first saw the Fuchsia-flowered Gooseberry, he also discovered "a humble but lovely plant" he came to call "the harbinger of California spring," the Baby-blue-eyes, now named *Nemophila menziesii* after its collector. In the Hartnell household it was called *azulejita,* the "little blue one."

No long-distance travel permit had yet been issued by mid-February 1831, but Hartnell did receive permission to take Douglas with him on a business trip to Mission San Juan Bautista early that month. In the foothills near that settlement Douglas rediscovered a pine with gray-green foliage and large cones that he named in honor of his English mentor Joseph Sabine (now *Pinus sabiniana,* the Foothill Pine). His previous specimen, collected in 1826 in southern Oregon, had been lost when he had to swim a river. In his dispatch to Sabine, which included seeds, cones, and needles as well as a description of the new pine, Douglas expressed the hope that Sabine's pine would "exist and flourish when we shall cease to be when we shall be gone forever." After San Juan Bautista, Douglas with an armed mission escort was conducted up to the Santa Clara mission. The party then traveled southwest toward Santa Cruz, where Douglas encountered the Redwoods, "the great beauty of California," remembering that Menzies had seen them there too. For Professor Hooker he collected Redwood specimens and seeds, but these were later apparently lost. Douglas estimated that some of the trees he saw were more than three hundred feet high, one measuring thirty-two feet in circumference three feet above the ground.

When Douglas returned to Monterey in early March, he found a new guest at the Hartnell residence, a woman at least in her forties dressed in a gray nun's habit. This was none other than Doña Concepción, who in 1806 at the San Francisco presidio had met and fallen in love with Count Rezanov. Still unaware of Rezanov's death and hence still awaiting his return, she had never married. Instead she had taken religious vows and had most recently been residing in Santa Barbara. She was godmother to the Hartnells' first son and now had come to Monterey to help Mrs. Hartnell with the birth of another child, eventually one among twenty-five! Doña Concepción Argüello—"Chonita," as she was affectionately called—and Douglas quickly struck up a friendship, the serious Scotsman enjoying Chonita's teasing ways. It was she who gave the Hartnell-household name of "Saint Francis" to Douglas because of his love for flowers.

About mid-April 1831, Douglas finally received his long-distance "passport," giving him permission to stay and travel in northern and central Alta California for six months, a period later extended to a year. But he was warned not to make any sketches or plans of Mexican fortifications. The naturalist was not tempted. After all, the fort at the Monterey presidio had only seven old cannons on a rotten platform.

Douglas quickly gathered together his astronomical instruments and set off southward on El Camino Real, the road connecting the chain of Alta California missions. This collecting trip would take him, in a month's time, as far as Santa Barbara. Traveling on horseback with a small caravan of packhorses and extra mounts and guided by mission dragoons, Douglas proceeded to Mission Nuestra Señora de la Soledad, followed by a delightful week in the Santa Lucia foothills at Mission San Antonio de Padua with Father Pedro Cabot, a sophisticated and thoughtful host. Cabot had served at San Antonio for almost thirty years, including when the mission had been severely damaged in the 1812 earthquake. He showed Douglas around the pleasant San Antonio River valley with its backdrop of coastal mountains, some of the wooded peaks rising to over a mile in elevation. There were extensive orchards and agricultural fields tended by mission Indians; and scattered throughout the oak-savannah pastureland were some six thousand cattle, ten thousand sheep, and one thousand horses. Cabot likely told Douglas that the special incense aroma permeating the mission buildings was pitch from a unique conifer—*Incensio*—collected from high in the surrounding mountains.

After San Antonio, Douglas went on to Mission San Miguel Arcángel. Monterey merchant Hartnell early in his trading career had contracted for all of that mission's tallow and hides. Douglas continued south to the missions of San Luis Obispo, Purísima, and Santa Ines, finally arriving at Santa Barbara in mid-May. Here he was entertained by Hartnell's father-in-law, Don José de la Guerra y Noriega, who was commandant of the presidio as well as father of four beautiful daughters. Douglas simply had to write to a friend that in Santa Barbara "the ladies are handsome," brunettes with "dark fine eyes" and good teeth who were best of all "very amiable." Attending to botanical business, Douglas remembered that the London Horticultural Society had directed him to travel into Baja California, and he himself hoped to visit the Colorado River desert, but Mexican officials informed him that he could continue no farther south.

Douglas was back in Monterey by late June 1831, having discovered something about the vernal climate: "Nothing can exceed the beauty of a California Spring," but "the plants bloom here only for a day," and by June "every bit of herbage is dried to a cinder." In late July he set off from Monterey again, this time for Northern California, with the optimistic intent of continuing on to the Umpqua country of southern Oregon, thus completing his West Coast collecting circuit that he had begun in 1826. He skirted San Francisco Bay, going on to the last of the northern missions, San Francisco Solano in Sonoma, where he asked Padre Buenaventura Fortuni if he could obtain soldiers and guides for his projected trip to Oregon. Padre Fortuni, demurring, sent him on to the Russian outpost at Fort Ross, where he spent only one day. At last realizing that about three hundred miles of unexplored wilderness still separated him from his destination in Oregon, Douglas reluctantly turned back toward Monterey by way of Mission Dolores, past fields of wild European oats around the Santa Clara mission, and across to San Jose.

Early August was very hot in San Jose, at the southern end of San Francisco Bay. Despite the silk handkerchiefs draped around his face and neck, the noontime siestas, and the chilled water from his canvas waterbags, Douglas suffered, although his recording of a temperature of 129 degrees was probably exaggerated. In a letter to Professor Hooker he complained, "How I lament the want of such majestic rivers as the Columbia." Douglas was impressed by Padre Narciso Durán at the prosperous San Jose mission, which was becoming famous for its clear brandy distilled from the extensive vineyard's grapes. Douglas had arrived during the annual slaughtering of hundreds of cattle for their hides and tallow, and he and Durán rode their horses atop a hillside where they could watch the vaqueros at work.

Durán supposedly arranged for Douglas to join a party of soldiers and converts on a foray into the hills to capture runaway and wild Indians. The naturalist tarried behind at Mount Diablo for some plant collecting, and among his specimens was a new species of mariposa lily, the rare

Mount Diablo Fairy Lantern, restricted in its distribution to that mountain area and today the logo for the Jepson Herbarium at the University
of California, Berkeley.

Douglas was again in Monterey by the last week in August. During
the coming autumn two pleasant surprises awaited him. On November
23, 1831, Douglas excitedly wrote to Professor Hooker that he had received a communication two days earlier from the Russian governor of
Alaska, Baron Wrangel, informing him that his proposed journey across
Siberia had been approved: "What a glorious prospect!" Douglas discussed the announcement with Cyril Klebnikoff, head of the Russian-
American Fur Company, who occasionally visited Hartnell, and received
further encouragement. The only problem was that Douglas must reach
the Russian-American Fur Company in Sitka, Alaska, by May 1832,
where he would be picked up by a Russian warship and transported to
the Siberian coastline. Douglas anxiously awaited a Sitka-bound vessel
to put in at Monterey, only to find that the regular Sitka transports never
proceeded farther south than Fort Ross. He then decided to take any ship
heading north, perhaps to Fort Vancouver on the Columbia River,
where he would be closer to Alaska.

In anticipation of his imminent departure, the naturalist determined
to pursue his collecting close to Monterey. He was now such a familiar
figure on the Monterey Peninsula that he was known as Don David El
Botánico. Also, ever since he had successfully set the broken arm of
young William Davis (who had fallen into the hold of the *Louisa* when
returning from Honolulu), Douglas had been called upon for doctoring.
As the weeks went by, he finally rationalized that "I continued to consider California *as still new to me,* and set to work a second time finding
new plants, and drying better specimens of those which I formerly possessed." For his old colleague Dr. Scouler he gathered together a collection of bones and an intact Sea Otter. Out of curiosity, he began searching again for the nest of the giant California Condor to determine the
color of the great bird's eggs.

The second surprise of the fall was encountering a fellow British
botanist, Dr. Thomas Coulter, late in November. Coulter had come into
Monterey Bay by ship about the beginning of October 1831, from Mexico, where he had initially been employed by the Real del Monte mining
company as a medical attendant and then as a mine manager. But in his
spare moments he was at work on a Mexican flora. Now he was planning to collect plants in Alta California, especially in the southeastern
desert country, for the great Swiss botanist Augustin de Candolle, with

whom he had spent several very productive years in Geneva. Shortly after his arrival in Monterey, Coulter had undertaken a fast inland trip as far south as Mission San Gabriel, to plan for his extended excursion into the desert in spring.

After meeting the Irish botanist, Douglas excitedly wrote to Professor Hooker that "I do assure you, from my heart, it is a terrible pleasure to me thus to meet a really good man, and one with whom I can talk of plants." Coulter and Douglas were together in the Monterey area into March 1832. Undoubtedly during the winter months the two often went out in the field together. They enjoyed hunting and fishing excursions, and Douglas considered Coulter to be an unsurpassed fisherman, adding, "besides being a beautiful shot with a rifle, nearly as successful as myself." As Coulter discussed his plans for the extended trip to the Colorado River and the Alta California desert, Douglas, remembering his promise to the London Horticultural Society, initially considered going along.

But Mexican affairs of state intervened. In January 1832, there was a threatened revolt against the provincial government in Monterey. Douglas and Coulter joined Hartnell's La Companía Extranjera (The Company of Foreigners) together with about fifty other Monterey residents, under command of loyalist captain Agustín Zamorano, to provide guard duty for the capital city.

By late March Coulter commenced his desert excursion, not returning until mid-July. Since no northbound ship had yet arrived, Douglas decided to travel south himself, into the wilderness of the Santa Lucia Mountains. He likely went down El Camino Real from Monterey and into the hills surrounding Mission San Antonio de Padua, more than a fifty-mile journey. There he at last saw an *Incensio,* the Bristlecone Fir, the rarest and most beautiful of American firs, and in the same vicinity the Sugar Pine, which he had first discovered in central Oregon's Umpqua River country. He noted that the fir was "never seen at a lower elevation than six thousand feet above the level of the sea [Douglas used his barometer], in latitude 36 degrees, where it is not uncommon." Today these conifers—together with the heavy-coned Coulter Pine, which Douglas also collected—still occur sparingly on secluded higher slopes and canyons in mountainous country north and west of the present-day Fort Hunter Liggett Military Reservation. Douglas collected the type specimen of the fir, and Coulter the type specimen of the pine, but at the time neither collected the seeds.

Sometime during early 1832 another naturalist showed up in Monterey, this time a German from the Berlin Museum. Ferdinand Deppe was in western North America for about six years, much of the time in Mexico. But he apparently made numerous trips north into Alta California, on one of them arriving in Monterey overland with a pack train from Baja California. Although he collected a few plants, Deppe was primarily interested in obtaining mammals and birds for the museum. His most noteworthy specimen, one that the Berlin Museum had especially desired, was a California Condor, eventually mounted and exhibited, becoming the fifth condor specimen in Europe. When Martin Lichtenstein prepared his scientific report on the bird, an engraved portrait of Deppe's specimen was included. Sometime before Douglas departed Monterey for the last time in August 1832, he met Deppe, noting that "the man seems to have been a backslider as far as botany was concerned." Dr. Coulter also met Deppe and reported that the German seemed to be dividing his time between natural history and mercantile pursuits. Deppe returned to the California coast several times after Douglas departed, and in the Sandwich (Hawaiian) Islands he would meet zoologist John Kirk Townsend, who had accompanied Thomas Nuttall across North America in 1834. While zoologist Deppe's condor received considerable publicity in Europe, botanist Douglas's serendipitous find at this time went essentially unnoticed until much later. In the roots of a Douglas-fir that Douglas collected in the Santa Lucias and packed to send to England were flakes of California gold.

Douglas finally gave up waiting for a northbound ship and instead sailed out of Monterey in August 1832 in a small American vessel bound for the Sandwich Islands. He had been in California for twenty months. As was his custom, he had kept a detailed diary for William Hooker. Unfortunately, on June 13, 1833, on the Fraser River in Canada, Douglas's canoe was demolished in rock-festooned rapids, resulting in the loss of all his belongings including the irreplaceable diary for the California interlude. But most of his California collections, except for a few sent through Mexico, successfully made their way to England. He had wisely divided shipments of plants into two equal bundles, transported separately.

At the end of his first twelve months Douglas calculated that despite drought conditions he had collected about five hundred species. Subsequently about 150 more were added, especially some of the annuals, and eventually more than two hundred species were growing in the London Horticultural Society garden. Among the collections there proved to be at least sixty plants new to science, including some unique beauties: the yellow-flowered Bush Poppy, the Mount Diablo Fairy Lantern and other mariposa lilies, the eye-catching penstemon called Scarlet-bugler, the golden-yellow Lindley Blazing-star, and the unusual Coast Silk-tassel, *Garrya elliptica,* a shrub with tough evergreen leaves and long, pendant male inflorescences. This last plant represented a new western American plant family. Douglas named it after Nicholas Garry, deputy governor of the Hudson's Bay Company, in appreciation for his "kind assistance." When a Coast Silk-tassel finally bloomed at the London Horticultural Society garden in 1834, it created quite a floral stir. Raved Professor Hooker about these shipments from Douglas in Alta California: "I think I scarcely ever in a collection of such an extent saw so much that is new and rare." Today, there are more than sixty California natives named in honor of Douglas.

In the Sandwich Islands Douglas learned that the London Horticul-tural Society had temporarily fallen on evil times as a result of extrava-gances and poor management, with debts exceeding assets by over three thousand pounds. Even King George IV was in arrears by five hundred guineas in his promised support of the society's garden. Joseph Sabine, criticized for administrative negligence, had resigned as secretary, and in part as a consequence Douglas resigned his own position on September 9, 1832. After only two months in Hawaii he returned to the Pacific Northwest, where as a free agent he continued his fieldwork, weather permitting, and began seriously pursuing plans for a return to Europe by way of Siberia. On March 20, 1833, he actually commenced the first leg of this enterprising journey, heading northwest through Canada. But by August, having been beset by very discouraging travel experiences—hos-tile natives, a challenge to a duel, ill health including failing eyesight, and finally the canoe wreck on the Fraser River—he returned to Fort Van-couver and then back to the Sandwich Islands, with a final visit to Cali-fornia along the way.

Douglas took passage again on the *Dryad,* leaving the mouth of the Columbia River on October 18, 1833. There were tempestuous autum-nal storms southward along the coast, and the *Dryad* arrived in San Francisco Bay on November 4 "nearly a wreck." Bad weather prevented a speedy departure, so Douglas set up his tent on Yerba Buena Hill (now Telegraph Hill). He provided to the new governor, José Figueroa, who was currently visiting the settlement, a list of the latitude and longitude of all the California missions he had visited, and made a trip north across the bay to Whaler's Harbor to examine an old Indian site with its huge midden of mollusk shells. There was no opportunity for botanizing here

Figure 14. Entrance to the David Douglas Garden, Perth, Scotland, with the Douglas-fir inscription slab and representative Douglas conifers and shrubs. (Photo by R. G. Beidleman.)

nor time to revisit Monterey, so Douglas wrote to the Hartnells, thanking them again for their "multitude of kindnesses" and sending greetings to his many Monterey friends, "to whom I have the distinguished honor of being known to."

As the *Dryad* sailed out of San Francisco Bay bound for the Sandwich Islands at the end of November 1833, David Douglas bade his final farewell to Alta California. On January 28, 1834, following in the footsteps of Archibald Menzies, he stood atop the summit of Hawaii's Mauna Loa at 13,679 feet. Writing his last letter to Professor Hooker in May, he concluded, "May God grant me a safe return to England." On July 12, while on a fern-collecting trip high on the slope of Mauna Kea, Douglas stumbled into a deep bull-pit trap and was gored to death by a trapped bullock. Billy, his Scottish terrier, was found standing guard at his owner's bundle of possessions nearby. Billy would be returned alone to England. When the tombstone for Douglas at Honolulu's Kawaiahoa Native Church was later engraved, the final Latin words were *Victima scientia* (victim of science).

In Perth, Scotland, nigh to the Tay River and not far from the towering Douglas-fir at Scone Palace grown from a David Douglas seed in 1826, is the landscaped David Douglas Garden, established in 1962 under the auspices of the Royal Horticultural Society of Perthshire with the permission and cooperation of the Perth Town Council to form a living memorial to David Douglas. Herein, on a Douglas-fir slab with a living backdrop of Douglas conifers and shrubs, is a large plaque (figure 14) reading, in part,

> Erected by the lovers of botany in Europe in memory of David Douglas. A native of this parish who, from an ardent love of science and desire to promote the improvement of botany visited the unexplored regions on the banks of the Columbia and southward to California whence he transmitted a great variety of the seeds of valuable trees and flowering plants, adapted to the climate of Great Britain and who, after devoting ten years of the prime of his life in adding to the arboretum and flora of Europe, suffered an accidental and lamented death in one of the Sandwich Islands on the 12th of July, 1834, in the 35th year of his age.

On an adjacent plaque are reminders of Douglas floral introductions from the New World: Coast Silk-tassel, Black Hawthorn, Douglas-fir, Salmonberry, Big-leaf Maple, and Oregon-grape, as well as the Pink-flower Currant, Salal, Noble Fir, Pacific Ponderosa Pine, red-barked Pacific Madrone, the lavender-and-white Bird's Eyes, Swamp Lupine, the Baby-blue-eyes representing Douglas's harbinger of spring, and the state flower of the Golden State, the California Poppy.

Thomas Coulter

First to the California Desert

Eminently calculated to work, full of zeal, very amiable, and I
hope may do much good to science . . .
—David Douglas to William Hooker (1831)

While a young man, Thomas Coulter developed an obsession with
lizards and snakes, keeping them as pets, carrying live ones in his pock-
ets, and holding them in his palm as he whistled to keep them motion-
less. An Irishman, he regretted the lack of snakes in the Emerald Isle and
threatened to ship a crate of writhing ones back to his homeland. With
such an interest in reptiles, Coulter seemed the ideal naturalist to be first
to explore the natural history of the California desert.

Born in County Louth, Ireland, in 1793, Coulter attended the University of Dublin (Trinity College), where he showed great promise in the
natural sciences: physiology, chemistry, and more particularly botany
and entomology. Indeed, as an undergraduate his collection of mosses
and insects would have made a professional envious. His specialty then
was bees, and his rapport with them was such that reportedly he was
never stung. Graduating in 1817, he went on to obtain both a Master of
Arts and a Bachelor of Medicine degree in 1820. Coulter then spent several summers at the Jardin des plantes in Paris. With a recommendation
from the British Museum's famous Robert Brown, he moved to Switzerland in 1822 to work with the eminent Augustin de Candolle. In 1816
Candolle had relinquished his university botany position at Montpellier,
France, home of the oldest botanic garden in Europe, to develop the natural history program at the Collège de Genève. Eventually he created the
natural history museum there, refurbished its botanic garden, and established one of the world's greatest herbaria. At Geneva Coulter not

only built up his own herbarium of European plants but under Candolle's guidance wrote and presented an excellent *mémoire* on teasels, a distant relative of sunflowers whose dried heads were used in carding wool.

At this time Candolle was involved with his monumental *Prodromus,* in which he hoped to include all the plants of the world. When Coulter departed in September 1823 from Geneva for England and then overseas, Condolle, already having had a genus of Brazilian legumes named in his honor, would certainly have encouraged his energetic protégé to collect far afield for the *Prodromus.* Coulter did indeed plan an ambitious collecting trip: across the Atlantic to Argentina, over the southern Andes to Chile, north through Bolivia and Peru, on to Mexico and then Alta California and perhaps even Canada. But upon reaching the New World, Coulter, instead of pursuing his enterprising collecting plan, signed a three-year contract in 1825 with the British-owned Real del Monte silver mining company in Mexico. He worked first as a medical attendant but eventually became a mine manager.

In his spare time Coulter was working on a Mexican flora, and in 1828 he shipped seventy species of live cacti to the botanical garden at Trinity College and fifty-seven to Candolle, forty-seven of the latter presumed to be new species. Coulter eventually moved on to western Mexico, where four robberies, a shipwreck, and the climate turned his interest from business back to botany. In early August 1831 he was in Guaymas across from Baja California, ready to travel north to Mexican-controlled Alta California by the hazardous overland route. Instead, with the rainy season in the offing, he embarked for Monterey about August 14 aboard an American brig.

After arriving in Monterey about October 1, Coulter made a quick trip south on El Camino Real to Mission San Gabriel near Los Angeles to plan for a trip to the Colorado River after the rainy season. In early November the despondent botanist wrote to his sister in London from San Gabriel that during his first month in Alta California no botanizing had been possible. It is likely that on this excursion he took a side trip into the Santa Lucia Mountains to Mission San Antonio. Near there he collected specimens but no seeds of the beautiful Bristlecone Fir and the Coulter Pine, probably at Cone Peak or the higher Junípero Serra Peak (5,862 feet), within hiking distance of San Antonio. Within a week of Coulter's return by ship to Monterey in mid-November 1831, he had a chance encounter with the London Horticultural Society's plant collector, David Douglas. The two certainly shared mutual interests, including a desire to collect plants in California's southeastern desert country.

But political unrest in the province discouraged them from making an immediate departure together.

The rainy season ended late in February 1832. On March 20, when dry weather would presumably make travel possible, Coulter embarked by himself on the desert trip, with compass, sextant, barometer, thermometer, and two chronometers. About the same time Douglas also traveled south, but only on a brief excursion into the Santa Lucia Mountains. Coulter's journey took him again toward Mission San Gabriel. At the end of sixteen days he reached Santa Barbara, having traveled along the Salinas River and then west to the ocean at San Luis Obispo. On one of his several visits there, Coulter discovered the new Bishop Pine. The major hazard on the current journey involved crossing rivers. Winter floodwaters had subsided, but wet sandbars—"quicksand"—awaited unwary travelers. When Coulter reached the Guadalupe River (now the Santa Maria), the bed looked treacherous, but he finally followed some bear tracks safely across. Past Santa Barbara he continued along the coast and then inland to Mission San Gabriel. From there he followed the track established by Juan Bautista de Anza in 1774–76 to the small settlement of La Pala.

La Pala is located inland in the beautiful San Luis Rey River valley surrounded by mountains, including Mount Palomar, the "pigeon-roost." Presumably in this area Coulter collected the spectacular Coulter's Matilija Poppy, its stems up to eight feet tall, with huge, papery, white-petaled blossoms and a dense central cluster of yellow-tipped stamens (figure 15). The poppy was given its scientific name—*Romneya coulteri*—in 1845 by William Harvey, keeper of the Trinity College Herbarium, who introduced many of Coulter's California plants to European and American botanists. Harvey had originally wanted to name the new genus after its discoverer, but another genus, a small tree from Mexico, had already been named after Coulter by Candolle. The Mexican tree's name was soon changed to *Caesalpinia conzattii,* one of the poincianas, but too late for Harvey to use Coulter's name for the poppy genus. At least Harvey's species name as well as the modern common name do honor Coulter. The generic name indirectly honors him as well, since *Romneya* refers to Coulter's close astronomer friend Reverend T. Romney Robinson of Ulster's Armagh Observatory. The reason Harvey could not use the astronomer's last name was that Robinson had already been used as a genus, honoring the shipwrecked Robinson Crusoe! How fitting for stargazer Robinson that Coulter's Matilija Poppy's type locality is close to the modern Palomar astronomical observatory.

Figure 15. Coulter's Matilija Poppy. (Photo by R. G. Beidleman.)

When the poppy was first put on public display in London, many years after Coulter's death, it created a sensation. Indeed, one admirer sat and looked at the flower for an entire hour. A popular English garden book of the time, William Robinson's *English Flower Garden,* called Coulter's Matilija Poppy "the fairest plant that ever came to our land from that country of flowers, California." The name *Matilija,* its Chumash Indian meaning unknown, comes from Matilija Canyon northwest of Ojai, where the poppy has grown in profusion. Some locals assumed that the canyon was named after the poppy.

From La Pala Coulter turned southeast, heading through the mountains and onto San Felipe Creek, passing southwest of what today is Anza-Borrego State Park and approximately along part of the route later to be used by the Butterfield Stage and many future naturalists. He traveled around the southern end of the Vallecito Mountains onto Carrizo Creek and into the Carrizo badlands, crossing de Anza's 1774 route when de Anza was on his way to Mission San Gabriel. Then for a hundred miles Coulter proceeded across the sandy, rock-studded, and barren Colorado Desert, "totally destitute of pasture," except that on the next-to-last day of the trip an alkaline wet spot called Agua Sola was encountered, with some rushes and reeds that the mules ate but the horses refused. Throughout the journey, to keep track of the distance, Coulter carried his chronometer on a belt, tightly cinched against his abdomen.

At the end of the first week of May 1832, he reached a crossing of the Colorado River south of where the Gila River entered it from the east (near present-day Yuma, Arizona). Thus Coulter became the first natural scientist to encounter both the lower Colorado River and what later would become the state of Arizona. It would be nearly a decade before the next naturalist, a teenager from Philadelphia, would traverse California's desert country.

At the edge of the hot, arid Colorado Desert of southeastern California (now recognized as a portion of the Sonoran Desert to the east), Coulter discovered another new genus, *Lyrocarpa* in the Mustard Family, growing on Creosote Bush alluvial fans. When William Hooker and Harvey later described this genus, they gave Coulter's name to the species he had collected, *Lyrocarpa coulteri*. In sandier sites Coulter found a distinctive twin-podded mustard, aptly called spectacle-pod, that Harvey named *Dithyrea californica*, the generic name meaning "two shields" in Greek.

Coulter spent more than a week exploring and botanizing in the vicinity of the lower Colorado River. For some days the wind blew strongly out of the southwest. When it finally ceased, on several occasions Coulter, while holding his shaded thermometer a few feet above the ground, recorded a temperature of 140 degrees. The day he left for his return to the coast, May 17, there was nary a breeze. Coulter reminisced later that it "was one of the most painful days I have spent." In a letter to Candolle he wrote of this desert region, "This is truly the Kingdom of Desolation."

Coulter retraced his route to Monterey, but with some collecting side trips. He reached La Pala on May 27 and was at the San Gabriel mission on June 15, but did not get to Santa Barbara until July 5, sixteen days longer, because of side trips, than when he came down from Monterey. Coulter's personal travel map indicated his side trips from San Gabriel to San Pedro and to San Diego by way of Mission San Luis Rey. He finally reached Monterey on July 19, 1832, after an absence of seventeen weeks. For Coulter the Colorado River excursion represented "the most interesting, the longest, and by far the most laborious of those I made in California."

When Coulter returned to the Monterey area he met another naturalist, German Ferdinand Deppe, who had arrived there from San Diego. Deppe had collected some plants, but his primary interest was in bird and mammal specimens for the Berlin Museum. David Douglas was still there and would remain until leaving on August 18, 1832, for Hawaii. Coulter and Douglas enjoyed each other's company again. Coulter told Douglas that he expected to return to England about a year later, in the autumn of 1833, and Douglas gave Coulter a letter of introduction to Professor William Hooker. As it turned out, Coulter and Douglas never met again.

Coulter went back to Mexico, where in the spring of 1833 he again became involved in a mining operation. Coulter was by no means the first naturalist to be lured into mining in Mexico. More than a decade earlier William Bullock—who developed famous museums in Great Britain and collected birds in Mexico, including Bullock's Oriole and Swainson's Hawk—had become involved in silver mining in the San Luis Potosí region. Coulter's mine activity was at Guanajuato, south of Potosí. He justified his return to mining by reporting that he was "not avaricious, but really I don't like to be always poor." Not only was this venture a failure, but the community was beset by a cholera outbreak, so Coulter volunteered his medical services. Also, another of Mexico's frequent revolutions broke out, with Coulter's residence in the line of fire.

Coulter finally returned to Europe sometime in 1834, reputedly having garnered a collection of about fifty thousand herbarium specimens and the wood of about one thousand additional specimens. Much of his collected material went initially to Candolle in Geneva but later was sent to London and thence to Dublin, where Coulter began its organization. Unfortunately, when the material was transported from London to Dublin, Coulter's botanical manuscripts and his American diaries were lost, including the diary for the California interlude. It is unfortunate that the California journals of both Coulter and Douglas have disappeared. Coulter was not only the first naturalist to explore the California desert, but also the first to travel the inland route all the way from Monterey to San Diego. Yet there is no evidence that he collected any of his favorites, the reptiles.

Back home in Ireland, Coulter began sending papers describing California to the London Geographical Society. For the first evening meeting of the Royal Dublin Society he exhibited cones of his Coulter Pine and the Sugar Pine along with a specimen of the Redwood, as well as describing the best method of bringing seeds from California to Ireland. In 1837 he obtained his professional medical degree from the University of Dublin, and in June 1840 he was appointed curator of Trinity College's herbarium.

Coulter, though described by the Irish astronomer Reverend Robinson as being of "a noble and commanding person; great stature, strength and dexterity in the use of arms," had throughout his adult life not always been in the best of health. Now he was suffering from gout and rheumatism. One April day in 1843, he missed his transportation in Dublin and had to walk a considerable distance in heavy rain, an exposure from which he never completely recovered. He died at his university on November 26, 1843, the college bell tolling for five days in his honor. The funeral procession, headed by the provost, included the fellows, the scholars, and all of the undergraduates. Coulter's eulogy was presented at a meeting of the Royal Irish Academy on March 16, 1844, by his longtime friend Reverend Robinson. Robinson emphasized that if the botanist's collections and notes had appeared in a single publication, it "would have formed probably the most valuable contribution to North American botany ever issued."

Work with Coulter's American specimens was continued by the new curator of the herbarium at Trinity College, as well as by Professor Harvey and by David Don at King's College, London; their work resulted in scattered publications. Many of the duplicate specimens became widely dispersed, the first set going to Kew. Asa Gray in the United States received specimens, some of which he sent on to John Torrey in New York. Indeed, it was Gray who coined the present scientific name for the striking lavender-flowered *Phacelia grandiflora* that Coulter may have first found in recently burned chaparral along the Southern California coast. Another beautiful waterleaf (now called *Phacelia minor*), collected on Coulter's Southern California visit, was named by Professor Harvey. Altogether there are about two dozen California plants whose species names honor Coulter. The most famous probably is the Coulter Pine, named *Pinus coulteri* by David Don, based on Coulter's specimen from the Santa Lucia Mountains near Mission San Antonio.

After their last parting, David Douglas added a postscript about his fellow naturalist, in a sense his competitor, in a letter to William Hooker dated October 23, 1832: "I left in California my friend Dr. Coulter, who will not, I trust, quit the country till he has accomplished every thing, for he is zealous and very talented."

A Frontier Naturalist
and His Protégé Visit
California Separately

THOMAS NUTTALL: "OLD CURIOUS"

I now arrived on the shores of California, at Monterey.
                                        —Thomas Nuttall

In St. Louis at Henry Shaw's old garden, now the Missouri Botanical Garden, there is an obelisk to the memory of Thomas Nuttall, inscribed "born in England, 1786, Honor to Him the Zealous and Successful Naturalist, the Father of Western American Botany."

After a childhood in the Ribble River Valley of Lancashire, England, Thomas Nuttall became an apprentice in his uncle's printing business in downtown Liverpool, although he was more interested in hiking the hills and vales of the Ribblesdale countryside looking at flowers. When the term of Nuttall's apprenticeship ended early in 1807, Uncle Jonas expected his nephew to become a printer, while Thomas was increasingly drawn to the pursuit of natural history. Over family objections, he finally decided to leave England for America in March 1808, hoping to explore a new natural world while subsisting as a journeyman printer.

Life for Nuttall on his first day in Philadelphia took a major detour when he visited Professor Benjamin Smith Barton of the University of Pennsylvania to have an unfamiliar plant identified. Barton at the moment was looking for an assistant to discover new plants, and to find cures for the gout that afflicted him. Before Nuttall knew it, he was out in the field as Barton's plant collector, on excursions first along the eastern seaboard and then, on April 12, 1810, off in a stagecoach from Philadelphia bound for Canada's far-distant Lake of the Woods in what was then called the Old Northwest.

When Nuttall chatted with a fellow passenger in the coach, the pursuit of western natural history accelerated remarkably. Fur trader Manuel Lisa regaled the neophyte naturalist with tales of the Missouri River hinterland and John Jacob Astor's overland expedition bound for the Pacific Coast. When Nuttall reached Michilimackinac between Lake Huron and Lake Michigan, he learned that there were major obstacles to reaching Lake of the Woods. On the other hand, Astor's party was about to leave for St. Louis, thence to proceed up the Missouri River and on westward. When the leader, Wilson Price Hunt, invited Nuttall to come along, the plans of Professor Barton were quickly dismissed.

At St. Louis, Nuttall again encountered not only Manuel Lisa, who had his fur house right by the Mississippi River boat landing, but also botanist John Bradbury, whom Nuttall had known at the Liverpool Botanic Garden. In March 1811 the two plant collectors were on their way up the Missouri River with Hunt's party. That fall Nuttall returned from the upstream Mandan Indian village with a panoply of new plants and in the company of Lisa and two of his partners in the Missouri Fur Company, Andrew Henry and Reuben Lewis, brother of Meriwether.

In England during the War of 1812, Nuttall later returned to Philadelphia, quietly telling of his wild Missouri River experiences at parties in Caspar Wistar's home near Independence Square. Nuttall rewarded his host by naming the genus of a newly collected North Carolina vine *Wisteria*. In mid-July 1818, at work with the Academy of Natural Sciences of Philadelphia, Nuttall published his classic *Genera of North American Plants*, including his many new species from the East and Middle West. To economize, he had set much of the type himself. In early October of the same year, Nuttall was off to become the first naturalist to travel up the Arkansas River and into the wilds of the Oklahoma Territory. Aside from his valuable plant collections on that trip, his published diary, *A Journal of Travels into the Arkansa Territory,* became a readable model of exploration writing for the period.

In 1823 Nuttall was invited by Harvard to become head of its botanic garden and teach an occasional botany course. Before Nuttall had completed his first decade at Harvard, he complained about the poor botany library and digressed to commence writing the first field guide to American birds. It proved to be so good that Ralph Waldo Emerson recommended it as a book "that everyone who lives in the country ought to read."

While working on the guide, Nuttall enjoyed watching birds at Fresh Pond near Cambridge. There was a hotel nearby, and Nuttall became friends with the proprietor's son, Nathaniel Jarvis Wyeth, who likewise observed the birds along with chopping ice from the lake for a living. In 1832 thirty-year-old Wyeth approached Nuttall with an intriguing suggestion: to join him as naturalist on an amphibious wagon expedition to the Oregon coast. The professor thought the idea bizarre, as indeed did some Harvard students, who coined the names *Amphibiums* and *Natwyethiums*. Nuttall declined but asked Wyeth to collect plants if the expedition materialized. The expedition did proceed, although the Amphibiums were abandoned by the time St. Louis was reached. Wyeth brought back plant specimens, including the only iris in the Rocky Mountains, which Nuttall later named *Iris missouriensis* in honor of the Louisiana Purchase.

In 1834, when Wyeth invited Nuttall to accompany him on another West Coast expedition, Nuttall—fed up with Harvard and academic life—resigned and accepted (figure 16). He proceeded to the Philadelphia Academy, where he began planning for the trip, including trying to find a young, strong, healthy assistant. The ideal candidate showed up at an evening academy meeting, twenty-four-year-old zoologist John Kirk Townsend. When the two left Philadelphia by coach on March 13, 1834, for Pittsburgh, eventually to join Wyeth's party at Independence, Missouri, there was one important item Nuttall had not mentioned to the zoologist: namely, that Nuttall had made out his will, anticipating that the two might not survive the trip.

By the autumn of 1834 Nuttall and Townsend had become the first experienced naturalists to complete a journey of observation and collection from the Atlantic Ocean to the Pacific Ocean across the breadth of what would eventually become the United States. Nuttall's new plant species would fill future floras for the Great Plains, the Rocky Mountains, and the Northwest, while the two naturalists' bird specimens would permit John James Audubon to add western birds to his monumental *Birds of America*.

After the cross-country trek and a brief collecting stint in the Northwest, the naturalists sailed to the Sandwich (Hawaiian) Islands for the winter of 1834–35. Here Nuttall gathered Hawaiian seashells and spectacular lobelia flowers, while Townsend collected the distinctive native birds, many of which are extinct today. Then in springtime it was back

Figure 16. Portrait of Thomas Nuttall. (From *Gardener's Monthly*, vol. 2, frontispiece [1860].)

to the Pacific Northwest. At the end of the 1835 summer, Townsend and Nuttall parted company. Townsend would stay in the Northwest, taking over hospital duties at Fort Vancouver until relieved by Dr. Tolmie, while Nuttall and the fort's ailing surgeon, Meredith Gairdner, boarded the Hudson's Bay Company's bark *Ganymede* bound for Honolulu. Townsend anticipated that Nuttall would soon visit California, and "either return to the Columbia by the next ship, and take the route across the mountains, or double Cape Horn to reach his home."

During the second winter in Hawaii, Nuttall continued his work on plants, seashells, and marine crustaceans. When he was able to visit "Owhyee" (the Big Island), he made a collection of ferns, which were dispatched to the Liverpool Botanic Garden, and he collected fresh basalt (lava) for his mineral cabinet from that active volcanic island. In Honolulu, where David Douglas was buried, and on the Big Island, dominated by Mauna Kea where Douglas had died, Nuttall could not have failed to remember his 1823 acquaintance, whose tragic death Nuttall and Townsend had learned about in November 1834.

By the springtime of 1836, Nuttall had decided to sail back to Boston with a stop at Alta California, rather than return to the Northwest and proceed overland. During the winters in Hawaii he had become acquainted in Honolulu with the American consul John Coffin Jones Jr., who had been a student at Harvard before Nuttall's time and on occasion provided Nuttall and Townsend with comfortable Honolulu quarters. Nuttall had also met Captain William Hinckley, who had taken Nuttall and Townsend to Kauai for an extended collecting trip, where King Kamehameha III provided his own home as headquarters for the naturalists. Captain Hinckley's partners were Nathaniel Spear and Jacob Leese, merchants in Monterey who in 1836 opened an establishment in Yerba Buena. Hinckley, Spear, and Leese reputedly laid the commercial foundation for San Francisco. Arrangements were made for Nuttall to travel from the Sandwich Islands to California on Spear's ship, the *Isabella,* arriving at Monterey Bay in March 1836. The *Isabella* may have stopped first at Yerba Buena, because one of the plants in Nuttall's collection, the Seaside Woolly Sunflower, was from there.

Nuttall certainly was delighted with his brief visit to the Monterey Peninsula, "one vast wilderness, neglected and uncultivated." The spring countryside was "spread out with its varied carpet of flowers; all of them had to me the charm of novelty, and many were adorned with the most brilliant and varied hues." He liked the abundant oak with its hollylike leaves (the Coast Live Oak), "a thorny gooseberry . . . with pendulous flowers as brilliant as those of a Fuchsia" (Fuchsia-flowered Gooseberry), and "a Ceanothus with sky-blue withered flowers" (Blue-blossom, named by Eschscholtz). Nuttall was distressed that the Blueblossom bush had been discarded on a woodpile. Altogether he gathered seventeen species of plants during his brief time ashore, but he was well aware that earlier botanists had collected many more plants around the Monterey Peninsula.

A tree in bloom along the border of a creek near Monterey, the White Alder, proved to be a new species, and Nuttall gave the modern name to the California Buckeye. He also found three new shrubs: Spiny Redberry, Monterey Ceanothus, and Sandmat Manzanita, as well as a new composite, Everlasting Neststraw. He took particular notice of the California Quail, and "the cheerful mocking-bird sent forth his varied melody." While beachcombing along the shore, Nuttall collected a new species of seashell. The naturalist left Monterey before the arrival of the new governor, Mariano Chico, whose rascally public behavior and licentious personal relations nearly precipitated a local rebellion.

Nuttall boarded the *Pilgrim,* a small three-masted bark belonging to the Boston company of Bryant and Sturgis, bound for San Diego. A trio of Bryant and Sturgis hide ships, the *Pilgrim, California,* and *Alert,* currently plied California coastal waters, collecting cattle hides at several stops along the coast. Because of the generally drier and hotter climate of San Diego to the south, most of the green (fresh) hides were taken down there to be dried in the sun before being packed for shipment around Cape Horn to the East Coast. Free passage for Nuttall on the *Pilgrim* had probably been arranged by his friend Captain William Hinckley, who was associated with the shipping company.

After Monterey, the next port for the *Pilgrim* in late March was the village of Santa Barbara, where the planted groves of olive trees were in full blossom, reminding Nuttall of flowering stands of willows. The *Pilgrim* remained here until late April, which pleased Nuttall immensely. His exciting bird find, chattering in the branches of a Coast Live Oak, seemed to be a new species of magpie, like the one he knew in England— but this California species had a conspicuous yellow rather than black bill. Eventually John James Audubon named the species from Nuttall's specimen in Nuttall's honor (now *Pica nuttalli*), and included its portrait in his Elephant Folio. The distinctive Yellow-billed Magpie is restricted in its distribution to certain portions of California. Nuttall was lucky to collect it in Santa Barbara, where it is uncommon. One wonders why Nuttall should remember the European magpie and yet apparently forget that with the Wyeth Expedition in 1834 he and Townsend came through the Rocky Mountain West, where the black-billed American Magpie, a species very similar to that in Eurasia, abounds.

There were plenty of new plants to be encountered around Santa Barbara, and Nuttall collected sixty-five species. A real find, spreading "its wide arms over the dried up rivulets," was the Western Sycamore, seen in California by numerous earlier naturalists but not recognized as a new species. The "very elegant evergreen tree" that caught Nuttall's attention was the Wax-myrtle, which Chamisso first discovered at San Francisco and gave its name. Another attractive native tree was the Holly-leafed Cherry, which Nuttall initially named. He also collected what he felt constituted several new genera of plants for Alta California: the crimson-flowered Chaparral Pea, which he named *Pickeringia montana* after Charles Pickering, his academy friend and naturalist with the United States Exploring Expedition; *Psilocarphus*, a sunflower genus commonly called woolly-heads; and *Leptotaenia*, a member of the Parsley Family, now included in the genus *Lomatium*. "On the rocks near the sea" Nuttall found Telegraphweed; a bush sunflower, *Encelia californica*, on a dry hillside; in the chaparral a new species of peony, eventually to be named *Paeonia californica* by Torrey and Gray; as well as several species of *Ceanothus* and *Rhus*.

Nuttall found a new shrubby evergreen oak forming thickets in the foothills east of Santa Barbara that he named *Quercus dumosa* (meaning "bushy"), the Nuttall's Scrub Oak of Southern California and northern Baja, now becoming rare because of habitat destruction. The new Birch-leaf Mountain-mahogany bush he collected on a chaparral hillside and named reminded him of a species he had seen just west of the Continental Divide in Idaho in 1834.

Santa Barbara was almost as good for birds as it was for plants. There had been, of course, the Yellow-billed Magpies, everywhere in April completing their stick-ball nests, tucked away among the branches of the Coast Live Oaks. There was much calling and chattering among the birds, but after Nuttall fired his gun, it was almost impossible to get within gunshot of them. The naturalist was quartered at a nicely gardened residence close to Mission Santa Barbara. Nearby he came upon the nest of California's familiar Black Phoebe; and there was an Anna's Hummingbird with a rough-looking, diminutive nest containing two tiny eggs on the branchlet of a Toyon. Nuttall actually captured the female with his hat while being attacked by the male. In the mountain thickets he encountered numerous Chestnut-backed Chickadees, the first specimens of this new species having been collected by Townsend at Fort Vancouver. There were flocks of Western Bluebirds and several subspecies of White-crowned Sparrows, one of which would eventually be named after Nuttall. Around Santa Barbara marshes he saw blackbirds resembling Red-winged Blackbirds but with very conspicuous white wing coverts. Nuttall collected the new species, the Tricolored Blackbird, which Audubon would name. When Nuttall tired of plants and birds, he headed for the Santa Barbara beach where he added twenty-six species of seashells to his growing mollusk collection.

The next stop for Nuttall and the hide ship *Pilgrim* was San Pedro, the main port for Los Angeles. Although the stay was brief, Nuttall collected two species of plants at the shoreline, the tarweed *Hemizonia fasciculata* and a new species of gumplant now included in *Grindelia camporum*. Nuttall's sailing aboard the *Pilgrim* ended when the ship passed Point Loma and entered San Diego harbor on April 15, 1836. But the naturalist remained aboard during a three-week wait for a Bryant and Sturgis ship to take him back to Boston. This section of Alta California was a natural history mecca, displaying a diversity of new animals and plants with a southwestern affinity. Only four naturalists—Menzies, Botta, Coulter, and Deppe—had been here earlier than Nuttall, and then only briefly.

Nuttall had a field day with the flora during his San Diego stay. In all, he discovered what at the time were nine new genera of plants. One of them, Desert Christmas Tree, *Pholisma arenarium* (known to the locals as Gamote de los Médanos), was not to be rediscovered for four decades, and then again on dunes at San Diego. Nuttall was indeed fortunate to find this bizarre parasitic plant of restricted range, since later in the spring it dies back to the surface of the soil or below. Nuttall's other new genera were *Streptanthus* (jewelflower); the robust, yellow-flowered *Isomeris*, with conspicuous bladderpod fruit; *Apiastrum*, a wild celery Nuttall probably collected in coastal scrub; *Nemacladus* of the Bellflower Family; and three new genera in the Sunflower Family, *Pentachaeta*, *Uropappus* (silverpuffs), and a new genus and species Nuttall named *Rafinesquia californica*, California Chicory, after the quixotic naturalist Constantine Rafinesque. Nuttall had never met Rafinesque but felt indebted to him because Constantine, usually a curmudgeon, had written a relatively favorable review of Nuttall's *Flora*. In all, Nuttall collected forty-four species of plants, many of them restricted to that southwestern section of Alta California.

The naturalist also had a great time picking up shells around San Diego as he awaited the hide ship *Alert* to take him home. There were various beaches at hand, and Nuttall ended up with twenty-one species of mollusks. It was during one of Nuttall's last beachline walks, within view of the newly arrived *Alert,* that a young sailor named Richard Henry Dana spotted the elderly beachcomber trudging along the shore, barefooted, with his pants rolled up to his knees, wearing a sailor's peacoat and atop his head a wide straw hat, picking up shells and stones. Earlier, when Dana had been talking to the second mate of the *Pilgrim,* Nuttall's ship from Monterey, the *Pilgrim* mate mentioned that aboard there was "sort of an oldish man"—he could not recollect his name—but definitely with white hair, who spent a lot of time out in the bush and along the beach, "picking up flowers and shells and such truck, and had a dozen boxes and barrels full of them." Now Dana suddenly realized that he was looking at his old Harvard professor, "though I should hardly have been more surprised to have seen the Old South steeple [of Boston] shoot up from the hide-house." Shortly Professor Nuttall remembered Dana, now a sunburned and stalwart sailor. Nuttall had not seen Dana since the lad had dropped out of college in 1833 because of eye trouble resulting from measles.

The *Alert* sailed from San Diego on May 8, 1836. Nuttall and Dana saw little of each other during the long voyage around South America to Boston. Occasionally, however, on a calm night at sea when Dana was at the helm, the naturalist would come aft "and hold a short yarn with me," though doing so was against the captain's rules. Nuttall kept belowdecks much of the time, and most of the crew had almost forgotten that they had such a passenger aboard. "Old Curious," they called him, "from his zeal for curiosities." When the *Alert* rounded Staten Island at Cape Horn, the tip of South America, and the cry "Land-ho" was shouted, Nuttall emerged on deck "like a butterfly, and was hopping round as bright as a bird." He asked Captain Alpheus Thompson, a man with a notorious lack of good humor, to put him ashore to collect, where probably no one had ever collected before. The captain's retort could have been anticipated, "that he would see the island—specimens and all—in another place, before he would get out a boat to delay the ship one moment."

The *Alert* at last came to berth in Boston Harbor on September 20, 1836. Nuttall's arrival was actually announced in one of the Boston newspapers. When Nuttall went to the office of Bryant and Sturgis to pay for his voyage from Alta California to Massachusetts, his payment was refused, with the emphatic declaration that they would not accept a penny of passage money from Professor Nuttall, because "you travel for the benefit of mankind."

In *Two Years before the Mast*, Richard Henry Dana's colorful account of his California experiences, published in 1840, the reason for the wandering and wondering of naturalists was shrewdly evaluated by one of the *Alert*'s old salts: "O, 'vast there! . . . I've seen them colleges and know the ropes. They keep all such things for cur'osities, and study 'em, and have men a purpose to go and get 'em. This old chap [Nuttall] knows what he's about. . . . He'll carry all these things to the college, and if they are better than any they have had before he'll be head of the college. Then, by and by, somebody else will go after some more, and if they beat him he'll have to go again, or else give up his berth. That's the way they do it."

Although Thomas Nuttall's sojourn in California was relatively short, the "Father of Western American Botany" was not to be forgotten by the Golden State. On Sunday, October 20, 1935, in the afternoon shade of the Willis Jepson memorial laurel at Spring Valley Lakes, San Mateo County, there was a celebration for the "centennial" of Thomas Nuttall's visit to California. In organizing the event, Professor Willis Jepson of Berkeley, who by then could certainly be called the "Father of California Botany," promised the gathered crowd, "To help keep fast the day in the sweet lanes of memory we will observe with winged words the anniversary of the visit to California of Thomas Nuttall, botanist, zoologist, explorer and quiet priest of the wilderness places." With W. W. Mackie of the University of California as "Master of the Day," the "winged words" included Adeline Frederick of the Berkeley Garden Club giving a talk called "Trees for Loyalties and Memories." Other speakers and their talks were University of California forester Emanuel Fritz, "Cornus Nuttallii, Its Place in the Forest"; Cornelia Pringle of the Cooper Ornithological Club, "Pica Nuttallii, the Yellow-billed Magpie"; E. Raymond Hall of the University of California Museum of Vertebrate Zoology, "Nuttall as a Mammalogist and Ornithologist"; and Jepson, "The Long Trail of Thomas Nuttall across the Continent in 1834 and 1835," with concluding remarks by Frank Dodd and the master of the day. No one at the celebration seemed too concerned that on October 20, 1835, Nuttall had been ready to pick up some more Hieroglyphic Venus seashells along the shore of Pearl Harbor in Oahu; he, of course, would not reach California until March 1836.

WILLIAM GAMBEL: THE FIRST
NATURALIST ACROSS THE SOUTHWEST

Am in California, on the banks of the broad Pacific
Ocean . . .
                                    —William Gambel

Today it is hard to believe that there could be twelve thousand applications for tickets for a series of botany lectures. This occurred in Boston in early 1840, the second season of the popular Lowell Institute programs. The lectures were delivered at the Odeon Theater and Concert Hall, which accommodated an audience of fifteen hundred. The speaker was the famous naturalist Thomas Nuttall, the intrepid explorer of the western American frontier for two and a half decades and the author of landmark flora, ornithology, and travel books. He was a fellow of the Linnean Society of London and member of the American Philosophical Society, an instructor of natural history and botany and director of the Botanic Garden at Harvard, vignetted in books by Washington Irving and James Fenimore Cooper, and soon to be portrayed in Richard Henry Dana's forthcoming *Two Years before the Mast.*

Nuttall's lecture series got off to a slow start when people in the audience had difficulty hearing the soft-spoken accents of the Englishman. In fact, Nuttall lectured for five minutes before anyone realized he had begun, but then came shouts of "Louder, louder, louder." During subsequent lectures he did speak up. After the series ended on April 14, Nuttall, accompanied by his field assistant, William Gambel, headed to Maine for field research on a supplement to François André Michaux's *Sylva* and to do a bit of mineral collecting, in which both Nuttall and Gambel were interested.

This would be Gambel's second year as Nuttall's assistant, having accompanied him to the southeastern United States for botanical and geological fieldwork from late 1838 until the fall of 1839, although Gambel was only sixteen (some claim eighteen) years old at the time. However, this young Philadelphian had already shown interest in and talent for natural history. He had begun a "cabinet of minerals" and presented a gold nugget he had found in North Carolina to the Philadelphia Academy of Natural Sciences for its collection. Nuttall was a willing mentor to his young assistant, having himself been befriended as a youngster by botanist John Windsor in England, and having later served as assistant for University of Pennsylvania professor Benjamin Smith Barton. In 1834 Nuttall had nurtured and depended on his assistant, zoologist John Kirk Townsend, during their excursion across the continent.

In 1841, because of restrictions in his late uncle Jonas's will, Nuttall felt obligated to leave North America and return to a sedentary life in England. But he realized that there remained one major wilderness region of the continent that no naturalist had explored, namely the American Southwest. Nuttall proffered the challenge to young Gambel, and he readily accepted. Nuttall, delighted, wrote to Dr. Engelmann in St. Louis that Gambel "has thus, at the risk of life and health, as well as considerable expense, undertaken to augment our knowledge of a region hitherto scarcely at all explored."

In March 1841, with good-byes to his mother and two sisters, and especially to his sweetheart, Catherine Towson, the seventeen-year-old was on his way west to Independence, Missouri, in the 1834 tracks of Nuttall and Townsend. At Independence Gambel joined the annual caravan departing in May for Santa Fe, with about eighty men, mostly merchants, and forty merchandise-loaded wagons. They arrived in Santa Fe on July 2, nothing very exciting having happened, according to a letter from Gambel to his mother. He did mention, though, that the party had been surrounded in western Kansas by five hundred Arapaho Indians, who were appeased with gifts, and that near Wagon Mound (New Mexico) about four hundred angry Utes had used the party for target practice throughout an entire afternoon. Also, he had celebrated his eighteenth birthday.

From Santa Fe, Gambel probably shipped back to Dr. George Engelmann in St. Louis, among other natural history collections, the type specimen of Gambel's Oak, found near Santa Fe and later named in his honor by Nuttall. On September 1 he joined William Workman and John Rowland and their company of twenty-five adventurers bound for Los Angeles. They departed from the village of Abiquiu northwest of Santa Fe rather than Santa Fe because there were fewer fleshpots and like distractions. Among the party were two others, named Mead and Lyman, who reportedly possessed "scientific proclivities."

The party proceeded up the Chama River for a short distance, across on the Spanish Trail toward Salt Lake City, then diagonally southwest through the Sevier River Basin and over the Virgin River, the southwestern tip of Nevada, through the upper Mohave Desert in Alta California and across Cajon Pass, finally arriving at the San Gabriel mission near Los Angeles on November 5, 1841. Early during the trip Gambel collected the type specimen of the Mountain Chickadee, which he named *Parus montana* (changed to *Parus gambeli* in 1886 by zoologist Robert Ridgway to honor its young collector and now called *Poecile gambeli*).

Somewhere in the desert shrubland covered with species of saltbush, he encountered small flocks of a new quail with distinctive guttural calls and a conspicuous crest displayed in flight. While he was skinning what he thought were the males, with black belly patches, the remaining specimens, presumably females, were carried away by a pair of thieving ravens. When Nuttall later read Gambel's manuscript describing the new bird, he recommended that its name honor the youthful collector, Gambel's Quail, now *Callipepla gambelii*. The official type locality is southern Nevada, although it seems possible, based on Gambel's travel schedule, that they were collected in Alta California's Mohave Desert. William

Gambel was the first naturalist to cross the Southwest from Santa Fe to Los Angeles, through deserts, as Gambel wrote, "worse than those of Arabia."

Gambel was excited about the "fine rich country" along the Pacific Coast. Traveling over the Transverse Ranges into the coastal basin he had seen a shrike with a pure white breast, a Loggerhead Shrike. For about two winter months in the pueblo of Los Angeles, Gambel lived in a house with five other members of the expedition. Here Anna's Hummingbirds were everywhere in the gardens. Gambel had to chuckle when he remembered Nuttall's description of the females possessing yellow head feathers; they had actually been covered with pollen from dunking into blossoms. Gambel's special find was a small black-and-white, ladder-backed woodpecker that he shot in a willow thicket near town. He described it and named it *Picoides nuttallii* (Nuttall's Woodpecker) after his mentor. Around Los Angeles there were flocks of wintering Gambel's White-crowned Sparrows, first collected near Fort Walla Walla by Townsend and Nuttall and originally named as a new species (now a subspecies) after Gambel. Mixed in with the sparrows was an occasional wintering Hermit Thrush, first collected by Nuttall and Townsend at Fort Vancouver, the species (now subspecies) later named by John James Audubon based on their specimen.

On January 14, 1842, Gambel wrote to his mother from Los Angeles that he had decided to spend some months collecting in California until late summer, then return to Philadelphia by ship around Cape Horn or by crossing Panama and visiting the West Indies en route. A week later he began his Alta California excursions, departing from San Pedro for Monterey aboard Captain William Phelps's ship, *Alert,* and returning to collect plants around San Pedro in early February.

Gambel became the first naturalist to visit the island of Santa Catalina, only twenty miles southwest of San Pedro. In mid-February on Santa Catalina he found a Bald Eagle nest with eggs high atop a rocky crag, with Ospreys nesting nearby. And on another visit in April he heard the Orange-crowned Warblers singing their "simple but lively song." He collected eight flowering plants there, including an eight-foot-tall snapdragon with scarlet flowers, originally named *Gambelia speciosa* by Nuttall (now *Galvezia speciosa*), the Showy Island Snapdragon. This rare shrub was not rediscovered until botanist Edward Palmer found it on San Clemente Island in 1875, its range being the islands of Santa Catalina, San Clemente, and Guadalupe off Baja. During April and May 1842 he visited and collected at both Santa Barbara and San Simeon.

By midsummer Gambel was in dire financial straits. He had entrusted all his funds to Workman and Rowland, who by now had disbanded their party, leaving Gambel penniless. Making his way north again to Monterey in September, he was, like David Douglas, befriended by English trader and rancher William Hartnell, who envisioned educating the naturalist to become a Catholic priest. But international events intervened. On October 19, 1842, two warships of the United States Pacific Squadron, flagship USS *United States* and USS *Cyane,* rounded Point Pinos and that afternoon anchored off the presidio. Commodore Thomas Ap Catesby Jones, acting on two rumors—that a Mexican newspaper had reported that the United States and Mexico were at war, and that the British were planning to take California—seized the capital of Alta California without firing a shot, running up the Stars and Stripes the next morning at eleven. When Jones's private secretary that afternoon determined that there was no war, the commodore apologized; but the next morning, when Jones's midshipman son was ordered to take down the American flag, he refused, got drunk on whiskey, fell off a nearby cliff, and nearly killed himself.

Meanwhile, Jones received a mild reprimand; sailors and the flagship band came ashore, the former "hunting wild Deer or dancing with tame Dear"; there were balls for all; and Gambel's own fortunes improved. He met the *United States'* assistant surgeon, R. T. Maxwell, in Monterey, and Maxwell "got the Commodore to take him on board as his clerk." So Gambel left Hartnell's home and took up quarters aboard the flagship. Maxwell being the ship's chief sportsman, he and Gambel were soon off on hunting trips, including after bear, finding the game so abundant one could knock a duck down with an oar. The two rode their mares across to the San Carlos mission, along the way shooting quail that an old Mexican woman stewed for their supper with red peppers, with tortilla corn cakes on the side. On one excursion over to Point Pinos, the two were set upon by a band of drunken Indians led by the "brutal" military commander José Castro, and they did not achieve their escape to the ship until nearly midnight. It was coincidental that during this same period there was another visiting naturalist at Monterey, "an odd fellow" who was "collecting curiosities for the German princes." When this chap (unnamed, but perhaps Ferdinand Deppe again, some twenty years after meeting Douglas and Coulter) was attacked in the woods by Indians, it was Maxwell who pulled the arrow out.

During his California sojourn, Gambel was primarily collecting plants for Nuttall, while his real interest was birds. He was the first to record Roseate Spoonbills for California, in small flocks reportedly as far north as San Francisco Bay. These birds are rarely seen today in the state. Gambel shot three White-tailed Kites near San Juan Bautista in one day. In Santa Barbara during mid-spring he noted that "the woods were ringing" with the inimitable songs of Northern Mockingbirds, imitating both Bullock's Orioles and Western Bluebirds. At almost every mission Gambel visited, he encountered Barn Owls, often a number under one roof, and he heard many a complaint from the mission padres about these birds imbibing sacred oil out of the church lamps. Along the Pacific coastline he made interesting observations on a variety of shorebirds, including the beautiful American Avocet, the red-billed Black Oystercatcher, and the Black Turnstone.

In the oak woodlands around Monterey Gambel collected the first specimen of the Oak Titmouse, a diminutive gray, crested relative of Gambel's Mountain Chickadee. In the Monterey Pine forests he encountered noisy, stub-tailed Pygmy Nuthatches in the type locality where they had first been collected by Alexander Collie in 1827. It was Gambel who discovered a mouse-colored wrenlike bird, secretive but with a penetrating song, skulking in low thickets and brushy places around Monterey, the Wrentit. This bird named by Gambel constituted a new genus, and for many decades it was considered to represent a new bird family restricted to North America, but it is now classified with the Old World Babblers. Another bird he collected from Monterey chaparral was the California Thrasher, *Toxostoma redivivum*. He named it *redivivum* because he believed it had not been reported since it was first seen and painted by the La Pérouse Expedition in 1786, and hence Gambel had "rediscovered" it. For a relative novice Gambel was surprisingly familiar with the ornithological literature, but at the time he did not know that another early painting of the California Thrasher had been done by José Cardero of the Malaspina Expedition in 1791 at Monterey. Gambel was aware that the Greater Roadrunner he collected was not a new species, but he felt that his description was better. Philadelphia Academy ornithologist John Cassin reported that this "gem of his collection is a most superb specimen . . . a beautiful cuckoo-like bird that walks on the ground."

Gambel, with broad natural history interests, also captured reptiles that he encountered. The old generic name *Gambelia* honored William Gambel for the species of southwestern Leopard Lizard he collected (now *Crotaphytus*). And he was definitely interested in the Alta California mammals. In the Los Angeles area he discovered and named the Pacific Kangaroo Rat, and at Monterey he collected and named a new species of mouse, the largest of the state's deer mice, now known as the California Mouse or, more descriptively, the Parasitic Mouse because it builds its nest within the stick nests of packrats. Another new species of deer mouse collected at Monterey was named in Gambel's honor by Spencer Baird of the Smithsonian Institution, now the subspecies *Peromyscus maniculatus gambelii*.

Gambel's vertebrate specimens eventually made their way back to the Philadelphia Academy, where curator John Cassin expressed great excitement about the specimens representing new species, as well as species already familiar. But Cassin was piqued when Gambel named what he thought was a new auklet, a sea bird from California, in Cassin's honor, *Mergulus cassinii*. Cassin discovered that the auklet had been given an earlier name back in 1811 and chided that "it is already very difficult to get him [Gambel] to examine birds that he has concluded are new—concluded, I mean in the woods of California without books—and the most doubtful bird, too, probably at least, he has called after me." But Cassin should not have been too upset. The common name for Gambel's seabird is still Cassin's Auklet.

Before the USS *United States* departed for Hawaii on November 22, Jones transferred his pennant to the *Cyane*. On December 11, then, the *Cyane,* with Commodore Jones aboard and Gambel as his personal clerk, was off for Honolulu, the youthful naturalist experiencing his first naval sea duty. The United States Pacific Squadron was back in California in late December, with an invitation from Mariano Vallejo for a lavish party in Sonoma. Then came the grandiose New Year's ball hosted at Monterey's Government House by the commodore, busy patching up American-Mexican relations. The *Cyane* and the *United States* invited the town's residents aboard, with cakes and pies prepared by the ships' stewards, whiskey in place of the exhausted Madeira wine supply, and continuous dancing. A fortnight later Jones's warships were in San Pedro, where Mexican governor Manuel Micheltorena staged a resplendent celebration for Jones and his naval staff, complete with military escort, dinner at Abel Stearns's mansion, and on the next day toasts, speeches, and caustic Mexican reprimands at the governor's quarters, and finally a ball until sunrise. The commodore danced through the night with Stearns's wife, Arcadia Bandini, and her beautiful sister.

On the evening of January 21, 1843, the *Cyane* was under way to

Mazatlán, Mexico, where it joined the USS *United States* and the sloop *Yorktown*. For months thereafter, amid rumors that Jones had been recalled and the British had seized the Hawaiian Islands, Jones's Pacific Squadron circled the South Pacific Ocean: Valparaiso (Chile), Callao (Peru), Hawaii (where the USS *Constitution* was encountered), the Marquesas, Tahiti, and back to Callao. Small wonder Gambel wrote to his mother that he had "sailed farther than twice around the world." Finally, on January 21, 1844, in Callao, Commodore Jones handed over his log, boarded the USS *Constitution,* and headed for home. Gambel was not sorry to see him go. For some time their relationship had been "unfriendly," as he had written to his mentor Nuttall in England, who replied that naturalist John Kirk Townsend had earlier had a similar experience with Commodore Jones.

Gambel was transferred to the USS *Savannah,* where he was surprised to find in the ship's library the journal containing his article on new California birds. On July 6, 1844, from Callao, he shipped land snail specimens back to the Philadelphia Academy by his old frigate *United States.* Although frequently the naturalist had considered making the navy a career, in March 1845 he was homeward bound from Valparaiso for Philadelphia, where he arrived in mid-August. Wrote academy curator Cassin, "Eureka! Gambel is here with his California birds and others— not very many, but some of the most magnificent specimens I ever saw."

Back home, Gambel began publishing more of his zoological findings, his plant specimens having been sent to Nuttall in England. Nuttall invited the young man to join him at Nutgrove Hall, near Liverpool, to work together in publishing Gambel's journals and selling some of his specimens. But instead Gambel decided to pursue a medical degree at the University of Pennsylvania. He was disappointed at not being named new curator of the academy, the position going instead to comparative anatomist Joseph Leidy on the recommendation of closet naturalist Cassin. In March 1848, Gambel received his M.D., then married his sweetheart in October. After unsuccessfully attempting to obtain a medical practice in Philadelphia, he was drawn back to California, with its flourishing gold rush.

By March 1849, seventeen thousand people had already departed from the East for California by ship. Naturalist Gambel decided instead to travel overland, shipping his medical books and instruments by sea. His wife, Kate, was to join him after he had become settled on the West Coast. On April 5, Gambel and another young man, Isaac Wistar, who one day would become president of the Philadelphia Academy, left by

Figure 17. Above Rose's Bar along the Yuba River, where William Gambel died in 1849. (Photo by R. G. Beidleman.)

train, then took a coach, and finally caught a steamer to St. Louis. Thence they met with other young men waiting at Independence, Missouri, to form a party that would pursue a northern route to the goldfields. Wistar's fast-moving group of sixteen, including Gambel, departed on April 25. But at Grand Island, Nebraska, the naturalist, tired of the fast pace, joined a slower-moving ox train with eighty men led by Captain Boone of Kentucky. Wistar had enjoyed Gambel, "an amiable excellent fellow and very pleasant in conversation," but was not really sorry at their parting, commenting that Gambel "is averse to camp duty and hard work, and fond of taking things easy." They expected to see each other again in the goldfields.

　　The Boone party was ill-fated. It did not reach the eastern flank of the Sierra until October, having by that time lost most of the livestock and wagons in Nevada's Humboldt Desert. Encountering early snows in the mountains, on the same route earlier recommended to the Donner party, the group delayed for better weather, which did not materialize. Gambel and some others finally made it across to the western slope of the

Sierra. At Rose's Bar, an active gold-mining camp on the Yuba River, there was an epidemic of typhoid fever. Dr. Gambel paused to minister to the ill, contracted the fever, and died on December 13, 1849, at the age of twenty-six. He was buried beneath the bole of a giant Pacific Ponderosa Pine above the river (figure 17), but within a short time his grave was destroyed by feverish hydraulic placer miners. Meanwhile, the ship carrying his personal belongings crashed into one of the Farallon Islands off the Golden Gate and sank.

Many years later, Benjamin D. Wilson, a distinguished Southern California citizen who was one of Gambel's companions on the 1841 Workman-Rowland expedition, paid tribute to the naturalist: "He was a very young man at that time and made himself, by his collection on that trip and in California, quite an enviable reputation." And in the late naturalist's hometown newspaper: "He has departed early, but not unhonored. Philadelphia owes to his memory a lasting tribute of respect for his science, virtue, worth, talent and energy." William Gambel lived too short a time to achieve his full promise as a natural scientist.

Karl Theodor Hartweg

All the way from London to look after weeds . . .
                    —Karl Theodor Hartweg

Karl Theodor Hartweg, a German naturalist, came to California on a British warship to discover that he had disembarked into a war between the United States and Mexico. Hartweg was another of the London Horticultural Society's collectors, according to that society's report a "steady, well-informed and zealous young man" sent out to the Americas more than a decade after the time of David Douglas. Born the same year as Charles Darwin (1812), he was one of a number of young Germans who left their homeland in the 1830s and 1840s to escape the growing unrest there. Another of these expatriates was Dr. George Engelmann, who would make St. Louis the scientific capital of the American hinterland.

Hartweg came from an old-time, well-known horticultural family in Karlsruhe, on the Rhine south of Mannheim in the foothills of the Black Forest. His father at one time served as the court gardener there. Like Thomas Coulter, Hartweg spent time at the Jardin des plantes in Paris before working as a clerk at the London Horticultural Society's Chiswick Garden, where his industry and know-how came to be appreciated. By 1836 he had been appointed a plant collector and dispatched to Mexico to search for hardy plants that might do well in England, with special attention to orchids, which were all the rage.

The seven years in Mexico, and later Central and South America, as well as Jamaica, were for Hartweg an adventure in themselves, often having little to do with the natural sciences. Bands of robbers were proliferating, as Hartweg himself experienced; foreigners were being stoned;

live plants he collected would only randomly be transported back to England, and then often arrived dead. A Central American revolution was brewing, the French were blockading, and finally France and Mexico went to war with each other. On the other hand, Hartweg did succeed in sending back 140 species of orchids, Mexican cacti, and conifers, as well as enough plant seeds to divide into innumerable packets for members of the London Horticultural Society. Then he returned to England, where he remained for several years.

The London Horticultural Society, pleased with Hartweg's success in the New World, decided to send him back, this time specifically to visit Alta California. They anticipated that he should "reap a rich harvest of hardy plants, especially of the beautiful Zauschneria and the evergreen Castanea, to which his attention has been especially directed." The latter was the Giant Chinquapin, first discovered by Douglas in the Multnomah Valley of central Oregon in 1826; later in California Douglas had been served some of the "chestnut" seeds as a dessert.

Hartweg departed from England on October 2, 1845, aboard a Royal Mail steam packet, arriving at the harbor of Santa Cruz on the east coast of Mexico on November 13. He traveled across Mexico to the port of San Blas, collecting along the way, then north by boat to Mazatlán to await any merchant ship destined for Alta California. Unfortunately, there had been none for the past six months and none were expected. Commodore John Sloat of the United States Navy was in the harbor with the USS *Portsmouth,* under John Montgomery's command, but Sloat emphatically turned down the botanist's request for passage north. Anticipating imminent war with Mexico, Sloat was leery of allowing an unknown foreigner aboard who might reveal "secrets" (actually common knowledge, particularly that Monterey was the determined destination of the *Portsmouth*).

Several days later when two British warships sailed into the harbor, Hartweg was more successful with Rear Admiral George F. Seymour, who permitted the plant collector to sail north to Monterey on HMS *Juno,* leaving the morning of May 12, 1846. The *Juno* arrived in Monterey Bay on Sunday, June 7, to discover that the United States had actually declared war against Mexico on May 13. The *Portsmouth,* meanwhile, having been in Monterey, had sailed on to San Francisco Bay six days after the declaration.

Not overly concerned about the war, Hartweg delivered his letters of introduction and settled down the next day "in the quiet little town of Monterey," a pleasant change from the drier Mexican coast. Here the open woodland of Coast Live Oak covered the lower slopes, while up to the skyline was a forest of tall pines he recognized as the conifer "collected by the late Dr. Coulter" and named by David Don. The Monterey Pines extended down to Point Pinos, where Hartweg observed their closest approach to the shoreline. The German naturalist recognized a number of plants around Monterey at least to genus: the very common Blueblossom, silk-tassel, Yellow Bush Lupine, honeysuckle, coffeeberry, Western Poison Oak, elderberry, and Bush Monkeyflower. On one California Buckeye he counted over four hundred open flowers and blossoms on one flowering "spike."

On June 22 Hartweg undertook a seven-hour ride, with an American gentleman who provided the horses, around the bay to Mission Santa Cruz, then up into the nearby hills where he collected "the beautiful Zauschneria" and encountered dense forests of the Redwood, or "bastard cedar" as some of the locals called it. A particularly large Redwood named "the giant of the forest" was 270 feet high with a circumference of fifty-five feet. Tall Douglas-firs were thinly interspersed among them. Returning to Monterey on July 2 aboard an American bark, Hartweg found that Commodore Sloat's flagship *Savannah* and two sloops of war had arrived. On July 7, 1846, with news of the defeat of the Mexicans at Matamoros, the Yankees landed a party of sailors and marines who raised the American flag at Monterey without protest.

Additional American naval ships were arriving in the bay, but Hartweg was more concerned with the rapidly drying vegetation and the fact that venturing far from Monterey had become difficult inasmuch as the Mexican general, José Castro, had appropriated all available horses in hopes of repulsing the Yankee invaders. Hartweg decided to restrict his field trips to the immediate area, because he realized that Mexican settlers in the surrounding countryside viewed with high suspicion this plant collector and his "weeds, which in their opinion are not worth picking up." He did walk cross-country to Carmel Bay, a leisurely two-hour stroll through pine and oak woodland, his particular find being the Monterey Cypress, which he originally named. His specimen became the type specimen of this rare cypress, reminding him of "a full-grown cedar of Lebanon." Spreading across the seashore dunes, the sand verbenas were "delightfully scenting the air with their perfume towards evening."

Hartweg next set off for Carmel Valley, passing by the old Carmel mission, which was "fast falling to ruin, and the lands nearly in the same wilderness as the first settlers found them." He proceeded up the Carmel

Valley a considerable distance, through a beautiful river bottom of Arroyo Willow and unidentified sycamores and alders, finally crossing northeast to ascend the summit of El Toro in the old Estrada land grant. Though unwooded on top, the peak did have an isolated stand of small Foothill Pines in a north-facing ravine.

On August 23 Hartweg was invited to sail to San Francisco Bay aboard the bark *Joven Guipuzcoana*. Then he hoped to proceed up the Sacramento River and return to Monterey on horseback. The boat stopped first at Santa Cruz, giving the plant collector an opportunity again to visit the mountains above the town, but from a different direction. Avoiding the abundant and "justly dreaded" *yedra* (Western Poison Oak), Hartweg came upon "a beautiful pine-wood," many of the trees more than one hundred feet high. He tentatively named the conifer *Pinus benthamiana* after George Bentham, late secretary of the London Horticultural Society, but these were actually Pacific Ponderosa Pines, named earlier by David Douglas. A short distance beyond was a stand of scraggly pines with clusters of knobby, closed cones. This new species, Knobcone Pine, would be introduced into England from seeds collected by Hartweg and named nearly fifty years later by John Lemmon.

The *Joven* attempted to leave Santa Cruz on August 28, 1846, but ran into a northwesterly tempest that forced the little vessel back into harbor for repairs. When the journey resumed, Hartweg noted that the coastline up to San Francisco Bay was mainly treeless except for Año Nuevo, where there appeared to be some cypresses or pines. The Santa Cruz Cypress does grow inland among the Redwoods in this area, but what Hartweg observed was the isolated, northernmost stand of the Monterey Pine on the West Coast.

At the customhouse in Yerba Buena, now in American hands, Hartweg encountered a problem. He had had shipped to him from London two large containers that had come by way of the Sandwich (Hawaiian) Islands. Unfortunately, neither contents nor ownership of the boxes had been marked on them. The Mexican officials, suspecting that they contained contraband, perhaps silk stockings or printed calicos, had torn them open, only to discover two small Wardian greenhouse cases, plus some garden seeds, nails, and so on. Hartweg had no papers to prove he was the owner, so he had to petition the American captain of the port and prove before the magistrate that the boxes were his.

Unimpressed by the sparse local vegetation, on September 10 Hartweg crossed the bay to Sausalito, planning a collecting trip up the

Sacramento River. Proceeding to Mission San Rafael he encountered General Mariano Vallejo, an important Mexican sympathetic to the U.S. cause in Alta California. News from Vallejo was not encouraging. Supposedly, a huge force of angry Walla Walla Indians was gathering near Sutter's Fort intent on hassling settlers, a rumor later proving to involve only forty peaceable Indian families. Nevertheless, at that moment American and Mexican volunteers were being sent up the Sacramento River, and Sonoma was in an uproar. The general immediately left for Sonoma, and Hartweg preferred to tag along on the forced march under a boiling sun rather than be abandoned in unfamiliar countryside.

The warlike state of Sonoma convinced the naturalist to abandon his Sacramento River plans and embark instead on a cross-country collecting trip to the Russian Fort Ross on the coast, which had just been purchased by John Sutter. Hartweg's route by way of Rancho San Miguel was relatively unexciting from the standpoint of collecting, although he did record what he thought was a burning bush (now *Euonymus occidentalis*), procured seeds from an unidentified sweet-shrub (now *Calycanthus occidentalis*), and collected a specimen of a new oak, the California Black Oak.

Hartweg returned to Yerba Buena by the end of September and on October 7, 1846, was back in Monterey, having sailed again on the *Joven*. He expected to leave soon aboard that bark for San Diego, where he would commence an overland return trip to Monterey. But again the Mexican War intervened. Since mid-August Los Angeles had been under United States control, but at the end of September Mexican insurgents forced an embarrassing surrender of the American forces. The second week of October, with the war at full boil, the Americans tried unsuccessfully to retake Los Angeles. Hartweg wisely decided to remain in Monterey. The rainy season arrived in early November and continued until the end of March. There was especially heavy precipitation during January and February, causing flooding of the region's rivers and again forcing Hartweg to postpone his excursion to San Diego.

After the wet, dreary winter, Hartweg was cheered by the appearance of spring on the Monterey Peninsula, marked by the early flowering of some shrubs: silk-tassel, Fuchsia-flowered Gooseberry, Pinkflower Currant, Oregon-grape, Western Chokecherry, Evergreen Huckleberry, and four species of manzanita. Also some nonwoody plants flowered as early as February, including shooting stars covering the grassy hillsides, as well as a yellow violet, Baby-blue-eyes, and California Poppy. Hartweg continued his local field trips, discovering on hills several miles from the seashore a new stunted cypress (Gowen Cypress) he later introduced into English gardens. On this same outing he found in a shady dell a rhododendron in bud (California Rosebay) and to his excitement a shrubby form of the Giant Chinquapin, also in bud—the very genus from which the London Horticultural Society especially wanted seeds. Considering the temperate environment of the peninsula, he was convinced that this shrub could grow in Great Britain.

Finally deciding that it might be safe to visit the Sacramento Valley, "where the settlers are all foreigners," Hartweg boarded an American bark bound for Yerba Buena on March 8, 1847. He waited there for five days, hoping to find some transport up the Sacramento River, all the while looking for new plants. One of the beauties he found was a scarlet columbine. By a great stroke of luck Hartweg ran into a friendly fellow German who had just brought a launch full of produce to Yerba Buena and was shortly to cruise back up the Sacramento River to his "farm" near the edge of the foothills of the Sierra Nevada. He invited the naturalist to accompany him and stay at his home.

The portly, jovial rancher was Theodor Cordua, originally from Mecklenburg north of Berlin, who had been a personal friend in Europe of none other than John Sutter. Sutter had corresponded with Cordua in Germany, inviting him to California. When Cordua arrived in 1841, Sutter leased him a parcel of his own holdings at the junction of the Yuba and Feather rivers. Here Cordua developed a farm, becoming the first settler up the Sacramento Valley. Two years later he received an immense Mexican land grant of eleven Spanish leagues (forty-nine thousand acres) north of the Yuba River, which he used as a stock ranch. Cordua's original farm, called New Mecklenburg (now within the city of Marysville), became a popular stopping place for travelers, including John Charles Frémont in March 1846. Frémont mentioned that Cordua's holdings had about three thousand cattle and some sheep, but his farm was primarily devoted to growing wheat and other grains and vegetables, with an Indian village on the ranch providing workers. Cordua boated his produce to a market in Yerba Buena.

The Cordua launch left Yerba Buena on March 23, 1847, with Hartweg aboard. They passed by Corte de Madera, went through the Carquinez Strait into Suisun Bay, and entered the Sacramento River on the twenty-sixth, making some stops for botanizing along the way. On the last day of March, Hartweg and his host disembarked at the mouth of the American River and proceeded upstream six miles to Sutter's establishment. Over the next two days, the two rode twenty miles on horseback north to Cordua's farm. Spring had arrived about a month earlier here compared with Yerba Buena, but as Hartweg noted with respect to the flora, he recognized "many old acquaintances."

During the next month and a half Hartweg traveled in country new and appealing to natural scientists. The only individual who had botanized in the Sierra Nevada countryside prior to Hartweg was John Charles Frémont, currently fighting in California, an enthusiastic amateur but lacking Hartweg's qualifications and experience. Thus Hartweg became the first "professional" naturalist methodically to botanize the western flank of the northern Sierra Nevada.

From the farm Cordua and Hartweg traveled north along the Feather River on a four-day trip upstream through Cordua's stock ranch on his Spanish grant land. For the first twenty-five miles the river wound through an open woodland of evergreen and deciduous oaks. The two left the river and crossed a prairie for twenty miles, the wildflowers forming "brilliant-colored patches," especially the California Poppy, which "produced a splendid effect," and a buttercup, *Ranunculus canus,* not to be collected again for decades. Hartweg observed that in the many small dry creekbeds there were plants very different from the surrounding vegetation, their seeds likely having been brought downstream by floods.

At Cordua's foothills rancho Hartweg seized the opportunity to accompany a group of settlers heading into the Sierra Nevada to build a sawmill. Leaving on April 13, the party traveled through oak woodland with scattered Foothill Pines, camping in the evening by a little mountain rivulet. The next day they rode through chaparral composed mainly of ceanothus, and finally into "a noble Pine forest" including Pacific Ponderosa Pine, Sugar Pine, and most likely White Fir and Incense Cedar, with an understory of Mountain Dogwood and a new species of *Ceanothus,* Mahala-mat, to be named by George Bentham based on Hartweg's specimen. Perhaps Hartweg's most beautiful find of this excursion was the Scarlet Fritillary, which he collected during a noon stop at the edge of the pine forest. His specimen had just a few pendant, bell-shaped blossoms, but some plants have been found in California with nearly three dozen flowers on a single stalk.

By the end of May 1847, when Hartweg returned to Cordua's farm in the valley, he observed that where several weeks earlier there had been a "carpet of flowers," now "a yellow, sickly tinge pervades the whole."

Apprehensive over the rapidly approaching hot, dry California summer, he "lost no time in collecting such seeds as were worth taking." He visited the "Butes" (Sutter Buttes) in the Sacramento Valley. Among Hartweg's finds there were a reddish-purple milkweed, a small-flowered flax, and a lemon-yellow owl's-clover.

His next excursion was back into the mountains, this time along the "Chuba" (Yuba) River. From the farm he traveled fifteen miles across parched prairie to the foothills, then up into a chaparral of ceanothus with scattered live oaks and Foothill Pines. Along a small rivulet he found several different mints, a *Collinsia* with whorls of whitish-yellow flowers, and an interesting Baby-blue-eyes. He collected seeds from the last, anticipating that if they germinated, "it will prove a great acquisition to that handsome genus." On this trip he hoped to penetrate higher into the mountains, but the Yuba River became so wild that he was forced to turn back. Later in the month he returned to the *Collinsia* to collect seeds. Finding his hands stained yellow from glands located on the fruit of the plant, he named it *Collinsia tinctoria,* essentially meaning "used for dyeing."

Toward the end of June Cordua invited Hartweg to join him and an Indian on another Sierra trip, this time across to the Bear River, which parallels the larger Yuba River to the north. The Bear River flows in the canyon (beside today's Interstate 80) that rises eastward to Donner Summit. Many were the flowering plants along the trail to delight Hartweg: mariposa lily, violet, monkeyflower, onion, wild-ginger, lotus, and others. The travelers' destination was Bear Valley, then a beautiful, green, grassy valley, about half a mile square. But at six thousand feet only a few flowers were in bloom this early, including a distinctive peony, *Paeonia brownii,* with pendant maroon-petaled flowers, and a clumped waterleaf with many funnel-shaped white blossoms, *Hesperochiron californicus.* In the surrounding mountain ravines and on the higher slopes there were still drifts of snow several feet deep. Blocking the upper end of the meadow was a granite massif, "larger than the largest building of Europe," and off to the north was a chasm eight hundred feet deep wherein cascaded the Yuba River. The tranquil Bear Valley was soon to be completely transformed by the placer mining of the gold rush, and today much of the uppermost portion where Hartweg botanized is nearly impenetrable.

After a final trip into the mountains to collect seeds he had missed, it was time to pack up and return to Yerba Buena. On June 30, 1847, with an American guide, he traveled by horseback down the Feather River to the Sacramento, where the baggage including collections was put in a canoe while the two men swam their horses three hundred yards across the river. The horsemen continued down the north shore to Carquinez Strait, where they crossed by ferry and rode on through San Jose to Monterey. By this time Hartweg, suffering from what he called "a kind of tertian fever" (malaria), could barely stay on his horse's back. When the two finally arrived in Monterey on July 8, the German botanist temporarily cured himself from his own little medicine chest and was soon out collecting more seeds.

Revisiting Santa Cruz at the end of the month, he had a recurrence of the ague, but ignored it in his excitement at discovering a Giant Chinquapin "with ripe fruit." Then the malaria returned, this time for a month. Finally, on September 6, he was able to return to Santa Cruz to collect seeds from the cones of Douglas-fir, Pacific Ponderosa Pine, and Knobcone Pine. Hartweg's last extended excursion in California, starting on September 20, took him with a guide south along the route of the missions, eventually to San Luis Obispo beyond the Santa Lucia Mountains. He encountered, as had Douglas and Coulter, the Coulter Pine, Bishop Pine, and Bristlecone Fir. The desirable cones of the fir were only at the crown of the more than fifty-foot-tall trees. Hartweg cut several of the trees down, only to find that their cones were stunted and frostbitten, with no seeds. Hiking up through thickets of Blue-blossom and manzanita west of Mission San Antonio de Padua he discovered a stand of Sugar Pines, but the cones had already dropped their seeds. Luckily he found some cones on the ground still containing nuts. Continuing on westward to the coast, he reached the isolated, southernmost stand of Monterey Pines (at Cambria). One of the abundant shrubs he discovered was a *Prunus* with hollylike leaves, whose red cherry-plum fruits, so Hartweg heard, were roasted and eaten as a favorite dish by the Indians. A decade earlier Nuttall had first come across Holly-leafed Cherry on the Monterey Peninsula. At San Luis Obispo he purchased a gallon of edible pine nuts from some visiting Indians, which were similar to the pinyon nuts he had encountered in Mexico. His last trip out of Monterey, on October 25 south along the coastline, finally came to a halt at the precipitous coastal mountains, but the London Horticultural Society collector nevertheless returned with some fifteen-inch-long Coulter Pine cones.

At the beginning of November, Hartweg, anticipating the arrival of

the rainy season on the Monterey Peninsula, began packing his collections for his return to England. No vessel arrived to transport him, and
indeed he had to wait for three more frustrating months. Meanwhile,
during this period gold was discovered at Sutter's Mill, somewhat south
of where Hartweg had earlier been botanizing. Finally, on February 5,
he boarded a Hawaiian sailing ship bound for Mazatlán, whence he
planned to cross Mexico overland to the Atlantic. But he was warned
against this overland route, so he sailed on south to Guatemala. There
he got halfway across Guatemala before being warned off again. Finally
he made a successful crossing of Nicaragua, found a ship bound for Europe, and reached Southampton, England, on June 3, 1848. On this same
date the United States Senate ratified a treaty with Colombia assuring the
future of a canal that would bring California closer in travel time to the
capital of the United States.

Karl Theodor Hartweg may have been the ultimate plant and seed collector, almost to the exclusion of paying much attention to the rest of
California's diversified natural history. Yet on his return Hartweg seemed
unappreciated by the London Horticultural Society and soon left its employment. Among other things, the society was irritated because he failed
to bring back seeds of the Bristlecone Fir from California. His many
specimens of plants were primarily studied by George Bentham of the
Linnean Society and published in *Plantae Hartwegianae,* including about
four hundred California species and varieties, of which eighty-one were
described as new, some of them in new genera. In addition, a number of
American plant species and a genus honor Hartweg. When he returned
to Europe, he was appointed inspector of the Duke of Baden's gardens
at Schwetzingen, where he pursued botany until his death on February
3, 1871.

Despite Hartweg's diligence as a field botanist in the face of war,
weather, and ague, despite his forays into Sierran terrain where no naturalists had ever trod, despite his trove of New World seeds and plants
collected for an Old World audience, his name seldom appears in our
western naturalist annals, unlike those of Douglas, Nuttall, and Menzies,
to name a few. Could his lack of acceptance be related to his having collected the type specimen of Mountain-misery, that bane of the Sierran
hiker?

PART THREE

The Overland Expeditions and Their Naturalists

LEWIS AND CLARK IN 1804–06 set the pattern for the overland expeditions, which were charged both with exploration and with assessing the country's natural resources. Commencing with Lieutenant Frémont's first of four California trips in 1843 and continuing through the Mexican Boundary Surveys, the U.S. Army Corps of Topographical Engineers spearheaded these expeditions during peace and conflict. Many topographical corps officers, including their surgeons, were interested in natural history, and included in each expedition's staff was one or more naturalists: invariably a botanist, occasionally a geologist, less frequently a zoologist or an artist, and rarely an ethnologist. The scientific collections went to institutions and individuals in the East and in Europe, and extensive reports appeared in government (and sometimes in private) publications. The naturalists' findings and accounts of their frontier adventures were well received by both scientists and the general public, especially Frémont's narratives.

John Charles Frémont

In Pursuit of the California Flora

The blue fields of nemophyla and this golden poppy represent
fairly the skies and gold of California.

—John Charles Frémont

THE FIRST EXPEDITION TO CALIFORNIA, 1843–1844

It may seem strange that John Charles Frémont—the intrepid explorer, the
"Pathfinder"—was concerned about flora. But actually his interest in natural history went back to teenage years at Charleston College in South Carolina, where he had shown aptitude as a student in the sciences, though easily distracted by feminine allures. Frémont was born in Savannah, Georgia,
on January 21, 1813, but grew up in Charleston. As a young man he caught
the attention of Joel Poinsett, the nation's first minister to Mexico. Poinsett
had strong interests in agriculture and botany and lent early support to the
establishment of a national scientific institution. It was he who introduced
the now-familiar Poinsettia into the United States from Mexico in 1828.
Poinsett facilitated Frémont's appointment as mathematics instructor on a
naval vessel, then as an officer serving with the U.S. Army Corps of Topographical Engineers, and culminating as assistant to French scientist Joseph
Nicollet on his Midwest expeditions. During the latter assignment Lieutenant Frémont, already versed in surveying, worked with Nicollet's young
German botanist, Charles Geyer, and gained expertise in field botany.

By 1842, when Frémont commanded his own government expedition
bound for the Rocky Mountains, he was definitely enthusiastic about
natural history. In Kansas near the Vermillion River he noted boulders
of compact reddish sandstone (Dakota sandstone). He observed that a
"red butterfly" (Monarch Butterfly) was always closely associated with

the Butterflyweed milkweed (food for the Monarch caterpillar). And in August when scaling what he thought was the highest peak in the Rockies, he caught a bumblebee, "certainly the highest known flight of that insect," later pressing it in a large book among flowers gathered on the expedition.

Frémont's trip was a topographical corps exploring expedition, not a Rocky Mountain flora excursion. Yet he collected, dried, and pressed a notable diversity of plants, and sent them without warning to Professor John Torrey, then at Princeton College, for examination. At first Torrey was bemused. But in March 1843, after completing the lengthy catalogue of the expedition's plants, Torrey described Frémont's first major plant collection as "a very interesting contribution to North American Botany."

About the time that Torrey's catalogue of Frémont's plants was being published with the official report of Frémont's First Expedition, Frémont was getting ready for his second army expedition westward, again departing from Kansas. The plan was to cross the Rockies over South Pass in southern Wyoming and continue to the Pacific Coast, connecting with the surveys of the 1841 United States Exploring Expedition in the Columbia River country. Frémont thought that he might then go south into California, through the Southwest, across the Rocky Mountains in Colorado, and back to the eastern seaboard.

When Professor Torrey learned about Frémont's new trip, he provided helpful suggestions about collecting and preserving plant specimens. In return, Frémont promised to pay attention to the botany, which, as Torrey wrote botanist Asa Gray at Harvard, "we, of course consider the main object of the expedition." Frémont seemed to take his role as plant collector seriously, because when he arrived at St. Louis in early May 1843, he sought out Dr. George Engelmann, a botanist and physician who was becoming increasingly important as a personal "clearinghouse" in St. Louis for scientists traveling into the West. Engelmann spent nearly two weeks with Frémont, providing "instruction for geological & botanical researches and collections." Engelmann bemoaned the fact that Frémont had not been authorized to take along an official botanist, but Frémont would again have with him Charles Preuss, the extremely able, though dour, German cartographer of his 1842 expedition. Also, Christopher Carson would once more serve as guide; Kit would be especially invaluable on this trip because he had already been to the West Coast with trapper Ewing Young's 1829 expedition. When Colonel John J. Abert, head of the Army Corps of Topographical Engineers, learned that Frémont intended to take along a twelve-pound howitzer cannon, he expressed considerable concern that in the eyes of Indians the expedition might seem more like a military than a scientific affair.

By early June the expedition with its complement of thirty-nine men was under way from eastern Kansas. A month later, as they approached the Colorado Rockies, Frémont was excited by the "many beautiful flowers, which we had not hitherto met," just a few of those Frémont would collect before autumn brought an end to the floral display of the West. Unfortunately, only a handful of these specimens was sent back from Fort Hall (Idaho) along the Snake River in mid-September. On October 25 the expedition reached the Columbia River at the Hudson's Bay Company Nez Percé fort (Fort Walla Walla), and a month later, with winter upon them, the travelers started south along the Deschutes River toward Alta California. At the Klamath marshes the party swung southeast into northwestern Nevada, camping on New Year's Eve 1843 at the western edge of Nevada's Black Rock Desert.

Their route now lay southward through the Great Basin along the eastern edge of the Sierra in Nevada, passing the Granite Range and then Pyramid Lake, of which a sketch was made and later published in the official report. They forded the Truckee and Carson rivers, with their cottonwoods and willows now leafless. In his journal Frémont commented that Big Sagebrush dominated the high desert, except in the alkaline sites where the common shrub was Greasewood. There was also a scattering of Mormon tea among the bushes of the slopes, and on the rocky lava foothills an open woodland of an apparently new species of pine with large nuts and single needles, some now covered with winter snow.

After a twenty-six-mile forced march in a heavy, chilling snowstorm, led by their new young Indian guide, the men made a high-desert camp about thirty miles southeast of Lake Tahoe on the last day of January 1844. They heard optimistic accounts from Indians that within "six sleeps" to the west was "where the whites lived." But the Indians also reported that the mountains would be impassable now because of the deep snow, and they urged the party to remain on the plains for the winter. Frémont, however, had already made his decision to continue. Although a few baggage mules had been lost since the Columbia River, all but two of the sixty-nine horses were with the party. Everyone was eager to travel, and on February 1, 1844, the expedition proceeded westward into Alta California.

On February 6, Frémont, mountain man Thomas "Broken Hand" Fitzpatrick, and Kit Carson, "with a reconnoitring party, on snow shoes," climbed a Sierran summit and saw to the west the broad, snow-less Sacramento Valley, beyond which was the barely visible outline of the Coast Ranges. Carson, looking through the telescope, thought he recognized a high point in the distance (Mount Diablo?) as one he had seen about fifteen years earlier when with Young's band of trappers. Surrounding the campsite of February 10 were many tall conifers of "noble appearance" that Frémont judged to be some new variety of cedar; this was the first written recognition of the Incense Cedar. Frémont also noted magnificent white pines (Sugar Pine), Mountain Hemlock, White Fir, and the red pine (Pacific Ponderosa Pine). On the morning of the fourteenth, Frémont and Preuss, fortified by the evening's starvation repast of "pea soup, mule, and dog" (the expedition's little dog, Tlamath), scaled a high peak from which they had "a beautiful view of a mountain lake at our feet" (Lake Tahoe).

Ten days later, Frémont's expedition was near present-day Placerville, California. The men had now descended about five thousand feet in elevation since the afternoon of the twentieth, when they had gone over the pass in the dividing range of the Sierra Nevada south of Lake Tahoe. This was indeed the "snowy mountain range," as its name suggests, and many a day's journey had been exceedingly difficult, to say the least (figure 18). The snow had lain deep among the timber; the open slopes were steep and icy; overhead there had been snowflakes, thunder in the distance, and squalls of rain; and rivulets of water gathered into roaring river torrents. Yet despite the aura of winter, on February 21 Frémont reported his first Sierra plant, "the common blue flax" (*Linum lewisii*, after Meriwether Lewis). As the track continued downward to the west, patches of green grass began to appear. The coniferous forest opened into a white-granite valley where conifers and live oaks intermingled, mountain birds were calling, and a sweet, almost summerlike breeze sprang up.

At the evening encampment of February 24, the half-starved horses were turned loose into the riverine rushes, except for one sacrificed to provide supper. In Frémont's mind, the "difficulties of the road" seemed at an end, and he decided to make a dash with a select party including Kit Carson and the best horses to Captain Sutter's settlement of Nueva Helvetia (Sutter's Fort). Meanwhile, Fitzpatrick was left to bring down the rest of the men, their exhausted horses and pack mules, and all of the baggage, at a much more leisurely pace. Frémont would later come back up to meet them with provisions and fresh animals.

The next morning Frémont's small group continued down the valley

Figure 18. "Pass in the Sierra Nevada of California" (near Carson Pass), by Charles Preuss with the John Charles Frémont Expedition, February 1844. (Courtesy California Historical Society, FN-30518.)

of the American River. En route Frémont took time to measure the circumference of a Sugar Pine, over twenty-eight feet at four feet up, and the following day he noted the first abundance of flowers in bloom, including a striking scarlet-colored gilia. As brushy understory began to increase, Frémont found a "new and singular shrub" with smooth, chocolate-colored branches, his first manzanita. On March 1, butterflies were swarming, and "numerous bugs are creeping out, wakened from their winter's sleep." There were manzanitas twenty feet high and a diversity of forest flowers were coming into bloom, dominated by slopes of shooting-stars.

Two days later the mixed coniferous forest gave way to a live oak woodland, but the main concern now was for Charles Preuss, who had missed the previous night's campsite. One man was sent back to look for him. The others continued, coming upon a trio of Indian women gathering basketfuls of Redstem Filaree, *Erodium cicutarium*, that Frémont recognized as a European weed. Finally, toward evening on March 5, Preuss showed up. Oblivious of the trail, he had become lost. Without any food, he had dug up bulbs, stuck his hand into an anthill and licked off the ants, which had "an agreeable acid taste," and appeased his craving for tobacco by smoking oak leaves. Thinking he heard dogs, he had succeeded in whistling up two wolves. When he encountered Indian men roasting acorns, he scared them away by offering his knife in exchange for some acorns. Finally, when he came upon the three Indian women, he noticed horse tracks and remains of the expedition's evening campfire, and rushed to rejoin the group. The next day, after passing banks of California Poppies in the beautiful oak savannah, the detachment reached Sutter's Fort. Frémont and Carson on horseback were greeted by the proprietor, Sutter himself, who had been strolling outside the fort wall.

The following morning Frémont led a relief party back to meet Fitzpatrick's larger group, with fresh horses and one of Sutter's packhorses loaded with provisions. Frémont was shocked when he encountered the group just below the main forks of the American River, declaring that "a more forlorn and pitiable sight than they presented cannot be well imagined," each "weak and emaciated man" leading an equally weak and emaciated horse or mule. Fitzpatrick reported that the "difficulties of the road" had actually increased over the past ten days. The lower foothills had proved to be made up of interminable steep spurs, and side ravines reverberated with churning water. The narrow trail for the pack animals and exhausted men on foot was sometimes hundreds of feet above the roaring river. To make matters worse, it had started raining heavily on the second day. The rain and melting snow made the mountain track so slippery that many of the animals, some with packs, had lost their footing and tumbled into the chasm. Frémont was stunned to learn that one of the mules that fell had been carrying almost all of the bales of plants they had collected between the Rocky Mountains and the eastern edge of the Sierra, over more than fifteen hundred miles of wilderness travel. The only plants saved from that long journey—a few from along the Snake River and the northwestern edge of the Great Basin—had apparently been in a separate plant press. But the bounty of the California spring and summer flora still lay ahead.

As John Sutter later wrote, he received Frémont "politely and his company likewise, as if an old acquaintance." At the fort some days were spent in replenishing supplies and obtaining new horses, mules, and some cattle. They had flour ground at the mill, got horseshoes and bridles from the blacksmith's shop, and acquired additional packsaddles, ropes, and bridles. Frémont observed that his party was "furnished in a princely manner" by Sutter, with repasts of fresh trout and salmon, beef, ham, venison, and even bear, along with fresh vegetables and fruit from the fort's garden, capped by white Rhine wine.

At Sutter's Fort, Frémont discharged five men, and another simply wandered away. Frémont met John Bidwell, a recent emigrant from the East, who had been in charge of Sutter's hock farm (vineyard) on the Feather River and would shortly become Sutter's assistant. Sutter provided Frémont with a young Indian vaquero to help manage the expedition's livestock, now 130 horses and mules and about thirty cattle, including five milk cows.

On March 24, 1844, Frémont's expedition headed south through the San Joaquin Valley with Frémont riding Sacramento, a beautiful gray saddle horse, Sutter's own favorite. Early spring was the perfect time for a march along the lower edge of the Sierra through grasslands and woodlands of Valley Oak and Interior Live Oak. There were numerous large rivers out of the mountains to cross: the Cosumnes, Mokelumne, Calaveras, Stanislaus, Merced, and Kern. Extensive thickets of twelve-foot-tall *Lupinus albifrons* (Silver Lupine) were crowned with spikes of blue flowers. In his journal Frémont noted great quantities of Common Soap Plant near the rivers, mats of Yerba-buena, and carpets of golden California Poppies: "A lover of natural beauty can imagine with what pleasure we rode among these flowering groves, which filled the air with a light and delicate fragrance." The party saw Tule Elk and bands of Pronghorns, wild horses and even fresh tracks of Grizzly Bears. There were ducks, geese, quail, and a Bald Eagle on the lookout for river salmon.

The San Joaquin Valley was a vast wetland as the snows that still extended low on the Sierra Nevada were beginning to melt, and on occasion rafts had to be built to cross the waterways. There were ponds and lakes crowded with sedges and rushes and Red-winged Blackbirds. One campsite was in the shade of Western Sycamores. Frémont remembered that such trees had been mentioned in the 1826 *Blossom* expedition report but was unaware that the type specimen had not been collected until a decade later, by Thomas Nuttall. On April 10 along the White River, Frémont took special note of the large cottonwoods, whose seed vessels were about to burst, and decided that they differed from any he had read about in François André Michaux's *Sylva,* which was being revised by Nuttall to include the trees of western America. This dominant riparian tree did turn out to be new, and it would be named Fremont Cottonwood, *Populus fremontii.* Frémont's incomplete specimens of the cottonwood survived this trip, but the actual type specimen was collected in 1846 during his next expedition. The following day along Poso Creek Frémont saw a beautiful magenta-flowered shrubby four-o'clock that reminded him of one he had once seen blooming in faraway Kansas. The Redstem Filaree, encountered repeatedly by Frémont from above Sutter's Fort in the Sierra to this southeastern part of the great valley, is today considered an indicator of misused land in the West, but in Alta California it was appreciated, at least before it went to seed, as a favorite forage for the vast herds of livestock that had undoubtedly facilitated its spread.

As the valley narrowed to the south, the countryside became noticeably drier, with a change in the vegetation. A Mormon tea, *Ephedra nevadensis,* that they had first encountered in the Great Basin near Pyramid Lake in Nevada became very abundant. Also, Coast Silk-tassel grew in the Kern River creek bottoms, in some sites actually replacing the Narrow-leaved Willow.

On April 14 Frémont wrote, "We here left the waters of the bay of San Francisco" (San Joaquin Valley), as the expedition began its climb eastward over the southern end of the Tehachapi Mountains. Frémont had come to realize that there was no Rio Buenaventura, no river running from the Rockies across the Sierra to the Pacific Ocean. Just the day before a Christian Indian, complete with long spurs and sombrero, speaking Spanish and riding a Spanish steed, had volunteered as a guide and recommended to Frémont that the expedition seek out the Spanish Trail across the desert rather than choose its own route. Following this guide, the expedition traveled alongside a flower-edged creek wooded with oaks, cottonwoods, willows, and sycamores. With "humming birds and other feathered friends of the traveller enlivening the serene spring air," the men proceeded over Oak Creek Pass, "a most beautiful pass covered with trees and flowers."

Joined by four more volunteer Indian *compañeros,* they descended through an expanse of fine bunchgrass, with at last no more filaree, then continued beyond a gushing foothills spring into a valley with "the sudden appearance of *yucca* trees." Frémont was less than kind in his description of one of his great discoveries, the Joshua Tree: "their stiff and ungrateful form makes them to the traveller the most repulsive tree in the vegetable kingdom." Climbing over a spur ridge, the expedition dropped into the Mohave Desert. At this point the main guide observed in Spanish, "No water, no grass, nothing . . . every animal that goes out upon them, dies." Preuss, as he looked at the arid landscape stretching into the distance, with its expanse of Creosote Bush (some up to ten feet tall), many varieties of cacti, and here and there a Joshua Tree, muttered, "We . . . must prepare for a miserable journey for the next months." Indeed, it was not long before Preuss was thrown off his mount, Polly, who had too close an encounter with a cactus.

Frémont's party traveled southeast across the desert toward another range, the San Gabriel Mountains, capped with snowy summits, through a land also dominated by the newly discovered "zygophyllaceous" shrub, Creosote Bush, initially to be named jointly by John Torrey and Frémont. Visible far ahead near the base of the foothills was a swath of rich orange Frémont mistook for variegated sandstone, but it turned out to be a brilliant field of California Poppies. This area today is in the vicinity of the Antelope Valley California Poppy State Reserve, northwest of Lancaster. At the edge of the mountains, while looking for pasturage for the horses and mules, Frémont found the single-needled nut pine again. His specimen from this spot became the type specimen of the Singleleaf Pinyon, unique among pines because its needles are individual, rather than in bundles of two to five.

On April 20 the party ran into the Spanish Trail, a relatively well-used route from Los Angeles to the Rocky Mountains. The jubilant Frémont wrote in his journal, "Once more, we felt like going homewards. A road to travel on, and the right course to go, were joyful consolations to us; and our animals enjoyed the beaten track like ourselves." At a fast pace for horses, mules, and men, the group reached the Mohave River in fifteen miles, just northwest of present-day Victorville. Here the stream of crystal water was sixty feet wide, but by the end of the next day only a dry bed remained. The expedition now moved rapidly east along the river, and then swung northeast to follow the Spanish Trail. Although Frémont asserted that the desert "afforded much to excite the curiosity of the botanist," he dared not spend the time to write at length, although he continued to collect.

For the next week they crossed "a miserable tract of sand and gravel" between "black, rocky ridges, bald and destitute of timber," but with "many beautiful plants and flowering shrubs." Some of those plants collected by Frémont and named by John Torrey were a mesquite called Screw Bean, with tightly coiled pods; the distinctive Desert Trumpet, with its inflated flowering stems; and the abundant purple-flowered *Psorothamnus fremontii*. Frémont's type specimen for this shrub, a member of the Pea Family, would be collected later, but not in California.

During this time some of the company's horses were stolen by Indians, and several men including Kit Carson set off to recover the animals. The next evening, after thirty hours and some hundred miles of reconnoitering, they returned with most of the horses and two Indian scalps, which shocked and disgusted Preuss but seemed to elate Frémont, although ordinarily he dealt benignly with Native Americans.

On April 30 Frémont's second western expedition left Alta California, southeast of Death Valley, with a goodly assortment of the new flora carefully stowed away, including a newly collected striking, tall-stalked, yellow-flowered mustard, Prince's Plume, which had been common for the past two days in alkaline sites. Frémont would see the same species in southeastern Colorado on his next expedition west. Thomas Nuttall first named this genus after a specimen collected on the Upper Missouri in 1811, describing it as "the most splendid plant in the Natural Order of Cruciferae."

Frémont was excited about the diversity of the Mohave Desert flora: "Throughout this nakedness of sand and gravel, were many beautiful plants and flowering shrubs, which occurred in many new species, and with greater variety than we had been accustomed to see in the most luxuriant prairie countries; this was a peculiarity of this desert." Frémont was the first really to sample the flora of Alta California's Mohave Desert. From the time he entered California near Lake Tahoe until he departed from it near Death Valley, his travels encompassed some six hundred miles.

His disappointment over having lost, in the wintry Sierra, the plants collected earlier in the expedition would be intensified during the evening of July 13, 1844, near the trip's end. In west-central Kansas a series of thunderstorms turned the Smoky Hill River into a raging monster more than five hundred yards across and, as Frémont wrote, "all our perishable collections [were] almost entirely ruined, and the hard labor of many months destroyed in a moment." Of about fourteen hundred specimens collected on the second expedition, many of them from California, about half were ruined or lost. Frémont tried to dry some of those soaked in the Kansas flood, but he could spare little time for such a task. Among those sent to Torrey were many unusable for identification, and Torrey was particularly dismayed since many of the plants had been collected "in regions not before explored by any botanist." Understandably, both Frémont and Torrey looked forward to another, more botanically successful expedition.

By the late summer of 1844, Frémont had begun preparing his expedition report in Washington, D.C., for Colonel Abert. Frémont and his young wife, Jessie, lived in his Missouri father-in-law Senator Thomas Hart Benton's commodious home with the family and developed a pleasant working schedule. Frémont organized his field data in the evening, then was up early the next morning for a breakfast of rolls and coffee. At nine Jessie would join him in the study, and they worked together on the report until one o'clock, with Frémont recollecting and relating general and specific information while Jessie recorded it, smoothing the presentation and making it more readable. As John Torrey provided identification of specimens collected, the scientific names often went into the report: "Here, among many new plants, a new and very remarkable species of eriogonum (*Eriogonum inflatum*, Tor. & Frem.) made its first appearance." After a light lunch, then, John and Jessie would take a leisurely stroll to the Potomac River.

The final document for the first California expedition was three times longer than the one prepared for Frémont's Rocky Mountain expedition. It was published late in 1845 as a U.S. Senate report and included John Torrey's monograph on the new genera and species of plants collected. So interesting was this report on Frémont's western exploration that it was published commercially in both the United States and Europe, soon becoming a best seller and a model for popular accounts of other explorations.

FRÉMONT'S RETURN TO CALIFORNIA, 1845–1847

An anticipated fish fry, scheduled for November 27, 1845, by the shore of Walker's Lake in Nevada never materialized. This was the designated rendezvous for the two parties of Captain Frémont's third topographical corps expedition (1845–47) into the West, his second to California. The parties had taken different routes across the western portion of the Great Basin and were now together again. But the fish fry was off because, as the few remaining Indians pointed out, the cutthroat trout were gone. There were other problems facing the expedition: provisions were almost exhausted, and it was already snowing in the mountains. Furthermore, Frémont's orders for exploration did not mention California. But the *Western Expositor* of Independence, Missouri, had correctly predicted their winter destination months earlier, namely, Alta California.

At Walker's Lake, Frémont once again divided his party of over sixty men, the majority armed, for the crossing of the Sierra Nevada. The main body was to travel south along the eastern side of the escarpment under the command of Theodore Talbot, guided by the famous Joseph Walker, who had thoroughly explored the country a decade earlier, and accompanied by Edward Kern, a reliable and talented cartographer. This group was to cross over the mountains toward the southern extent of the San Joaquin Valley and wait there for Frémont. Meanwhile, Frémont, with fifteen select men including Kit Carson and some of the Delaware

Indian hunters, would proceed directly westward over the mountains to Sutter's Fort. They hoped to obtain provisions from Sutter, then travel south down the San Joaquin Valley to rejoin the larger group near the headwaters of the present-day Kings River in the mountains east of Fresno.

Leaving the Great Basin at the end of November, Frémont and his small detachment headed west up the Truckee River, quickly crossing the emigrants' pass, which the next winter would become the most infamous route over the California mountains, Donner Pass. Here, Frémont collected his first plant in California, the Sierra Stonecrop. The succulent stonecrop in its rocky crevice was not in bloom on December 4, but on the other hand, to Frémont's relief, there was not even a powdering of snow on the pass. He had successfully avoided the Sierra winter. Before any fast-moving blizzards struck, the men continued down the Bear River toward Sacramento, "where the winter king already shrunk from the warm breath of spring."

Traveling down the western slope, Frémont's party left behind the forest of giant conifers—the Sugar Pines and Incense Cedars—and headed into oak woodland that included a new species with very long acorns, the Canyon Live Oak, which appeared to be a favorite of the local Indians. One village had wicker storage bins bulging with about two hundred bushels of the sweet acorns.

At Nueva Helvetia, John Sutter himself was away, but his right-hand man, John Bidwell, who had been left in charge, was delighted to see Frémont again, remembering him from 1844. Frémont officiously requested sixteen mules, six packsaddles, some flour, and other provisions, and wanted to use the smithy to shoe the mules. Supposedly there were no mules available and no coal for the forge. After some unpleasantries, including Frémont's suggestion that Sutter was siding with the Mexicans in not wanting to help the expedition, Bidwell managed to obtain fourteen mules and have them shod.

John Bidwell, a native New Yorker, had arrived in California toward the end of November 1841 with the Bartleson Party. Guided by Thomas Fitzpatrick, they crossed the Sierra Nevada south of Lake Tahoe. The emigrants were on the verge of starving, reduced to eating crows and even a Bobcat, and winter was approaching. When Bidwell was scouting for game, just at dark he stumbled upon a gigantic fallen log at what is today Calaveras Big Trees State Park. But his claim to have discovered the Giant Sequoias was preempted by Zenas Leonard's 1833 discovery farther south.

Frémont and Bidwell, despite present differences, had kindred interests. The two were shortly thereafter involved in the Bear Flag Revolt.
Bidwell later became a major horticulturist at his estate in Chico, with its
huge orchards, vineyards, and livestock herds; and he was active in agricultural experimentation. He and his wife would host, among others,
world-famous natural scientists Asa Gray, Joseph Hooker, John Muir,
and David Starr Jordan. Both Frémont and Bidwell during their respective careers served as military generals and ran as presidential candidates.

On December 14, 1845, refreshed and refurbished, Frémont and his
fifteen men were ready to travel south and rejoin the rest of his company.
The route was essentially that of the 1844 expedition, from Sutter's Fort
south through the San Joaquin Valley along the edge of the Sierra
foothills to the planned rendezvous with Talbot, Kern, and Walker. At
this season about their only collections were of oaks and Red Heather.
Noted again was the abundance of Redstem Filaree, or *alfalferia* as the
locals called it, which with another introduction, Black Medick, was favorite forage of the abundant livestock before spring grasses sprang up.

By January 1846 filarees, which Frémont called the "first valley flowers of the spring," and golden Douglas Violets had begun blooming
across many of the slopes. But what preoccupied Frémont was the
whereabouts of his other party, which he had sought in vain over Christmas and on New Year's Day up the Kings River and into the mountains.
On this excursion upstream at thirty-five hundred feet Frémont recorded
the intermixed pines and oaks on the ridges, with oaks, cottonwoods,
and sycamores in the canyon bottoms. At eleven thousand feet Frémont
and his party finally reached a granite divide, where he could look down
on the apparent headwaters of the San Joaquin River. A thousand feet
below, the streamlet dashed and sparkled, and Frémont reflected that
water was indeed "a jewel beyond price," and frequently a higher priority for an expedition than gold and silver and copper. For explorers,
he contemplated, among the most "prized objects of our daily life" was
"a clear cold spring of running water or a good camp, big game, or fossils imbedded in rock."

Returning down Kings River to the San Joaquin Valley, Frémont, ra-
tionalizing that Talbot's group had decided to travel slowly to keep their
animals in good condition, went back to Sutter's Fort and camped near
the mouth of the American River late on the evening of January 14. By
this time, in fact, Frémont's men were on foot and exhausted, having survived by eating their horses.

The next morning Sutter, out on a tour with two dignitaries and passing Frémont's American River encampment, ran into Kit Carson and asked to whom the camp belonged. Frémont himself was still asleep, but Sutter had him awakened, introduced him to his companions—William Leidesdorff, the vice-consul for the United States in Alta California, and William Hinckley, now captain of the port of Yerba Buena—and invited Frémont to dine at Nueva Helvetia. At Sutter's suggestion the expedition camp moved closer to the fort. When Sutter returned later, having visited Leidesdorff's huge land grant on the Feather River (Rancho de los Americanos), he ordered a seven-gun salute to honor the Frémont Expedition, and a sumptuous repast was prepared for all. Sutter warned Frémont that the Mexican authorities were increasingly uneasy about Frémont's visits, or any other Yankee visits at present. Sutter had appeased the "Californios," as the Mexicans were called, by reporting that Frémont was simply engaged in a geographical survey of Alta California. As it turned out, the Californios' misgivings about Frémont and Yankees in general were certainly justified.

Frémont obtained a passport from magistrate Sutter to visit Monterey. On January 19 Frémont boarded Sutter's launch with eight of his own men, Leidesdorff, Hinckley, and a newly arrived emigrant, William B. Ide, and sailed down the Sacramento River to Yerba Buena. It was this first visit to San Francisco Bay that inspired Frémont to name the ocean entrance to the immense, beautiful harbor the Golden Gate. At Yerba Buena, Frémont was entertained by former New Englander Hinckley and by Leidesdorff and his attractive, girlish Russian "wife," who lived in the finest house in Yerba Buena, an adobe bungalow with a wide porch facing the bay and a pretty garden. Apropos of his interest in geology, Frémont, together with Hinckley, took a trip south to the quicksilver (mercury) mine at New Almadén. Planned as a quick tour, initially by boat down the bay to the Alviso embarcadero near Santa Clara, the excursion fell apart as the boat was becalmed in the bay, then got lost. As a result the men had to spend a dismal, chilly night tied up among the rushes. But Frémont was rewarded at New Almadén by getting all the specimens of vermilion-colored cinnabar he could carry.

On January 24 Frémont set off from Yerba Buena in the company of Leidesdorff for Monterey. The two enjoyed chatting as they rode their horses south down the valley beyond San Jose. Frémont learned about Leidesdorff's prized garden and his intense interest in natural history, and was impressed by his knowledge of the country. Near Monterey, Frémont collected his first plant in three weeks, the Chamisso Bush Lupine.

The specific purpose of Frémont's Monterey trip was to obtain permission from the Mexican government for his expeditionary party to remain in Alta California, to refit and obtain supplies, fresh horses, and mules. He then planned to proceed north up the Sacramento Valley to the Columbia River by way of the Willamette Valley, thus completing in Frémont's view "from beginning to end *this road to Oregon,*" and linking with the 1841 surveys of Wilkes's U.S. Exploring Expedition. After meeting with U.S. Consul Thomas Larkin in Monterey on January 27, 1846, the next morning Frémont talked with Don José Castro, the Mexican commanding general, and other officials, informing them that his expedition was essentially engaged in surveying "the nearest route from the United States to the Pacific Ocean . . . the object of the survey was geographical . . . made in the interests of science and of commerce, and that the men composing the party were citizens and not soldiers." Without hesitation, General Castro granted the requested permission.

Frémont enjoyed the shrubs and other plants coming into bloom on the Monterey Peninsula. On January 31 he collected two more specimens: the penstemon *Keckiella breviflora* and Black Sage. Then he traveled north to Laguna Ranch, owned by William Fisher, thirteen miles south of San Jose, through fields of European wild oats three feet tall. Here Frémont's entire company was reunited, including Talbot's party, which Kit Carson had been sent to bring down from Sutter's Fort. Apparently there had been confusion over the designated meeting place at the southern end of the Sierra Nevada, with Talbot and Walker coming into the San Joaquin Valley over present-day Walker Pass about seventy-five miles farther south than Frémont expected. The most exciting zoological encounter at the San Jose encampment was when one of Frémont's men accidentally backed into a captive California Condor wounded by a Frémont nimrod, and received a sharp pinch and a ripped shirt for his carelessness.

Reorganized, the Frémont Expedition commenced its survey work again. They moved up into the nearby Santa Cruz Mountains the third week in February, where for Frémont especially "the great trees in the forest . . . had aroused my curiosity" and where, incidentally, he searched for a future homesite. He was impressed by the *palo colorado* (Redwoods), measuring many of them; by the handsome evergreen oaks; and especially by the Pacific Madrone, with its polished red trunk, one of them sixty feet tall and four feet in diameter. Many shrubs and flowers were in bloom, and Frémont was able to procure a variety, including Suncup (an evening-primrose), Pinkflower Currant, Hairy Manzanita, Blue Nightshade, Red Maids, and a shooting-star.

Frémont would not collect another plant (Buckbrush) until March 11, 1846, when he and his "party of highwaymen," as General Castro by then characterized them, were on their way across the San Joaquin Valley, having been ordered out of the territory after a controversial but somewhat humorous standoff between Frémont and the Mexicans in the vicinity of San Juan Bautista.

Now for the last time in Mexican California, Frémont had some free time and opportunity to pursue his exploration and natural history, completing his Oregon–Northern California survey of the 1843–44 expedition. By mid-March 1846, traveling east, Frémont had reached the Merced River in the San Joaquin Valley, "the country clothed in the floral beauty of spring." The expedition swung north across the Stanislaus River toward Sutter's Fort. On the twenty-second, camping at a favorite spot on Eliah Grimes's ranch near the fort, Frémont spent several days trying to obtain more horses and mules. Finally, on March 24, the expedition was on its way up the Sacramento Valley for thirty miles through undulating countryside covered by an unbroken blanket of yellow-and-white composites (Tidy-tips) and golden California Poppies, with Baby-blue-eyes on the streambanks. Many years later Frémont would reflect that "the blue fields of nemophyla and this golden poppy represent fairly the skies and gold of California." Fittingly, a subspecies of *Nemophila pulchella* is now named in Frémont's honor.

On March 26, near the junction of the Feather and Yuba rivers, Frémont paused at the fine rancho of German emigrant Theodor Cordua, where just a year later the London Horticultural Society's plant collector Karl Hartweg would sojourn. Frémont probably found the distinctive red, saprophytic Snow Plant up the Yuba River from here, and along the Feather River added the Yellow Mariposa Lily and Adobe-lily.

The expedition on March 30 reached the ranch of a Danish emigrant, Peter Lassen, near the mouth of Deer Creek, a tributary of the Sacramento River southeast of present-day Red Bluff. Lassen came to the United States in 1830 as a master blacksmith, settling for several years in Missouri, where he became acquainted with another emigrant named Sutter. A year after Sutter was bitten by the western bug, Lassen arrived in California. Through his friendship with Sutter, Lassen obtained Mexican citizenship by 1843 and a twenty-two-thousand-acre land grant as his ranch. Here Frémont found Lassen running cattle and growing wheat and cotton. Lassen had "lately planted a vineyard, for which the Sacramento valley is considered to be singularly well adapted." Frémont and Lassen quickly became congenial friends. Around Lassen's Ranch Fré-

mont collected a host of plants, seven of them new to science. The specimens ranged from the familiar Western Poison Oak to the type specimen
of the Fremont Cottonwood.

By the end of April the Frémont Expedition—having made a side trip
north toward Mount Shasta and back to Lassen's Ranch and traversed
rugged canyonland east along the Pit River branch of the Sacramento—
now moved northeast into Modoc Indian country. Here the landscape
reminded Frémont of the Great Basin, but more fertile and well watered.
They traveled through good grassland plains, north past the expanse of
black basalt flows making up present-day Lava Beds National Monument, and on May 1 camped at the edge of today's Clear Lake. The party
waited several days for hunter Auguste Archambeault, who had failed to
return from his game-hunting foray. Search parties failed to locate the
well-liked Canadian, but after three days of being lost and luckily avoiding a band of marauding Indians, he turned up. The company, having enjoyed a rest, shifted camp a short distance up a creek running into Clear
Lake, a move portrayed in a sketch by Edward Kern. The Indians of this
region were considered unfriendly, but the only noteworthy altercations
for the Frémont Expedition here involved encounters with bothersome
bears. During one of the bear bashes, Delaware Charley broke his nose
in a fall from his horse, and naturalist Frémont managed to set it in an
outline considered to be superior to the original.

On May 6, 1846, the expedition continued north up Lost River, left
Alta California, and headed toward the Klamath Lakes basin of Oregon.
Frémont's last collection on the border of what would soon become the
new state of California was a prostrate mustard that would be named
Smelowskia fremontii (now *Polyctenium fremontii*) in his honor by
Sereno Watson of Harvard.

The night encampment on May 8 was near the shore of Oregon's
Upper Klamath Lake. This was less than twenty miles from the Klamath
Marsh, where in December 1843 Frémont's previous expedition had
swung to the southeast into the Great Basin and eventually into Alta
California. Frémont was excited to link the tracks of his two expeditions,
but as he later wrote in his memoirs: "How fate pursues a man!" Standing by his campfire in the chill of this early spring evening, Frémont heard
faint hoofbeats, then saw two horsemen coming into the firelight. They
reported that some distance behind them rode Lieutenant Archibald
Gillespie of the U.S. Marine Corps with dispatches for Frémont from
Washington. Early the next morning, worried that Indians might ambush
the marine, Frémont with a few others galloped south a fatiguing forty-

five miles to rendezvous with Lieutenant Gillespie and his three companions. Gillespie's dispatch from Washington announced that the United States intended to take California from Mexico. Frémont was no longer simply the leader of a scientific surveying expedition, but now in addition an active officer of the American army.

The expedition hastened south toward Sutter's Landing, suffering periodic attacks by Klamath Indians in southern Oregon that killed three of Frémont's men. Despite these distractions, he still thought to name a new river after his botanical mentor John Torrey and in Alta California collected thirty additional plants before the end of the month, especially around Lassen's Ranch again. One unusual new genus would come to bear Frémont's name—*Fremontodendron* (Flannelbush)—the California species of which he collected on May 27, 1846, on Deer Creek (figure 19). By the first week in June Frémont's company was camped at Sutter Buttes, where many other Americans were congregating, and Frémont soon became a directing force in what came to be known as the Bear Flag Revolt. This controversial revolt evolved into the Mexican War, in which Frémont would be heavily involved. During the remainder of 1846 he collected only one more plant, the eye-catching Scarlet Monkeyflower, in early September somewhere north of Los Angeles. And throughout 1847 but one additional California specimen would be added.

On January 13, 1847, three days after the Articles of Capitulation were signed for the war in California, Frémont was appointed governor and commander in chief of the territory of California by U.S. Navy Commodore Robert F. Stockton. From this moment, Frémont's ability to pursue his natural history interests diminished. On August 22, 1847, Brigadier General Stephen Kearny issued an order from Fort Leavenworth to arrest his former family friend, Lieutenant Colonel Frémont, and have him transported to Washington for a court-martial, in part because Frémont—an army officer—had been taking orders from a U.S. Navy commodore. Remarkably, on the trip back to Washington, Frémont, with important matters on his mind, still collected a few plants.

After the court-martial trial early in 1848—where Frémont was charged with disobedience but not mutiny and the punishment was remitted—he resigned his commission and left the U.S. Army Corps of Topographical Engineers. In future years he would organize and lead two more expeditions to California, neither under the auspices of the federal government.

Figure 19. Flannelbush, *Fremontodendron californicum*. (Photo by R. G. Beidleman.)

FRÉMONT'S FINAL TWO CALIFORNIA EXPEDITIONS, 1848 AND 1853

Frémont penned a short note to Professor John Torrey from the nation's capital on November 1, 1847, when his general court-martial was imminent. He had planned a longer letter, but "accumulated business pressing upon deranged health" had intervened. Enclosing a little flower from his last expedition for identification, Frémont explained that when he thought of California and plants, he thought of Torrey and simply had to write. As he put it, the "pleasant work in California" was interrupted, but his floral collections had been sealed in airtight cases and would eventually reach Torrey by ship. Frémont's regret at the moment was that Torrey's generic name of *Fremontia* for Greasewood, collected during the first western expedition, had been replaced by *Sarcobatus*.

By the time of his next letter to Torrey at the end of February 1848, Frémont had, as he put it, "entirely withdrawn from the public service." But, he optimistically continued, "our work will still go on, and . . . I even hope to be able to give it a greater extent and beauty than we had before anticipated." Sure enough, only a week later Frémont wrote that he would leave for Missouri in April 1848 to lead a civilian exploring party directly to San Francisco. Frémont did not reveal that he hoped Congress would underwrite this expedition, which had multiple goals: to complete Frémont's West Coast survey, to strengthen ties between the restless Californians and the Union, and to scout for a railroad route across the headwaters of the Rio Grande in the southern Rocky Mountains. Frémont had long dreamed about surveying a central railroad route, and many businessmen in the Midwest, not to mention his father-in-law, Senator Benton of Missouri, were highly supportive of such a venture.

The plant cases with their floral treasures from Frémont's second expedition to California finally reached the Navy Yard in Brooklyn on June 29, 1848, aboard the *Erie,* and one of Frémont's men brought them to John Torrey's home. The two large shipping containers filled with many tin cases had been delivered to the *Erie* in California by mule. Each case was filled with dried specimens, then soldered closed and packed within a strong wood frame. A green cowhide was sewn snugly around each frame. What a scene it was beneath the trees on Torrey's lawn in New York City, with Torrey, his two young sons, and Frémont's man busily cutting off the tough cowhide coverings and finally getting the tins open. Some specimens were damaged, but Torrey enthusiastically concluded that "there must be a thousand species of plants in the collection," not to mention containers with pinecones, fruit, and seeds. Torrey fervently hoped that Frémont "would be going out again, & will doubtless be as active & zealous as ever in observing the Botany of his favorite regions." And, of course, "his plants will all fall into my hands, & I shall give account of them in a work on California which he intends writing."

Way behind schedule, Frémont was still in Washington in early September, now anticipating a winter trip that would be "a severe one." But he wrote to Torrey that he would try to collect such plants as the season might afford. Indeed, for this expedition Frémont, encouraged by Torrey and Asa Gray, hired a young German gardener-botanist, Frederick Creutzfeldt. The stoic German cartographer Charles Preuss from his previous expeditions would also be along again, as well as Edward Kern, who had earlier been left in command of Sutter's Fort after the American flag had been raised there, much to Sutter's irritation. In addition, Edward's two younger brothers were included, Dr. Benjamin Kern, to serve as surgeon and assistant naturalist, and Richard Kern, a talented landscape artist.

This expedition—Frémont's fourth into the West, and his third to California—was destined for disaster. Congress refused to support it, so it had to be financed privately. Frémont's wife, Jessie, and their two children accompanied him to St. Louis, where Frémont again spent a short time with botanist George Engelmann. The first of several misfortunes struck in October 1848 on the trip from St. Louis to the expedition's starting point near Westport on the Kansas River, with the death of the Frémonts' six-week-old baby boy. Frémont and his thirty-three men departed much too late in the year. They were caught in winter snow high in the southern Colorado Rockies, where they had been misled by their venerable mountain-man guide, "Old Bill" Williams. By the time what was left of the expedition reached New Mexico, ten of the members had perished, along with 120 mules. Their equipment, collections, and journals had been abandoned. The three Kerns, Preuss, and botanist Creutzfeldt did survive, but the latter, in a small relief party, came close to death and later was unjustly accused of cannibalism. When Benjamin Kern and Bill Williams went back from Taos in an attempt to recover the expedition's equipment, they were set upon by Ute Indians and killed.

A few of the men, including Creutzfeldt and Preuss, continued to California through the Southwest with Frémont. But the plants Frémont collected along the Gila River and in Sonora and turned over to Creutzfeldt—who professed, as Frémont noted, "to be a botanist"— were repeatedly soaked and "many are ruined & the rest he did not even label." Frémont nevertheless wrote Torrey that he was still optimistic about future collecting in Sonora and California, since "there are a great many new plants and my interest has rather increased than diminished in the pursuit." About sixty-five plants from the 1848–49 trip reached Professor Torrey. Fewer than half were collected in California, and of these only two were new species, including the attractive Desert-lavender, named *Hyptis emoryi* by Torrey.

Writing to John Torrey in the fall of 1850, Frémont promised that when he returned to California with his next expedition, he would have an employee "collect the Spring plants of our neighborhood," the Sierra foothills region around his Mariposa Grant—the estate he acquired in 1847. It was true that after Frémont reached California in the late spring of 1849 following the Colorado disaster, botanist Creutzfeldt and Frémont's old-time cartographer Preuss were employed as surveyors. But as Frémont planned for another private—and final—expedition to California in 1853, no botanist was included. This expedition again crossed the southern Rockies, just slightly north of the 1848 expedition route. At this same time the United States Boundary Commission was busy surveying the new Gadsden Purchase from Mexico, and the government had initiated its series of Pacific railroad surveys. The one on the 38th parallel led by John Williams Gunnison crossed Cochetopa Pass (Colorado) just a few months ahead of Frémont's party. Alas, Frémont's departure once more was delayed, and he barely avoided another major winter disaster, this time in the Utah mountains.

In 1853 John Torrey's "Plantae Frémontianae; or, Descriptions of Plants Collected by Col. J. C. Frémont in California" was published by the Smithsonian Institution. But Frémont's promise to Torrey about productive collecting on the fifth expedition essentially came to naught. Only about seven specimens were garnered, although a yellow-rayed composite probably collected in the California desert, *Syntrichopappus fremontii,* did represent the nineteenth new plant genus based on Frémont's western collections. After 1850, however, there was virtually no further correspondence between Frémont, erstwhile botanical collector and the new state of California's first senator, and Professor John Torrey, the botanical savant. John Charles Frémont's love of flora had waned.

Battles and Botany

Emory and the Army of the West,
1846–1847

... should be employed in collecting data which would give
the government some idea of the regions traversed . . .
—Emory's orders from Colonel J. J. Abert

On August 18, 1846, the Army of the West under General Stephen
Watts Kearny marched into the famous old town of Santa Fe (New
Mexico), and the Mexican War on the upper Rio Grande was concluded without a single shot being fired. In the late afternoon of September 25, at the temporary headquarters there, First Lieutenant
William Hemsley Emory, Kearney's chief topographical engineer, with
his advance guard surveying party, was ready to depart for Alta California with General Kearny and three hundred dragoons to join in the
fighting continuing there.

The red-haired, bearded, serious-minded lieutenant Emory was much
more than a topographical engineer with a talent for cartography. Appointed to West Point at the age of only eleven by Secretary of War John
C. Calhoun in 1823, he became a member of the West Point Class of
1831. As a cadet, nicknamed "Bold Emory," he was a classmate of
Speaker of the House Henry Clay's son and had a close boyhood friend
named Jefferson Davis. Emory's marriage to Matilda Bache made him
son-in-law to Alexander Bache, who was director of the U.S. Coast and
Geodetic Survey, while Matilda's great-grandfather was none other than
Benjamin Franklin. In addition to the influential Alexander Bache, young
Emory's circle of friends eventually expanded to include botanist John
Torrey, Harvard College botanist Asa Gray, zoologist Louis Agassiz, as-
tronomer William Bond, botanist George Engelmann of St. Louis, and
members of the staff at the new Smithsonian Institution including Sec-

retary Joseph Henry, a physicist; zoologist Spencer Baird; and herpetologist Dr. Charles Girard.

In 1838 Emory was recruited to serve as one of thirty-six officers for the new U.S. Army Corps of Topographical Engineers by its chief, Colonel John J. Abert. His first assignment involved harbor improvements along the Delaware River, while from 1844 to 1846 he received invaluable experience working on the Canadian boundary survey. In addition, using a variety of publications he produced a comprehensive map of the Republic of Texas. This so impressed Abert that Emory was assigned to General Kearny's Army of the West in 1846 to prepare a map of the proposed southwestern route to Alta California and give "the government some idea of the regions traversed," including the botany, zoology, and geology.

Emory's surveying team included another topographical engineer, First Lieutenant William H. Warner, who was charged with making the topographical sketches; landscape painter John Mix Stanley with his assistant, Dabney Eustis; Norman Bestor, a statistician, who as Emory's naturalist aide would be responsible for specimen collecting and sketching; Bestor's assistant, John Riley; plus six teamsters and drivers, and the private servants of Emory and Warner. In Washington Emory obtained from the navy two Gambey sextants and two excellent box chronometers, and at Fort Leavenworth a Bunton syphon barometer, used especially for measuring altitudes, all to be carried in the instrument wagon. On the way west from Washington the stagecoach with Emory and his staff crashed, thanks to its drunken driver, while crossing the Allegheny Mountains in Pennsylvania. Bestor proved his worth, having wisely carried the chronometers separately in a basket on his arm. Later he would also prove to be a source of amazement to southwestern Indians because of his spectacles.

In addition to instruments, Emory brought along a working library, including two copies of Frémont's report of his 1843–44 expedition to California; maps of Texas, Oregon, and California by Philadelphia geographer Samuel Augustus Mitchell; Emory's own map of Texas; two nautical almanacs; and Santa Fe trader Josiah Gregg's two-volume *Commerce of the Prairies* (1844). Between 1831 and 1839 Gregg had made four trips along the Santa Fe Trail to New Mexico, recording many observations on the natural history of the region that were included in his book. When Lieutenant Emory arrived in St. Louis, Missouri, he visited botanist George Engelmann at his home near the courthouse, where Dr. Engelmann, becoming used to such visitors, provided valuable natural history advice for the excursion into what was essentially a virgin region for exploration. But as Emory later reminded Professor Torrey, the trip was primarily an army concern, and those "intervals snatched from military duty" could not compare with the accomplishments of a strictly scientific expedition.

During the layover at Santa Fe in August and September 1846, Emory took time to design and supervise construction of Fort Marcy, which would overlook and protect the town, naming the fort after his friend Secretary of War William Marcy. On September 25 it was time to move out, following General Kearny's force south along the Rio Grande. The army was well equipped for the trip, hauling a number of howitzers and tents for all the men—but no tent poles!

On October 6 the Army of the West encountered Kit Carson in central New Mexico, appearing out of a huge dust devil stirred up by an accompanying band of horsemen. Carson was riding hell-for-leather east to Washington on an express trip from California. He told them that he carried the message that Alta California "had surrendered without a blow." Commodore John Sloat of the United States Navy, with no resistance, had raised the flag at Monterey on July 7, Mexican generals were supposedly in flight, and Frémont had probably been appointed military governor. In light of such optimistic news, General Kearny felt he could safely and successfully carry out his orders to establish a California government when he arrived on the coast, so he immediately sent back about two hundred of his three hundred troops to Santa Fe. Furthermore, since "Broken Hand" Fitzpatrick, Frémont's old guide who was now leading Kearny's First Dragoons, had never been farther into the Southwest before, Kearny tried to persuade Carson to replace him. After all, Carson had just traveled through the country where Kearny was bound. With much reluctance, Carson finally agreed and was joined by old-time trapper Antoine Robidoux, who was likewise familiar with Alta California, as scout and interpreter. Meanwhile, Fitzpatrick was sent on to Washington in Carson's place with the favorable war news. Emory, it seemed, could now complete his own reconnaissance and mapping for a wagon road to San Diego.

With Kit Carson, the large party circled west across the Black Mountain range, crossing the Mimbres River and past the historic copper mines at Santa Rita del Cobre. Skirting the Burro Mountains, the party on October 20 reached the upper Gila River, which bounced down out of the Mogollon Range from a wild land that in 1924 would become the first designated Wilderness Area in the United States, on the New Mexico–Arizona border, named in honor of conservationist Aldo Leopold.

For the next month the Army of the West, on the Gila Trail, followed the Gila River westward through the Sonoran Desert, passing haunts of the Apache, Pima, and Maricopa Indians. Emory, the first "naturalist" to traverse southwestern New Mexico and southern Arizona, astutely commented that "not one object in the whole view, animal, vegetable, or mineral, had any thing in common with the products of any State in the Union," save, he thought, for the cottonwood. This was actually a new species, too, the Fremont Cottonwood. Emory's observations on the Indian tribes were detailed, especially for the Pimas and Maricopas, and many exceptional new plants including cacti were collected. But by November 17 the desert landscape bordering the Gila River had become dreary "beyond description," as Emory wrote. The vegetation consisted of a scattering of dwarf Creosote Bush and desert mint bushes across a fragmented pavement of black volcanic basalt with interspersed stretches of sand.

On November 22, 1846, near the junction of the Gila and the Colorado rivers—a scene painted by Stanley (figure 20)—the expedition's optimism about the political security of California began to dissipate. It was apparent that a huge party of horsemen had been in the area recently; and from some Mexicans whom Emory managed to capture, it was learned that five hundred horses were being driven into Sonora from California for use by Mexican general José Castro's troops. The next day another Mexican was picked up who carried letters from Alta California reporting that a counterrevolution had taken place and "the detestable Anglo-Yankees" had been driven out of Los Angeles, Santa Barbara, and elsewhere. Evidently Alta California was not a part of the Union yet.

On the twenty-fifth, Kearny's Army of the West forded the Colorado River, fifteen hundred feet wide at the crossing near present-day Yuma, and touched the sandy desert shoreline of California. They pushed hastily on toward San Diego, the dragoons crossing the Colorado Desert's *jornada del muerto*—the Devil's Highroad—fifty-four miles in two arduous days. Among the manifold difficulties during this forced march were problems with newly captured wild horses, which had never been ridden before. Travel at first was along the edge of the immense Algodones Dunes, through a scattering of Mesquite and Mormon tea. Increasingly the desert became a landscape of flats covered with lava fragments, alternating with sand berms, and some horses and mules actually

Figure 20. Near the junction of the Gila and Colorado rivers, looking west toward Picacho Peak in California. Sketch by John Mix Stanley with the Emory Expedition, November 1846. (From William Emory, *The Notes of a Military Reconnaissance* . . . [1848].)

had to be pushed through the sand. Despite optimistic promises by the Mexican prisoners, there turned out to be no surface water, and water could be procured only by digging a deep pit—fifteen feet down—into a dry, sandy creek bottom. Emory reflected that "the desert was almost destitute of vegetation," with only an occasional Mormon tea, Creosote Bush, saltbush, or woody desert mint. The only luxuriant shrubs, growing along the dry creekbeds, were Mesquite, whose beans both soldiers and horses relished.

Although the Mexicans warned against it, the expedition, now traveling below sea level, detoured to the northwest past a stagnant, saline lake. As they approached the playa border, according to Emory's report, it turned into "a thick soapy quagmire," with the stench from dead shoreline animals unbearable and the water "wholly unfit for man or brute." Straggling troopers seeking to quench their thirst were sorry. In this geologically interesting country southwest of the present-day Salton Sea, Captain Abraham Johnston of the First Dragoons collected fossilized "mussel" shells buried in the basin of Salton Sea's predecessor, the late Ice Age Lake Cahuilla. Farther west in old marine sediments, the men found even more ancient fossilized clams. Years later a paleontologist called the region "one of the most remarkably complete sequences of animal life to be found anywhere in the world." At the entrance into the desert foothills along Carrizo Creek were reddish badland buttes of sand and clay with gypsum seams that had eroded into fantastic shapes, reminding Johnston of similar strata seen along the Canadian River in northeastern New Mexico.

A sudden morning fog out of the southwest, lasting for several hours, brought some relief from the heat, but many of the mules and horses perished before they reached a magnificent spring—Ojo Grande—at the source of Carrizo Creek in the granitic foothills. Emory's subsequent analysis of a handful of salt gathered from the spring's edge indicated sodium chloride (salt), calcium sulfate (gypsum), and magnesium sulfate (Epsom salt); small wonder that a number of the horses and mules became bloated and died from drinking too much of this water. In the crossing of the *jornada* the soldiers had not suffered as much as the animals, but the abrasive vegetation and rocky outcrops had torn their uniforms and boots to shreds. In the vicinity of Carrizo Creek, Emory collected two new plant species: a euphorbia, Broad-leaved Stillingia, a spiny-leafed annual; and Brittlebush, an abundant desert sunflower. On November 28 Emory found an unusual composite without ray flowers, *Encelia frutescens,* which Gray would describe as new. Later, botanists' confusion over southwestern hot springs (*aguas calientes* in Spanish) resulted in the Rayless Encelia's type locality being listed as southwestern Arizona's Agua Caliente Springs, which Emory never visited.

Out of the Colorado Desert and west of present-day Anza-Borrego State Park, the army traveled northwest up the Vallecito and San Felipe valleys (traversed years earlier by botanist Thomas Coulter), with an escort of harassing "wolves" (probably Coyotes). The diversity of the flora improved with the appearance of a new buckwheat, a possibly new *Plantago,* and extensive thickets of the "centennial plant," Agave, including one in flower. Many of the distinctive Ocotillos, as yet botanically undescribed, were in striking scarlet bloom. Clustered about rocky hillside springs (and first seen at Ojo Grande) were what the soldiers in the party familiar with Florida called "cabbage trees," stands of the California Fan Palm. These palms had been noted earlier by Spanish travelers in Baja California, including Fra Junípero Serra. Later there was speculation that these palms had been brought from the Mediterranean by the mission fathers. Indeed, when Serra first arrived in San Diego in 1769, he planted an Old World palm there. But the California Fan Palms are true natives of the California desert draws, and Lieutenant Emory made the first scientific recording.

Continuing up San Felipe Valley on December 1, 1846, against intermittent gale winds from the west, the Army of the West camped at the abandoned Indian village of San Felipe. Next morning the company proceeded over the ridge at the head of San Felipe Creek, about thirty-five hundred feet above sea level, between mountains with summits frosted by ice and snow and whose flanks bore woodlands of evergreen oaks and pines. In the vicinity a Holly-leafed Cherry specimen was collected. After dropping down to Jonathan Warner's rancheria in the beautiful grassy valley of old Agua Caliente (now called Warner Hot Springs), the first priority was food. At the evening's camp seven of Emory's men consumed an entire sheep at one sitting. Up the hillside a short distance was an impressive hot spring, discharging from a crack in the granite wall at a temperature of 137 degrees Fahrenheit and filling the air with a sulfurous aroma. Just above was another spring, but this one was cold, 45 degrees. Emory anticipated that the day would come when "the invalid and leisure seeking portion of the white race" would assemble "here to drink and bathe in these waters, ramble over the hills which surround it on all sides, and sit under the shade of the great live oaks that grow in the valley." In the future Warner Hot Springs was indeed destined to become a popular resort.

For the Army of the West there was little lingering in the shade of the evergreen oaks. After a much-needed two days of recuperation, the men moved on toward San Diego. By good fortune, on December 5 Kearny's Army of the West was augmented by a party of three dozen men from San Diego led by Captain Archibald Gillespie, the same marine who had brought news of imminent war with Mexico to Frémont the previous May. Gillespie was accompanied by navy lieutenant Edward Beale and by Frémont's veteran but youthful hunter and guide, Alexis Godey, of Frémont's Second and Third expeditions.

This countryside—as Kearny had been warned by Marshall, the young New Hampshire chap in charge at Warner's hot springs—was definitely "near the heart of the enemy's stronghold." Indeed, the next day farther southwest at San Pascual the resident "Californios" (Mexicans), on horseback and with nine-foot lances and pistols, attacked in superior force and with clever tactics, despite Carson's claim that they would not fight.

The so-called major battle of the Mexican War in California had erupted, evaporating for the moment Emory's consideration of natural history. Only one Mexican died, but eighteen Americans were killed in the skirmish and thirteen were injured. Four of them later died, there being only one army surgeon, Dr. Griffin, to administer to them. General Kearny, heavily beset, was saved by Emory but still sustained several deep lance wounds, and Emory's topographic assistant Lieutenant Warner was similarly wounded. Antoine Robidoux was so seriously injured that he was given up for dead, but during the night, when the temperature dipped below freezing, he regained consciousness and begged Emory for a cup of coffee. Although there had been no coffee for days, the cook actually found enough to brew a cup that Emory "poured into the waning body of our friend Robideaux." Antoine so revived that in appreciation he pressed into Emory's hand a dirty piece of brown-flour cake from his pocket, which proved to be laced with tiny flour beetles. Unfortunately, Godey and two companions, who had been dispatched to San Diego for help, were captured and held by the Mexicans. Today the site of what became known as the Battle of San Pasqual, east of Escondido, is preserved as a California State Historic Park. Because of Emory's "gallant and meritorious conduct," he was breveted a captain.

Killed in this battle was Captain Abraham R. Johnston of the First Dragoons, Kearny's aide-de-camp and Emory's closest friend. Like Emory, Johnston had strong interests in geology, archaeology, and contemporary Indians. His journal, kept from the time the expedition departed from Santa Fe on September 25, 1846, until two days before his death, was filled with detailed sketches of prehistoric Pueblo Indian pottery shards, Indian "wigwams" of willow branches, geological formations, and even the Saguaro in bloom, as well as notations of natural history interest. He mentioned Greasewood, growing in the alkali flats, using the old name *Fremontia*. Johnston's final entry on December 4, after he heard there might be a party of up to eighty Mexican soldiers in the vicinity, read, "too certain to make a dash on them in a dark, stormy night; so we slept till morning." Early that next day, Johnston led the advance charge into the enemy camp, and as Emory noted later in the day upon returning to the battle scene, "the first object which met my eye was the manly figure of Capt. Johnston. He was perfectly lifeless, a ball having passed directly through the centre of his head." He was one of only two soldiers killed by bullets.

The next day, five miles beyond, the Californios attacked again, at a site that fittingly became known as Mule Hill, keeping the Americans under siege for four days and forcing them to survive on mule meat. Carson, Lieutenant Beale (the same naval officer who would several years later carry news of California's gold strike back to Washington), and Beale's Indian servant volunteered to dash to San Diego, still in American hands, for help. During their precarious trip, the trio managed to sneak quietly past three rows of Mexican sentinels by taking off their shoes. They finally arrived in San Diego, still barefooted having lost their shoes, in such terrible condition that Carson suffered for several days and Beale never completely recovered from the ordeal. Commodore Stockton quickly sent 180 sailors and marines from his naval command to the rescue of the Army of the West, and the entire assemblage reached San Diego on December 12, 1846. Probably the last plant Emory collected, on the "plains of San Diego," was the common aromatic Dove Weed, a relative of the Poinsettia with stinging hairs, used by the local Indians to stun fish.

As the Army of the West came into San Diego, the rain was falling in a winter torrent. From one of the Mountain Men the first sight of the ocean elicited the exclamation, "Lord! there is a great prairie without a tree." Emory and his crew were assigned to a miserable one-room hut but preferred to stay instead in the muddy plaza, storing their notes, chronometers, and other instruments in the hut. Fortunately, the "indefatigable Mr. Bestor" was left on guard and prevented four burly fellows who broke into the hut from stealing any of the valuables. The next morning Captain Gillespie procured a house with two rooms for Emory, where the topographic party was able to work on the expedition collections and fieldwork notes. During the brief stay in San Diego, Emory and Gillespie became good friends, neither man anticipating that in a few years the marine would marry another great-granddaughter of Benjamin Franklin, as had Emory, and the couple would become the next-door neighbors in Washington of Emory and his wife, Matilda.

Emory's official topographical reconnaissance was at last completed, and on December 28, receiving orders to report to General Kearny for military duty, he was able to store his equipment and notes aboard the USS *Cyane* anchored in the harbor. During January 1847 as he served with Kearny there were several more armed encounters between the army and the Californios in this southwestern region, with the Americans the victors. The third week of January Emory received special orders from Kearny at Los Angeles to report back to San Diego, thence to sail on the captured brig *Maled Adhel* carrying dispatches for Washington, including a brief letter to Lieutenant Colonel Frémont that would play a role in Frémont's subsequent court-martial.

Emory departed from California on January 25, but not before he had penned the final entry for the reconnaissance, about California's flora. Emory regretted that the season of his stay "was unfavorable to obtaining a knowledge of its botany." The deciduous trees had lost their leaves, and neither flowers nor seeds could be collected. Much of the country was without trees, except for a few live oaks, sycamores, and pines, mostly in the mountains. The hillsides were covered with wild oats, mustards, "carrots," and immense herds of cattle. Optimistic residents enthusiastically assured Emory that this southern portion of Upper California was "unequalled for the quality of the grape and the wine expressed from it." As to the future of agriculture in the dry Southwest, Emory warned, as did Major Long decades before him and Major Powell decades after him, "In no part of this vast tract can the rains from Heaven be relied upon, to any extent for the cultivation of the soil."

In a letter dated October 8, 1847, to Albert Gallatin, who was writing an essay for the New York Ethnological Society dealing in part with Emory's observations on southwestern Indians, Emory regretted having departed from California in such haste, leaving behind many of his notes and collections, including the seeds of Pima Indian native cotton, a valuable long-fiber variety. But fortunately most of his preserved plant specimens had already gotten into the hands of Professor John Torrey, who was preparing an elaborate catalogue with drawings, the latter probably done by Isaac Sprague of Cambridge, Massachusetts. With considerable enthusiasm Torrey wrote to Emory that "the route which you passed over is extremely rich in botanical treasures."

Despite the difficult circumstances, Emory and Norman Bestor, his assistant, were able to collect some thirty species from Southern California, among which were at least eight new species, including Red Shank (a relative of Chamise), Southern Mule's-ears, *Eriogonum trichopes,* Broad-leaved Stillingia, Brittlebush, Rayless Encelia, and *Dicoria canescens,* from a new composite genus found "in the desert of drifting sands." Undoubtedly more new species were gathered, but Professor Torrey summed up the problem in his comment on "an apparently new species" of *Grindelia:* "the flowers had fallen from the heads, and one specimen is therefore scarcely sufficient for determination." Torrey did honor Emory by naming *Perityle emoryi,* a new rock-daisy, discovered in the San Felipe Valley. However, the abundant southwestern oak that honors Emory, *Quercus emoryi,* occurs in Arizona but not California.

Emory collected parts of, or had sketches made of, some dozen new cacti for George Engelmann's perusal. Although many of these do grow in the California desert, they were first encountered east of the Colorado River. One of the chollas was named by Engelmann after John Stanley, Emory's artist (*Opuntia stanlyi*, Devil Cholla). The most remarkable cactus was the Saguaro, some growing more than fifty feet tall; its type locality was somewhere along Arizona's Gila River. Engelmann initially named this world's largest cactus based solely on its first written description and sketch, in a letter from Emory.

Emory's experience with Kearny and the Army of the West was definitely not at an end when he arrived back in Washington in 1847. Dispatches Emory carried from Kearny outlined the current problem of command in California: the navy faction with Commodore Stockton and army officer Frémont was at odds with the army faction of General Kearny. In addition, Emory on the way home had written a lengthy letter, subsequently published in several newspapers, delineating the command problem and favoring Kearny's side. During the court-martial trial of Frémont, commencing on November 2, 1847, Emory as a witness supported Kearny's position, causing a major rift between Emory and both Frémont and his influential father-in-law, Senator Benton.

A decade later, in his own military career, William Emory commanded cavalry troops in the Indian Territory, where "Bold Emory" won distinction for never losing a soldier to capture or death during a tour extending into early 1861. He fought laudably on the Union side in the Civil War and finally retired from the army in 1876 with the rank of brigadier general after serving for forty-five years. William Emory died on December 1, 1887.

George Engelmann, when examining what appeared to be a new barrel cactus collected by Emory along the upper Gila River on October 25, 1846, paid appropriate tribute to this soldier-scientist: "I propose to name it after its zealous discoverer, who has, surmounting numberless difficulties, although occupied by severe and arduous duties, found leisure to do so much for the advancement of our knowledge of the wild countries traversed by him." Engelmann made this statement prior to Emory's next western assignment—one of even greater scientific import—associated with the Mexican War aftermath.

Emory certainly lived up to Colonel Abert's earlier expectations. During Emory's 1846 reconnaissance of the Southwest with Kearny, he drew definitive maps, was adept at astronomical observations, had representative flora collected, recorded the daily temperatures and weather conditions, calculated each day's mileage by the odometer attached to a howitzer wheel, picked up specimens of galena and copper, recorded the dips of the sandstone strata, and reflected on such geological processes as the role erosion played in landscaping the arid region. He even dabbled in archaeology and anthropology, being the first to examine thoroughly the Pueblo ruins at Pecos (New Mexico) and those at Casa Grande (Arizona), while contemplating the origin of the prehistoric builders and making observations on the contemporary Indians. From the time he left Fort Leavenworth, Emory's botanical collecting, under the circumstances, was prodigious, including many species new to science, and his collections would eventually go to John Torrey and be shared with Asa Gray. His magnificent assortment of cacti went to George Engelmann in St. Louis. The official 1848 congressional report, *Notes of a Military Reconnaissance, from Fort Leavenworth, in Missouri, to San Diego, in California,* included a catalogue of the plants by Torrey and of the cacti by Engelmann. Emory's report, according to a later historian, "ranks in importance with Frémont's volume on the road to Oregon." Emory's map of the Southwest, together with those of Frémont, Preuss, and Nicollet, made more complete and accurate the mapping of the West from the Great Plains to the Pacific. His draftsman, John Mix Stanley, completed 151 paintings during the trip, many of which were included in the final report. Unfortunately, a fire that destroyed the Smithsonian building in 1855 consumed all but five of Stanley's original paintings of the expedition.

The Mexican
Boundary Surveys

A graduate school for collectors . . .
—A. Hunter Dupree, *Asa Gray* (1959, 205)

THE OFTEN CONFUSING PRELIMINARIES

In mid-September 1847 the United States Marines had at last entered the "Halls of Montezuma" in Ciudad de México, and the major fighting in the Mexican War was over. In California the war had come to an end much earlier, with the military commandant of California, Lieutenant Colonel John Charles Frémont, adding his signature to the Articles of Capitulation on January 13, 1847. But it would still be more than a year before the Treaty of Guadalupe Hidalgo—signed on February 2, 1848, and ratified by the Senate on March 10—essentially brought a formal end to the war. Patriotically enough, the treaty went into effect on July 4, 1848. As reported in the diary of Henry W. Bigler, this was about six months after James Marshall, the "Boss of the Mill," found some kind of "mettle" that looked like gold up the American River at John Sutter's new sawmill in the Sierra Nevada (Rawls and Orsi 1999). Inevitably, these two momentous events would change the western North American scene forever.

A provision in the treaty called for establishment of a boundary commission "to run and mark the said boundary in its whole course" between Mexico and the United States and be responsible for the final boundary decision. The commission was to consist of an appointed United States commissioner and surveyor and their Mexican counterparts, each group with its own support staff. For the United States, the responsibility for the nation's first full-scale survey fell to the Army Corps of Topographical Engineers, to delineate the specific boundary between Mexico and the United States and to make a comprehensive evaluation of the semiarid and arid countryside through which the boundary line would extend.

Perceptive William Emory, by a letter of April 2, 1849, urged James Buchanan, the outgoing secretary of state, to negotiate a national boundary at 32 degrees north, extending roughly from Ensenada on the coast of Baja California eastward to just above Mexico's El Paso del Norte (present-day Ciudad Juárez). This recommendation was based on Emory's own experience in crossing the Southwest, and—of immense significance—could ensure that the United States would have an eventual route for a national railway and wagon road that avoided both the mountainous landscapes and the vicissitudes of winter weather farther north.

The well-intentioned Mexican Boundary Survey of 1849 quickly encountered problems—topographic, economic, social, and political—a few humorous, most of them vexing. Generally the treaty designated the Rio Grande as the boundary from the point where it enters the Gulf of Mexico northward to the "southern boundary of New Mexico," leaving within Mexico the old town of El Paso del Norte; thence the line would continue to the western New Mexico boundary, north to the Gila River and west to its junction with the Colorado River, then in a straight line to just below San Diego Bay. Three important questions requiring answers involved the exact location of the starting point on the coast just south of San Diego, what constituted the New Mexico border, and the exact location of the junction of the Gila and Colorado rivers.

It was now discovered that the main map the American and Mexican diplomats at war's end had initially depended on for determining the boundary was inaccurate. This was New Yorker John Disturnell's 1847 map of Mexico and Texas. Disturnell was mainly a publisher of guidebooks, and no one knew where he had obtained the inaccurate information for his map. The Americans definitely desired enough borderland for a wagon and railroad route, but the boundary designated in the treaty, based on Disturnell's map, headed west about fifty miles north of El Paso del Norte (thus eliminating much of what has today become the booming southern Arizona–New Mexico Sunbelt).

Beyond this major shortcoming, there ensued within six years the appointment, one after another, of six United States commissioners, each with differing politics and qualifications. Also, operational funds for the project were not readily obtained from Congress, and internecine squabbles erupted among the professional scientists and nonprofessionals over who would be the boundary survey naturalists. And there were unending minor incidents: a fight between army officers over a San Diego woman, a growing array of friends and relatives who became survey "baggage," bickering among survey staff over decisions, the saga by one commissioner of returning an attractive kidnapped Mexican señorita to her family, a squabble between Mexicans and Americans over the definition of a "marine league" measurement (which would determine how far south of San Diego the boundary line started), chief surveyors who became distracted by personal interests, and last but by no means least, the problem of gold. From the active placers and mines and those yet to be discovered came the call of Midas to members of the boundary survey and its military escort, as well as to the citizenry at large. Desertion for the goldfields was to burgeon like an epidemic.

President Polk's first choice for American commissioner was former Arkansas Democratic senator Ambrose H. Sevier, appointed on December 12, 1848. But Sevier died before the Senate could confirm him. Then on January 16, 1849, two months before Polk left office, the Democratic president named John B. Weller, a lawyer and Democratic congressman recently defeated for the governorship of Ohio. Weller had seen service in the Mexican War, but although he now resided in California he knew little about the Southwest. Meanwhile, the Mexicans appointed as their commissioner General García Condé, who proved to be a superb administrator, scholar, and gentleman. The group of United States "surveyors" was a mix of topographical engineers and nonmilitary professionals. Former Texas Ranger Andrew B. Gray, a civilian and loyal Democrat who had done surveying in upper Michigan and had worked on the Republic of Texas boundary survey in 1839, was named by Polk to be the chief American surveyor, while William Emory, now a major, was selected as both chief astronomer and commander of the escort.

From the very start the lines of command were not clear, and for some time, to his disgust, Emory was not even included in the commission's meetings and plans, while the less experienced Gray was. Fortunately, Emory had two able assistants. Lieutenant Amiel Weeks Whipple, a topographical engineer who had graduated from West Point in 1841, had already gained considerable knowledge of the Southwest during the Mexican War and exhibited a growing professional interest in the diversity of southwestern Indian tribes. The other assistant, a topographical engineer who had graduated fifth in his class of 1846 from the U.S.

Military Academy, was Lieutenant Edmund Hardcastle. He had previous surveying experience during the Mexican War and had been breveted for bravery especially at the Battle of Molino del Rey.

En route to California by way of Panama, Emory was already disillusioned by the organization—or lack thereof—of the commission. Among other things, he discovered that Weller had made no arrangements for transportation of the commission members from the western coast of Panama north to San Diego. And because of pressure from gold seekers for transportation out of Panama, the commission group had to delay from mid-March 1849 until mid-May before any were able to leave for California. Trust Emory, however, to make good use of the delay by setting up the surveying equipment and training his assistants, as well as determining the latitude and longitude of Panama City and recording other phenomena. His careful survey of Panama City eventually permitted an accurate map of the region to be produced.

The team of Americans who finally managed to convene in San Diego on June 1, 1849, just before the deadline, to commence the western boundary survey numbered thirty-nine people, about half of them civilians, plus an infantry and cavalry unit of some 105 soldiers. The Mexican delegation did not arrive until July 3.

Among the United States civilians was Charles Christopher Parry, appointed as assistant surgeon and chief botanist and geologist for the survey. Professor Torrey promoted Parry's appointment, not only because Parry was qualified but also because Parry would undoubtedly send his botanical collections to his friend Torrey. On the other hand, Asa Gray, now professionally prominent at Harvard and patronizing western collectors, urged the appointment of Charles Wright, a Yale graduate who had moved to Texas in 1837, where he occupied himself with land surveying and botanizing. In 1844 Wright commenced a correspondence with Gray and sent him a collection of east Texas plants. Impressed, Gray invited Wright to the Harvard Botanical Garden as curator during the winter of 1848–49, to work on western collections. But Wright soon became eager to be back in the field, and after conferring with Engelmann, Gray gained permission from the secretary of war for botanist Wright to join the eastern division of the boundary group, assembling near El Paso. Meanwhile, Torrey managed to have Parry included with the California division.

Parry was born near the Malvern Hills of England's Gloucestershire in 1823. His family immigrated to the United States and had settled north of New York's Catskills by the time Parry was ten. He graduated with honors from Union College (Schenectady) and in 1846 received his

M.D. degree from Columbia College, where he became well acquainted with John Torrey. Shortly thereafter he moved to what would become his lifelong home, Davenport, Iowa. In central Iowa he joined a small government land surveying party during the summer of 1847, and the next year he served with David Dale Owen's second important geological survey of Wisconsin, Iowa, and Minnesota.

The immediate decision on the West Coast at the end of the Mexican War involved the governance of the new territory of California. On April 13, 1849, General Bennet Riley, having just arrived in Monterey, assumed the office of both commander of the Tenth Military Department and military governor of California. In his dual position he made arrangements to send out exploring parties throughout the territory, for military purposes as well as to gather information about California's natural resources, especially geologic. In his opinion, expressed in a letter to the War Department, no other part of the western frontier required as complete an exploration as California. Lieutenant George Derby was sent to map the Sacramento Valley, followed by a reconnaissance south along the eastern San Joaquin Valley and southern Sierra together with Lieutenant A. J. Wilson. Captain William H. Warner, Emory's former assistant topographer, and Lieutenant Robert S. Williamson were sent into northeastern California with eleven civilian assistants and an old Hudson's Bay Company voyageur, François Bercier, as guide to search out a possible railroad route to the Great Basin. Unfortunately, Warner and his small field party were ambushed by two dozen Pit River Indians, and Warner and the guide were killed. Lieutenant Edward Ord was dispatched to examine Southern California's Cajon Pass area and look into possibilities for an eastward rail or wagon route extending beyond the Navaho country. And Riley requested that Major Emory with the Mexican Boundary Survey proceed to investigate the junction of the Colorado and Gila rivers.

These minor surveys were timely. Just the previous September (1848), according to a letter from John Henry Alexander to Joseph Henry at the Smithsonian, Philip Thomas Tyson of Baltimore—a chemist and geologist involved in mining—had expressed an interest in visiting California "upon a scientific exploration of the Country, for his own gratification and the prosecution of possible ulterior views." This he did on a private tour extending from June into October 1849. With recommendations from Washington, D.C., including one from Henry, he was able to travel around the state, often with Georgia's Whig congressman Thomas Butler King, who had been sent out by Whig president Taylor. The president not only was interested in California's natural resources but also hoped to encourage California to come into the Union as a "Whig free state" (without slavery). King and Tyson were abetted in their reconnaissance by the commander of Division of the Pacific, Colonel Persifer Smith. In October 1850, when King, defeated by Frémont and Gwen, failed to become one of California's new senators, he was appointed collector of the port of San Francisco.

This flurry of surveying activity between the Mexican War and the Boundary Survey reports resulted in a publication in 1851 by Tyson, *Geology and Industrial Resources of California,* including his observations and the scientific reports by the topographical corps, comprising geology, mineralogy, and botany.

LIEUTENANT WHIPPLE AND BOTANIST PARRY: THE MEXICAN BOUNDARY SURVEY, 1849

Despite the Mexican Boundary Commission's diversified problems, planning for the survey was already under way in San Diego, and the staff was divided into several parties. Chief Astronomer Emory, responsible for setting up headquarters and the astronomical observatory, named the encampment at the southern end of San Diego Bay after General Riley. Establishment of Camp Riley and the determination of the initial western boundary point on the Southern California coast were completed on June 9, 1849. Meanwhile, Chief Surveyor Andrew Gray was initially responsible for surveying the San Diego port. Emory assigned Lieutenant Edmund Hardcastle, assisted by civilian George Gardner and with an army escort of twenty-three men and Mexican engineer observers, the difficult task of surveying the 140-mile boundary line eastward through wilderness mountains and desert to the Colorado River. Emory devised a unique surveying technique that involved plotting gunpowder explosions at high points along the boundary. Remarkably, when checked later, the completed boundary length came within six inches of Emory's calculations!

Emory's other assistant, Lieutenant Whipple, with a cavalry escort under Lieutenant Cave J. Couts, was to travel from San Diego along much of the Army of the West's 1846 route to the junction of the Colorado and Gila rivers and determine the exact latitude and longitude of that important spot. Surgeon Parry was to accompany Whipple. On this trip to the Colorado River and back with Whipple, Parry hoped to make more detailed collections and scientific observations than earlier ones by Thomas Coulter and Emory on sections of almost the same route. Also

Parry, though primarily a botanist, was to prepare the final geological report.

Near the end of July 1849, when everyone was finally ready to go to work, Commissioner Weller dropped a bombshell. He informed Emory that the commission's appropriation was completely exhausted, forcing a delay in sending out the major surveying parties. Naturalist Parry thus busied himself during the summer months around San Diego in the diversified natural history haunts of Menzies, Botta, Coulter, Deppe, Nuttall, and Hinds. Like his naturalist predecessors, Parry enjoyed visiting Mission San Diego, relocated in 1774 to a new site five miles up the San Diego River's Mission Valley. Like the other California missions, this one had essentially been abandoned following its secularization by the Mexican government. After the Mexican War the U.S. Army had put the mission buildings to use as stables, but in the padres' old mission garden Parry still found exotic grapevines, olive and fig trees, and pomegranates, watered by an aqueduct from the old mission dam about six miles up the gorge.

Whipple's survey party finally left San Diego on September 11, 1849, for the Colorado River, there to set up a base camp on the California bank across from the mouth of the Gila River. The men started northeast across to the valley of the San Diego River, then headed upstream, passing the mission, and into the foothills of the Peninsular Ranges. Parry jotted down notes about the bluish igneous greenstone and granitic outcrops, the extensive open groves of evergreen oaks, the grassy basins, the fields of corn, wheat, and barley, and the fruit orchards around Don Miguel de Pederina's Santa Monica rancheria. At Santa Maria (near present-day Ramona) the expedition was hosted by sixty-eight-year-old Don José Ortega, the eldest of twenty-one siblings and with twenty-one children of his own. Here Whipple procured as guide and Indian interpreter Chief Tomaso, who apparently talked only of women and liquor. Lieutenant Couts called him "a great old rogue" and sarcastically noted that "Whipple and Dr. Parry (both being as modest as a young maiden) fear him very much because of his bawdy language."

In mid-September the party continued on to the beautiful stream valley of Santa Isabel, where, as Whipple wrote, "Nature appears more smiling." It is fitting that today this is the site of the California Fish and Game Department's delightful Boden Canyon Ecological Reserve. Parry recorded his first cottonwoods here, and there were rocky outcrops containing crystals of garnet and tourmaline. Later expeditions also found tourmaline in this region, associated with igneous intrusions of coarse-grained pegmatite. In fact, about forty miles to the northwest near the abandoned mission of Pala, one of world's most famous tourmaline mines, the Queen Tourmaline Mine, would be developed near the end of the century. Around Santa Isabel itself were irrigated fields of barley, corn, wheat, peaches, figs, melons, pears, and apples. During the evening Parry thought he felt an earthquake. From here the route lay due east over a granitic mountain range (the Volcan Mountains), on a road passing below conifer-clad peaks of five-thousand-foot altitude, and southeast into the San Felipe valley, where "groves of oaks are filled with the music of the lark and blackbird."

Whipple was now on General Kearny's 1846 route, currently crowded with "gold sot" emigrants. Past the old Indian village of San Felipe and dropping in elevation to the southeast, the countryside became "barren and desolate in the extreme," where "thorny cacti and arid shrubbery usurp the soil." Encountering the first Creosote Bushes and Ocotillos, Parry realized, as he wrote in preparing his report, that "at this point, indeed, we may say, that the desert proper commences." On September 20 the party stopped for the night at Vallecito, where "the camp is gay, mirthful, and merry" with the many groups of gold seekers. But warnings were received of the desert ahead and of "5000" hostile Indians congregating at the mouth of the Gila because of rumors (unfounded) that the American surveyors were spoiling for a fight. At Vallecito Whipple again told escort commander Couts in no uncertain terms that he, Whipple, "was to receive all the credit for the observations at the mouth of the Gila." Meanwhile, Couts put Whipple in "a terrible stew" by announcing that he, Couts, planned to march the survey party across the desert at night (really a good idea), commenting to himself that "Washington City dandies with white gloves, etc., [referring to topographer Whipple and Dr. Parry] don't like roughing it any more than having to get up early in the morning, saying nothing of losing a night's sleep."

With respect to traveling at night, Couts prevailed. The group departed from Vallecito on Monday, September 21, at 5:00 P.M., and Whipple had to admit that "the scenery here, by moonlight, was beautiful." Where the wagon road left Carrizo Creek and crossed eastward through the Carrizo Badlands with their clay banks "eroded into fantastic shapes," Parry discovered fossilized oyster shells in what he correctly called Tertiary marine formations, and recognized gypsum "in extensive beds." Never could Parry have anticipated that in just over one hundred years the U.S. Gypsum Company would begin developing the nation's largest gypsum strip mine north of this spot.

East beyond the mountains the desert land—"sandy hills behind, a dreary, desolate plain before us, far as the eye can see"—dipped toward the distant Colorado River, at one point actually below sea level. While the escort forged ahead, the wagons lagged behind. Morning temperatures reached 120 degrees Fahrenheit in the sun and 110 in the shade. Digging for water proved of no avail. In late afternoon on September 22 a "terrible sand hurricane" abruptly battered the party, followed by a violent cloudburst, with rain, hail, thunder, blinding lightning, and streamlets and pools of water everywhere. Shortly the expedition had an even greater surprise when they reached the edge of a broad expanse of shallow water—a quarter of a mile in length and one hundred yards wide—lying in an impermeable clay basin. The surprise lake, New Lake or Signal Lake as it came to be called, had just developed in the past months as a result of wet weather and overflow of the Colorado River into the basin. This explained why Kearny had not reported it when he crossed with the Army of the West in 1846.

Traversing the vast expanse of water, the mules sinking up their knees in muck, the expedition headed toward the only visible landmark, Signal Mountain five miles to the south. The storm was over, the stars were sparkling, and not a cloud was in the sky. Parry and Whipple had been traveling with the ambulance and other wagons when the storm erupted and had become completely lost. After wandering around, sometimes wading on foot, they returned the way they had come. The ambulance and wagons were finally spotted some two miles off the main track by Couts's servant, to the escort commander's immense amusement. Meanwhile, Couts reported having felt an earthquake during the cloudburst, accompanied by a "large opening break into the surface of the earth," a claim later questioned by both Whipple and Parry.

Finally, all the party—horses, mules, soldiers, surveyors, wagons—was reassembled southeast of the lake on September 23 at what Couts christened Camp Salvation, with his parting comment: "Poor Whipple! Times are hard on him (?). Ambulance, umbrella. Oh my!" The expedition had arrived at the edge of New River (near present-day Calexico), "the oasis of the desert, where sweet water and excellent grass are abundant." Whipple's group was not alone there, because on all sides were the white tents of emigrants; and Camp Salvation would serve for several months as a resting spot for many travelers on the southern emigrant trail. A two-day layover for Whipple's party provided time for a reconnaissance. American Coots were seen in the stream backwaters, and

Great Blue Herons dipped for fish near the shore. There were disk-shaped Planorbis and other live freshwater mollusks, similar to some of the geologically recent fossil shells Parry had dug up in the desert farther west. The naturalist decided on a junket by himself south toward some low mountains (Signal Mountain, situated just below the boundary line); these turned out to have a sprinkling of locustlike trees on the lower slopes but were destitute of vegetation on top. Meanwhile, during the heat of the day all at the encampment were provided with a shimmering mirage of riparian trees in the midst of the desert.

After the full party left Camp Salvation, the next night's camp was eight miles east at the "Lagoon." Departing at four o'clock the following morning by the light of Venus, the party dipped into Mexico on September 28, making that night's stop twelve miles from the Colorado River. At 8:15 P.M. there was an earthquake lasting two minutes, this one strong enough to shake tents, spill water from buckets, wake up those asleep, and frighten those awake. By the next evening the Whipple survey party had arrived at the Colorado River, surrounded by many emigrants and Indians, in Mexico below Pilot Knob and about twelve miles south of the Gila River mouth and Fort Yuma.

By October 2, 1849, Whipple had worked his way up the west side of the Colorado River, estimated by Parry to be five hundred feet wide, through the Mesquite riparian woodland. There he found a hill ideal for setting up his astronomical observatory, across from the mouth of the Gila. However, in his report Couts claimed to have supervised cutting the route through Mesquite north along the bank and then performed "the unpleasant duty of selecting his [Whipple's] point for observation, or rather the Camp, myself." Quoth the sarcastic Lieutenant Couts, on the verge of his twenty-eighth birthday: "Take him [Whipple] away from his books, and he is not worth a tinker's d—n for anything under God's heaven. I now doubt his capacity for determining the position of the mouth of the Gila. With the aid of Dr. Parry he may succeed."

For the local Yuma Indians, with their irrigated fields of maize, beans, melons, and pumpkins, early October was harvesttime. It was Colorado Desert autumn weather, the sky clear, air dry, and midday temperature 100 degrees in the shade! During the next two months here near Fort Yuma at Camp Calhoun , as it was called, Whipple determined the exact location of the Gila River in relation to the Colorado River, while Parry was charged with keeping the weather records and collecting autumn plants. Couts had to handle a variety of Indian affairs, some very touchy. He first dealt with Pablo, grand chief of the Yumas, who showed up on October 3 in full uniform: scarlet coat trimmed with gold lace, epaulettes of silver wire, impressive green goggles, but legs and feet bare. Couts supervised a ferry across the Colorado River for emigrants and spent much of his time dealing with the hordes of travelers and their myriad problems.

October 15, 1849, turned out to be a special day at Whipple's camp. There arrived a large and very interesting party of emigrants from the East, commanded by none other than John Woodhouse Audubon, "Johnny" as he was called, the younger son of the famous John James Audubon. Lieutenant Couts made available his sergeant's boat—nothing more than a caulked wagon body—and within two days all of Audubon's party and supplies were across the river. Audubon was received kindly by Whipple and enjoyed a beafsteak dinner with Lieutenant Couts, his first real meat repast in four weeks, Audubon reflecting that Couts was "most kind and hospitable." On the morning of the seventeenth Audubon proceeded westward across the desert with his entourage of forty-six men, only half of them mounted, with minimal rations and jaded mules, toward San Diego and eventually the Sierra goldfields in the mountains east of Stockton and Sacramento.

During this time of Audubon's emigrant company passing Fort Yuma, Colonel James C. Collier, the newly appointed collector of customs for California, arrived from the East, escorted by Captain Herman Thorne and thirty dragoons. Unfortunately, Thorne accidentally drowned the next day while transporting his large group across the river. His replacement, a young lieutenant named Edward Beckwith, later led one of the Pacific railroad surveys to California, after the death of its commanding officer, Captain John Williams Gunnison, in western Utah in 1853.

Later in October the hot, arid climate at Camp Calhoun began affecting some of the expedition's surveying equipment. The cover of one of the wood chronometer boxes actually cracked apart from dryness, and the rest of the boxes showed signs of breaking. Even the rim of Whipple's reading lens snapped into three pieces. The desert weather may in part have precipitated another quarrel between Whipple and Couts that later resulted in a reprimand from Emory, who instructed the two men to quit feuding and pay more attention to their respective jobs.

Finally, on December 1, with the surveying assignment completed, the camp at the Colorado River was struck, and Whipple and his crew headed back to San Diego. Parry was with the advance party led by Lieutenant Couts, who audaciously left Whipple and the wagons behind on the wagon road along Carrizo Creek and proceeded directly west over the high Laguna Mountains, "literally climbing up steep rocky slopes, or winding along rude ravines," by Parry's account, until twenty-five hundred feet above the desert. In this section of the southern Elsinore Fault Zone, the taller forested mountains rise almost four thousand feet higher. But once past the crest, the route was a gentle downward slope through beautiful, shaded valleys "watered by clear flowing streams." The group finally reached the coast at the Mexican village of Tijuana, just south of the tentative initial boundary point, on December 10, 1849. A fitting December welcome-back was a cloudburst followed by a flash flood on the San Diego River that sent a foaming body of water down the stream channel into the bay.

Meanwhile, at Camp Riley near San Diego, a prevailing aura of gloom had developed. The gold rush had increased the cost and decreased the availability of essentially everything. Horses, for example, were now two hundred dollars apiece, and no mules were to be had. The local salary of common laborers jumped to $150 a month. As a result of Eldorado Mania, farmers abandoned their crops, ranchers their livestock, so any available provisions were very expensive. Furthermore, the gold fever stimulated desertions by commission staff and members of the army escort. Commissioner Weller desperately canvassed for loans, while Emory optimistically solicited a requisition from the army through General Riley, based on his personal signature.

In mid-September 1849 it was discovered that Democratic commissioner Weller had actually been officially relieved of his duties back in June by the new Whig president, Zachary Taylor, the notification having been given to his replacement, John Charles Frémont, to deliver to Weller. Frémont, however, had not made the delivery until he was certain that he would become the first senator when California became a state. Meanwhile, none of Weller's money drafts was honored. Fortunately, the loan Emory did receive from the army permitted him to send out surveying parties in October. Lack of funds continued to plague the commission, even into 1850 when Emory's promised $50,000 deficiency appropriation from Congress was diverted to John Russell Bartlett, a prominent Whig from Providence, Rhode Island, whom President Taylor had appointed as the new boundary commissioner. Emory was understandably discouraged and began to consider resignation.

The 1849 Colorado-Gila survey itself proved in part a frustration. Everyone had finally become disgusted with the surveyor, Andrew Gray, who instead of completing his overview of the surveying with a journey to the Colorado River had encountered Custom Collector Collier's

group on the wagon road heading for the coast and immediately returned with it to San Diego. There he joined in a round of partying, his
surveying obligations forgotten. Furthermore, Gray's location of the
mouth of the Gila River, based on the unreliable Disturnell map of 1847,
had been found not to correspond to the detailed survey by Whipple at
the actual junction.

As for naturalist Parry, his three-month period with the Colorado
River survey had not been seasonally ideal for botanical endeavors. Nevertheless, he managed to accumulate a good collection of plants, all of
them new and interesting to him. Most unfortunately, these specimens
were later lost in a disastrous fire where they were stored in Panama,
awaiting transport to Washington. Parry had also collected geological
specimens that did survive, among them rose quartz, feldspar, a fossilized oyster, and San Diego County's famous black tourmaline.

With official surveying ended for the winter and no further expedition
funding, Parry began venturing away from San Diego on his own for extended periods. He went north in February 1850, past San Luis Rey and
through Santa Barbara. Continuing beyond San Luis Obispo to the pine
woods of the Monterey Peninsula, he wrote of the latter that "this vicinity is the pleasantest I have yet seen in California. . . . The convenience
of the harbor and abundance of pine timber give it an advantage." By
February shrubs were already blooming. Parry collected a new species of
California-lilac, Hoaryleaf Ceanothus, in the mountains south of Los
Angeles. During March he traveled and collected up and down the coast
again, in the oak woodlands around San Diego and Santa Barbara, at
Santa Ana and San Juan Capistrano and Monterey, along the river at San
Luis Rey, and south of San Luis Obispo at the huge Rancho Nipomo,
whose proprietor was Captain William Goodwin Dana, cousin of
Richard Henry Dana who wrote *Two Years before the Mast*. The ranch,
with its thirteen-room Casa de Dana adobe, still standing today, was a
favorite stopping place for travelers. At Dana's ranch Parry collected a
specimen of the Pale-yellow Layia, a California endemic annual, today
listed as rare and threatened. Along the San Diego bayshore he gathered
seeds of the showy Sea-dahlia, those very seeds introducing this favorite
yellow-flowered perennial into gardens, whereas in its native haunts it is
now designated as rare. In April 1850, Parry traveled possibly as far
north as San Francisco, collecting again at Santa Barbara and San Luis
Obispo, and on the hillsides and along the beach at Monterey. That
spring on a trip to Mission San Juan Bautista, Parry was accompanied
by the energetic English plant collector William Lobb. It may have been

after this experience that Parry enthusiastically wrote to Professor Torrey, "What an interesting work a Botany of California would be! to a
collector properly equiped and prepared to decide what is new." During
May he ranged once more between San Diego and Monterey, commenting that the hills around Monterey were yellow with the blossoms of
California Goldfields. In June, as the countryside dried up, he traveled
to the mountains east of Monterey, up to San Felipe northeast of San
Diego where he had been with Whipple the previous fall, and to nearby
Santa Isabel, whose fruiting orchards again impressed him.

In late June and through July, Parry botanized around San Diego,
where he collected along the seashore and in the salt marshes, and then
went east over the Laguna Mountains into the desert. In the mountains
he found the second specimen of a pincushion, *Chaenactis artemisiifo-
lia*, the first having been collected earlier by Thomas Coulter. He became
acquainted with entomologist John Lawrence LeConte from New York,
who in the midst of his decade of North American natural history wandering just happened to be in San Diego, where he was casually associated with the Mexican Boundary Survey. LeConte asked Parry about a
strange pine he had encountered up the coast. In mid-June, when Parry
was tracking down a rumor of a lignite vein in a shoreline promontory
near the mouth of the Soledad Valley, he discovered LeConte's conifer:
"a new species of pine growing in sheltered places about the bluff." Parry
quickly wrote to Professor Torrey that "if new I wish it with your permission to bear the name of *Pinus Torreyana* (n.sp.)." The new Torrey
Pine turned out to be quite rare, found only in its type locality and on
the channel island of Santa Rosa.

In August Parry traveled north again to Santa Barbara, and then he
spent the fall and winter of 1850–51 in residence at the Moorish-style
Mission San Luis Rey, long in disuse until taken over by the U.S. Army
during the Mexican War. Father Antonio Peyri's old garden, enclosed by
the mission buildings, probably attracted Parry's interest, especially the
Peruvian Pepper Tree that Peyri had supposedly planted in 1830 as a seed
from Peru given to him by a sea captain. Legend had it that the San Luis
Rey tree was historic, because not only was it California's first pepper
tree, but it was also the ancestor of all the rest. It is disconcerting to discover that Father Peyri had been forced to depart San Luis Rey in 1829,
a year before he supposedly planted the pepper tree seed.

Left to his own devices, Parry was certainly industrious with respect
to his collecting for John Torrey and George Engelmann. Torrey shared
some of Parry's plants with the "prince of systematic botanists," George

Bentham, longtime secretary of the London Horticultural Society who had worked on the Wilkes Expedition and Karl Hartweg plant collections. During Parry's peregrinations up and down the Southern California coast in 1850–51, he procured more than 350 specimens, including new species and genera, and often provided written observations. In a gravelly ravine near San Felipe in June Parry found a new relative of ceanothus, Lotebush, which Professor Torrey honored with Parry's name, *Ziziphus parryi*. A new species of the composite genus *Calycoseris* was named *parryi* by Asa Gray, while Parry discovered a new composite genus and species in the Laguna Mountains, which Torrey named *Hulsea californica*, San Diego Sunflower, in honor of Dr. Gilbert W. Hulse, a plant-collecting army surgeon.

During Parry's travels he obtained the first specimens of four important new cacti. Golden-spined Cereus, *Bergerocactus emoryi*, he found on a dry hill along the boundary line near the San Diego seashore, where it was "growing in thick masses and covering patches of 10 or 20 feet square." It is now uncommon in the region because of impacts by development, goats, and cactus collectors. The other three species of cacti were discovered in the mountains around San Felipe in June 1849: *Echinocereus engelmannii*, the California Barrel Cactus, and the Cane Cholla *(Opuntia parryi)*, the latter named by Engelmann. Parry took note of the use of plants by the Californios, including little children at San Luis Rey eating the berries of Douglas Nightshade, *Solanum douglasii*. A Mormon tea with the suggestive but now incorrect species name *Ephedra antisiphilitica*, found at San Diego, was reputedly used by natives as a remedy for gonorrhea. Of the yerba-santa shrub Parry wrote that people used an infusion of the leaves as a tonic, and then he aptly added that "the plant has a strong terebinthine [turpentine] taste and abounds in a resinous matter which sometimes exudes so copiously from the stalks and leaves that in drying, the specimens stick firmly to each other and to the paper." Along the ocean near San Diego he collected Cliff Spurge, today listed as rare in the area, and in the dry plains around San Diego now covered by suburbia, he found large tracts of tarweed, *Hemizonia fasciculata*, filling the air with "a strong balsamic" aroma. One of the flowering trees he encountered in Southern California near San Fernando was originally thought to be the same as the walnut collected on the Sacramento River in 1837 by Hinds. But eventually it was determined to be a new variety, Southern California Black Walnut, while Hinds's is the Northern California Black Walnut, *Juglans californica* var. *hindsii*.

Parry departed California in the late winter of 1851, returning to the East Coast to commence work for his report on the geology and botany of the California boundary section, in cooperation with John Torrey and New York geologist James Hall. Major Emory had preceded him, having departed for Washington in October to bring the financial and administrative plight of the Pacific boundary survey party to the direct attention of the government. For Emory the prospects quickly brightened. The new secretary of the interior, Alexander Stuart, was sympathetic, and shortly the monetary crisis was satisfactorily settled. Furthermore, Emory heard from Captain Hardcastle in San Diego that the survey work on the Pacific Coast had been concluded, and he so reported to Secretary Stuart, adding that the last of the permanent boundary markers had been installed. Finally, in September 1851, Emory was notified that he had been named chief astronomer and head of the Scientific Corps for the boundary survey's relatively new commissioner, John Bartlett. At Frontera, near El Paso, Texas, Major Emory and naturalist Charles Parry were later reunited, to continue boundary work with a new assortment of personnel. But Parry did not visit the Golden State again for a quarter of a century.

Two others associated with this first Mexican Boundary Commission unexpectedly maintained much more immediate relationships with California than Parry. Although John Weller's term of office with the commission was short, in 1851 he was elected senator from California, replacing the man who was supposed to have replaced him as commissioner, John Charles Frémont. In 1858 he became the fifth governor of the Golden State, following again in the footsteps of Frémont who in a sense had been her first governor. Later Weller served as minister to Mexico.

Cave Couts resigned from the army the same year Weller became senator. He shrewdly married well—to Ysidora, the daughter of Juan Bandini, whose adobe in Old Town San Diego had been General Kearny's headquarters in 1846. In addition to being one of Alta California's foremost citizens, Bandini improved the culture of San Diego by incorporating the waltz into San Diego society. Couts's wife was the sister-in-law of Massachusetts-born Abel Stearns, who had become the most important trader, rancher, and landowner in Southern California. As a wedding present, Stearns gave Rancho Gaujome (Ranch of the Big Frog), a large parcel of land near San Luis Rey, to Couts, and there Cave held forth as a cantankerous majordomo until his death in 1874.

BARTLETT, THURBER, AND WEBB TOUR CALIFORNIA, 1852

On June 19, 1850, John Russell Bartlett, who had just retired from the bookstore business in New York City, received his appointment as the fourth United States boundary commissioner for the Mexican Boundary Surveys. Bartlett, who was born in Providence, Rhode Island, but grew up in Canada, seemed a most appropriate choice. Besides being an amateur botanist, bibliophile, antiquarian, linguist, and ethnologist, he had established the American Ethnological Society with statesman Albert Gallatin and had long been a member of the Franklin Society of natural science in Providence. Also, he had an itch to travel, was a talented artist tutored at the age of fourteen by an itinerant German, and, perhaps most important, was a committed Whig, appointed by the new Whig president, Zachary Taylor.

The boundary commission had been adjourned until the first Monday in November 1850, at which time the new staff assembled by Commissioner Bartlett was to congregate at the eastern end of the boundary line near El Paso. Andrew Gray would continue as surveyor, while Lieutenant Colonel John McClellan, a Mexican War topographical corps veteran, replaced William Emory as astronomer and head of the scientific party, with Lieutenant Whipple as his assistant. Emory himself was back in Washington, having expressed no interest in serving on another "mixed commission." Although Commissioner Bartlett, in selecting his scientific staff, had solicited recommendations from President Francis Wayland of Brown University, ethnologist Henry Schoolcraft of the Smithsonian, and Alexander Bache of the U.S. Coast Survey, he showed a preference for his own favorites. Among his choices was Dr. Thomas Hopkins Webb, a physician and amateur scientist especially interested in geology and zoology, whom Bartlett had known for years and who would serve as commission secretary. Bartlett also added Louisiana Frenchman Augustus de Vaudricourt, a lithographer and art and piano teacher, recommended by Jefferson Davis, who had been working as a lithographer for the topographical corps. John Weyss would be responsible for the sketches of each boundary marker site, and Henry C. Pratt became the commission's official draftsman and artist. For the position of botanist, and also to serve as quartermaster and commissary, Bartlett invited his good friend George Thurber, who gave chemistry lectures at the Franklin Society in Providence. Thurber was an enterprising twenty-eight-year-old practicing pharmacist, locally famous for his work with herbal medicine. He had become interested in the chemistry of plants, would never take medicine unless he knew what was in it, and was developing an urge for western travel. Thurber also received a recommendation from West Point chemistry-geology professor Jacob Bailey, who among other talents happened to be the country's authority on single-celled "infusoria." When the genial New Englander Thurber arrived at the Texas meeting place he was formally dressed in a fashionable black woolen broadcloth suit. Meanwhile, John Torrey took it upon himself to advise Bartlett to retain Dr. Parry on the naturalist staff, and about the same time recommended to Joseph Henry of the Smithsonian that there should be a specifically designated scientific corps attached to the new commission.

Bartlett and his personal retinue had come from New York City to Texas by ship, then overland to the rendezvous near El Paso. It was already apparent that for all his admirable qualifications Bartlett was not a devotee of frontier hardships. He and Webb had traveled in a heavy-duty, enclosed, rockaway coach-and-four; in case of Indian attack it was supplied with a double-barreled shotgun, a Sharp's repeating rifle, and two six-shooters strapped to the coach doors, while the driver carried a pair of one-shot Derringer pistols and Bartlett and Webb each had his own Colt five-shooter. Rather than camp out with the other twenty-three sojourners, the pair whenever possible would overnight in a hotel. On November 13, 1850, Bartlett and his company arrived at the El Paso rendezvous, ten days late, but McClellan's scientific group as well as Condé's Mexican Boundary Commission contingent were even later. Surveyor Gray was still in Washington, suffering from smallpox.

It was not long before McClellan, who had been found competent except when (frequently) drunk, was replaced by Lieutenant Colonel James Graham, who had had experience with the Northeast Boundary Survey between Canada and Maine (where Emory had been his assistant), as well as resurveying the Mason-Dixon line. Of particular note, he had been a member of the United States' first western scientific exploring expedition under Major Stephen Long, to the southern Rocky Mountains in 1820. In April 1851 Bartlett, not enthusiastic about the stark El Paso area, decided to move survey headquarters to Santa Rita del Cobre (the Copper Mines, east of present-day Silver City) in the pleasant foothills of the Mimbres region of New Mexico, a comfortable distance away from the designated desert boundary line. This happened to be fine countryside for both geology and archaeology (especially Mimbres Indian artifacts), and Bartlett was in his element until he was thrown from his mule and kicked in the shoulder, which laid him up for several weeks.

On May 16, 1851, tired of waiting for Colonel Graham to arrive, Bartlett took off with a small party, heading south into Sonora, Mexico, for a fortnight. When Graham and Gray at last did show up in mid-July, manifold petty controversies erupted. Finally the boundary surveyors departed for the Gila River where Condé awaited them, taking along a young, attractive Mexican captive, Inez Gonzalez, whom Bartlett had "rescued." He was determined now to return her to her family in Mexico, so a major portion of the Boundary Commission—including Bartlett, Graham, Gray, Whipple, and artist Pratt—abandoned survey duties and disappeared south of the border with Inez for some months.

In September Emory, who was still in Washington, was informed that he was assigned to take the place of his old superior, Colonel Graham, as chief astronomer and head of the Scientific Corps for the Boundary Commission. In short order Emory was on his way to El Paso, accompanied by topographical engineer Lieutenant Nathaniel Michler and Professor Torrey's botanical protégé Arthur Schott. En route Emory wrote to Dr. Henry at the Smithsonian that he had hired Torrey, as well as several other prominent scientists, to work later on the specimens that the survey would collect. Torrey at this time was leaving his position as professor of chemistry and botany at the College of New Jersey (Princeton) and moving back to New York.

In November, when Emory reached El Paso, where botanist Parry presumably had been waiting, he was informed that Commissioner Bartlett was absent, and no one seemed to know where he had gone or when he would get back. Establishing headquarters at nearby Frontera, and without ever encountering Commissioner Bartlett, Emory put three survey parties into the field in Texas. By early October Gray had been relieved of his responsibilities, and Emory became not only the commission's astronomer but also its surveyor. In Washington dissatisfaction had been increasing over the Bartlett-Condé boundary compromise of December 1850, which supposedly had solved the Disturnell map discrepancy but in reality gave up six thousand square miles (in southern New Mexico and Arizona) rightly belonging to the United States. Emory himself increasingly felt that a new treaty must be effected before additional Mexican settlements were established in the area. To compound matters, on December 19, 1851, General Condé died of typhus in Arizpe, Sonora, and his position as Mexican commissioner was assumed by his chief surveyor, José Salazar.

Commissioner Bartlett, still wandering around in Mexico, was oblivious of all of these developments. It was probably not Bartlett's five

months away from the boundary duties, nor necessarily the wild festivities of the Fiesta of San Francisco in Magdalena, Mexico, on October 1,
1851, which were responsible; at any rate, Bartlett within a few days
after the October fiesta became too seriously ill to risk travel north to the
Gila River. Consequently, his party was sequestered in the village of
Imuris, south of Nogales in Sonora, and there Bartlett would remain,
ministered to by his own Dr. Webb, by a German and English physician,
and especially by Dr. J. W. Campbell, an American resident. Soon Dr.
Webb and the rest of the party, leaving Bartlett behind, departed north
to Fort Yuma and thence to San Diego. With Webb were, among others,
botanist Thurber, interpreter John C. Cremony, and artist Pratt.

By the end of 1851 Bartlett had recovered sufficiently to travel to the
Pacific Coast in Mexico. In early February he boarded a mail steamer,
the *Oregon,* for passage north from Acapulco. When the ship arrived in
San Diego on February 9, 1852, Bartlett found that the Gila River surveying parties under Whipple had been encamped nearby for about a
month, having totally exhausted their provisions during their survey.
Two days later Dr. Webb's group, including botanist Thurber, showed
up from their inland trek across the desert. Webb reported that they were
"all in good health and spirits, notwithstanding . . . ": he meant that they
were running out of food; had lost most of their animals, not to mention
their clothing and cooking utensils; had abandoned tents, bedding,
books, and portfolios; were without an escort in hostile Indian country;
and had survived "scorching heat" in the daytime and "cold searching
blasts and chilly atmosphere" at night. Webb traveled on foot for twelve
days of the journey, as did most of the others. Then there was the desert:
"It is doubtful if any tract of land can surpass the jornada which we
crossed." There was a bright spot at the western edge of California's Colorado Desert when Thurber was able to collect "a few scanty specimens"
that he preserved in his pocket notebook. Among these were a woolly annual of a new genus (to be named *Trichoptilium incisum* by Asa Gray)
with buttonlike blossoms, and a new prostrate desert star only two
inches high, *Monoptilon bellioides,* both plants desert composites that
bloom in early February.

Now in San Diego in mid-February 1852, all members of Bartlett's
western boundary survey group were finally together within the United
States again. In addition to other commission demands, not to mention so-
liciting a visiting band of "Diegeno" Indians for "a vocabulary of their language" with the Spanish equivalents, Bartlett laid plans for a trip to San
Francisco for supplies. On February 24, 1852, Bartlett, Webb, Thurber,

and Whipple departed aboard the *Sea Bird*. The steamer put in at San Pedro, Santa Barbara, and Monterey, but because of squally weather none of the men went ashore until they arrived in San Francisco on February 27.

The rainy winter weather continued into March, but this did not stop botanist Thurber from looking for early spring flowers. In a damp, rocky area near San Francisco he found a tall white poppy, *Meconella californica,* and along the shore Beach Knotweed. Bartlett, meanwhile, had heard about some "geysers" at the head of Napa Valley, so when the weather improved, he took off with Webb and Thurber on March 17 for Benicia, in a boat crowded with Chinese bound for the goldfields. Benicia was the headquarters for the army's Pacific Division, so Bartlett approached General Ethan Hitchcock for an escort to accompany one of his future survey parties planning to go from San Diego to the Pima villages on the Gila River, since all but five of Colonel Craig's escort had deserted for the gold country. When Bartlett mentioned that he would now like to visit some local geysers, the general was good enough to furnish horses and a pack mule for the trip. While Bartlett was conferring with General Hitchcock, Thurber continued his collecting around the fort, finding a Douglas Sandwort and a Douglas Meadowfoam, both named after botanist David Douglas.

On March 19, with a guide provided by the post's quartermaster, the men traveled west along San Pablo Bay and then up the Napa Valley, covered with fields of wild oats, planted barley, and immense herds of cattle. In the undisturbed areas there were beautiful wildflowers Thurber hastened to collect: a jewelflower, Hairy Fringepod, and Narrowleaf Fringepod, all mustards; in damp rocky sites Common Pussypaws and in the shade of the oak woodland expanses of Miner's-lettuce, both of them portulacas; the pink-and-white creeping Woodland Pea; up the shaded hillsides the bicolored Miniature Lupine; the white-flowered Hillside Gooseberry; the delicate saxifrage Hill Starflower; the eye-catching scarlet Indian-warrior clumped in the understory; the beautiful Coast Larkspur and Red Larkspur; a few Blue Skullcaps; and the abundant Baby-blue-eyes, *Nemophila menziesii,* named after Archibald Menzies. Among the phloxes he found the common Pink-lobed Linanthus and Bird's Eyes; and Dutchman's-pipe was just budding its unique flowers. On the slopes under the oaks were Ground Iris, and about ready to bloom was the Giant Trillium, the three leaves in a whorl reminding Thurber of the New England trilliums. Scattered everywhere on the grassy hillsides were the blue-violet heads of Blue Dicks. Thurber could not have traveled up the Napa Valley at a more appropriate season!

Bartlett, always good with words, would later write in his reminiscences about Napa that "if this romantic valley were transferred to the older countries of Europe, it would be taken for the domain of a prince or a nobleman." He noticed that in many areas there were few trees and shrubs, aside from the centuries-old evergreen oaks and the just-leafing Valley Oaks. Bartlett speculated that the lack of woodland was the result of the immense cattle herds, as well, perhaps, as early clearing by Indian fires.

The Bartlett party traveled northward up a new road lined "with brilliant wild flowers" in the center of the valley near Napa Creek, with new fenced farmland on either side. For the group's Rhode Islanders it was a delightful surprise to encounter farmers from their home state, as well as other New Englanders. The party was joined by merchant Joseph Osborne, whose large landholding they were passing through, and then by the valley's oldest white settler, George C. Yount, who had arrived in Southern California with the Wolfskill party in 1831. Upon arriving, Yount had hunted for Sea Otters among the Channel Islands off Santa Barbara, noting in his diary the first recorded observation of otters using rock platters on their chests to smash mollusks. Here in Napa Valley Yount had received a grant of twelve thousand acres from General Vallejo that became Rancho Caymus in 1836.

Away from the distraction of surveying, Bartlett, Webb, and Thurber thoroughly enjoyed viewing and discussing the natural history of this area new to them. As they traveled north, the side hills became more heavily wooded, and tall Redwoods could now be seen rising above the rest of the forest. They crossed the valley to examine some hot springs, from which wisps of steam were rising. Farther along, Mr. Fowler, a local resident, showed a specimen of local ore to Dr. Webb, who identified it as zinc but doubted that it could have come from the area. Fowler immediately led the group up a deep ravine to several outcropping ledges, which Webb identified as serpentine, not zinc. The man finally admitted that the specimen had been given to him by someone who had claimed that rich deposits of zinc abounded in the region, a classic mining hoax. Bartlett commented that such tricks of deception "even sometimes lead scientific men astray."

The travelers were just west of Mount St. Helena, and Bartlett took time to make a sketch of the panorama, "the finest alpine scenery" he had yet seen in California. Crossing a ridge they came into another valley, where there was a village of what Bartlett called "Digger Indians," with dome-shaped huts of wooden posts, thatched with tules (reeds) and a compressed layer of earth atop the roof. Bartlett simply could not resist obtaining a full vocabulary of their language from three of the Indians. On March 22, 1852—having spent the night with the MacDonalds, the last settlers until north of the distant Oregon border—the party set off for the so-called geysers, with MacDonald volunteering as guide. The journey took two more days in the saddle through ravines, across steep hills, over summits with vistas, and through "thorny chapporal." Dinner the second night was a feast of venison and bear meat, provided by some hunters they encountered.

Next morning, March 23, the men were up by sunrise. After a bath in the nearby stream and a breakfast of bear meat, venison, hard-bread, and coffee, they continued toward the geysers (or "volcanoes," as the guide called them), encumbered, as Bartlett wrote, "Dr. Webb, with his hammer and leather bags for minerals, and with boxes and bottles for small zoological specimens; Mr. Thurber, with his portfolio for plants; and I, with my sketch-book," plus, of course, pistols and rifles for emergencies. At the first stop, in an area of decomposed rock, there were sulfurous fumes and wisps of steam. Then, after a quarter-mile hike over steep hills, the principal geysers came into view in a narrow defile, "truly a grand prospect, and difficult to describe" (figure 21). There were columns of steam jetting forth on every hand from high, almost vertical banks. At the chasm bottom where ran Pluton (now Big Sulphur) Creek there were cavities in the bank filled with boiling water gurgling over egg-size rocks, presenting, as Bartlett observed, "a curious resemblance to a pot of boiling potatoes." And from beneath the hot earth there came an incessant, deafening roar. This "geyser field" had first been scientifically described a year earlier in *Silliman's Journal* by Professor Charles Shepard, a geologist at Western Reserve College. Bartlett presumed that the present excursion represented the second scientific visit.

While Webb, Thurber, and MacDonald continued along the canyon, where there proved to be more, but less impressive, geysers, Bartlett and Joseph Osborne stayed behind, Bartlett working his way onto a precarious cliff perch in order to make an up-gorge sketch of the hellish scene. With specimens of multicolored decomposed rock, sulfur crystals, and several sketches, the group finally headed south for their evening campsite. Next day, having heard rumors of some hills of "black flint" in the area, they discovered at the farm of a Mr. Kelly a rocky ridge of obsidian resembling "a newly made macadamized road."

With a goodly collection of obsidian fragments, and after a final pastoral sketch of the Napa Valley, the group camped for its last night, the

Figure 21. The "geysers" north of Napa Valley, by Commissioner Bartlett of the Mexican Boundary Surveys; pencil with a sepia wash. (From John Bartlett's *Personal Narrative . . .* , vol. 2 [1854].)

twenty-sixth, before returning to San Francisco by way of Benicia. The so-called geysers in the Mayacamas Mountains have become famous, because at this spot by 1989 the many commercially developed steam wells reportedly represented at the time the world's largest geothermal power site. The area is not a true geyser field, such as that at Yellowstone. Here a subterranean body of hot magma is generating steam from percolating surface waters, thus producing the so-called geysers, which are actually fumaroles, constituted essentially of steam, not water. The region is seismically active, with volcanic eruptions as recently as a thousand years ago. At the present time clusters of minor earthquakes are recorded here almost every day.

By the end of March 1852 a number of individuals in Washington and El Paso, and at the Copper Mines in New Mexico, must have wondered just exactly what the Mexican Boundary Commission was up to in the Southwest. And well people might wonder. At the geysers area the boundary commissioner was approximately six hundred miles north of the United States–Mexico border. But Bartlett was certainly gathering grist for his future *Personal Narrative.*

Bartlett and his group next decided on a trip out to the Farallon Islands from San Francisco, but this had to be called off because the steamer departed too late in the day. On April 2, Bartlett, Thurber, presumably Webb, and Dr. Andrew Randall—who had been with David Dale Owen's Old Northwest Survey and would shortly become the first president of the new California Academy of Natural Sciences—headed off south from San Francisco by stage, with a pause in San Jose to change coaches. Here Bartlett encountered an Indian woman of the San Luis Obispo tribe and "obtained a most satisfactory vocabulary" from her. Then in the afternoon the four departed in another stage, this time headed for the famous quicksilver mine at New Almadén. Three busy days were spent becoming acquainted both above and below ground with, as Bartlett pointed out, the only mercury mine in the United States and second only to its namesake in Spain. This was the same mine Frémont had visited in 1846.

The commission men were shown the reducing furnaces, and in an adjacent building the huge vats holding the pure quicksilver. To verify the high specific gravity of mercury, Bartlett sat on a wooden board atop the quicksilver, sinking hardly an iota. Then he plunged his bare arm down into the mercury, "a most singular and chilling sensation." Bartlett professed to be well aware of "the effects of the mercury on the system" and was informed that the men working the furnaces were replaced every four weeks and that twenty to thirty horses and mules used in hauling the mercury died each year.

Probably the high point of the New Almadén visit was the descent into the mine itself, led by engineer Bester and armed with special torches made up of a candle at the end of a stick. In some places the shaft took a downward turn and one had to descend on a miner's ladder—a single notched log—twelve feet or so, with one hand on the ladder and the other on the torch. Back above ground, the party collected specimens of cinnabar, and Thurber continued his floral quest, adding the California Blackberry and Hooker Onion.

At the end of the Almadén tour, Bartlett hoped to continue on to Monterey overland. But heavy rains had flooded rivers and knocked out bridges, so the party returned to San Francisco. While waiting six days at that port for a steamer bound for San Diego, Bartlett met and interviewed Captain Sutter, whom he described as intelligent, courteous, fluent in several languages, kind, hospitable, and generous to a fault. He no longer owned his immense estate in the Central Valley but seemingly remained cheerful, making "the most of his misfortunes."

Finally, on April 14, the Bartlett party departed in the steamer *Ohio* for Monterey. After San Francisco and its gold energy, Monterey seemed rather dull even though it had been the capital of Alta California. Bartlett obtained a horse and made the traditional ride across to Mission San Carlos in Carmel, while Thurber collected a specimen of two shrubs, the common Chaparral Currant, still in blossom with pale rose-colored flowers, and Twinberry, a honeysuckle with orangish blossoms and black fruit. Along the shore Thurber also obtained the yellow-flowered Pacific Cinquefoil and a California Buttercup, both of which he had also collected in San Francisco, and the California Poppy.

Bartlett arranged passage south from Monterey on the U.S. Revenue Cutter *Frolic,* under Captain Douglass Ottinger, and the boundary group was on its way on April 17, in fine weather with a fresh breeze from the northwest. Passing Point Conception, where even from sea the ubiquitous herds of cattle could be seen grazing on the green-hued shoreline, they finally arrived on the evening of April 21 at San Pedro. Bartlett took a stage inland to La Ciudad de los Angeles to purchase mules, at first skirting several lagoons filled with ducks and shorebirds, and then on the plains passing extensive patches of six-foot-tall mustard, for which the area was famous, as well as a scattering of brilliant flowers in tinges of purple, orange, and yellow. And, of course, there were vast herds of cattle and horses. Bartlett encountered a former mission Indian from whom, naturally, he extracted another vocabulary in the "Diegeno" language, but different from what he had obtained earlier in San Diego.

From Los Angeles, Bartlett proceeded to Mission San Gabriel, enjoyed a breakfast there, and was back on shipboard by evening. The next morning the *Frolic* sailed on south and into San Diego harbor. The first news the delinquent commissioner received there was that Gray had been replaced as surveyor by Major Emory, whereupon Bartlett turned the completion of the survey of the Gila River over to the trustworthy Whipple.

The time seemed to be approaching to head back to the Boundary Commission's headquarters in El Paso. Camp equipment was obtained from San Francisco, but mules were still at a premium. Also, the company was shorthanded because of many desertions. When Antoine Leroux showed up, already famous as a southwestern frontiersman and guide—he had just recently been serving with Brevet Captain Sitgreaves's expedition to the coast from the Zuni River—he was immediately put in charge of the animals and *arrieros* (muleteers).

But yet another diversion arose. Captain Ottinger of the *Frolic* invited a dozen of the boundary staff to join him on a quick cruise to the Coronado Islands, lying off Baja about twenty miles southwest of San Diego. The ship left on May 5, reaching the main large rocky island, with other islets nearby, in three hours. It was difficult to find a landing site, but eventually the men were able to get ashore and make their way to the summit. There were few shrubs, but wherever soil was present it was covered with grass and "a great abundance of wild flowers," whose collecting kept Thurber busy. He found three succulent plants Thomas Nuttall had first collected along the San Diego coast in 1836: the red-flowered, fleshy-leaved Seaside Calandrinia and two species of red stonecrops, Lance-leaved Dudleya and *Dudleya pulverulenta,* the latter a striking, chalky-white coastal-cliff plant. One of the smaller islands was crowded with dozing California Sea Lions, an invitation for some of the sailors to go ashore and kill several for mere sport by hitting them on the head with stones.

Back in San Diego Thurber added about three dozen more species of plants to his growing collection. Across the dry plains the tarweed *Hemizonia fasciculata* covered large areas, giving the air a balsamic pungency. On the dry hills the red-tinged buckwheat Fringed Spineflower grew in such profusion as to give the slopes a red appearance. Along the sandy coastline the minty bushes of White Sage were abundant. While Thurber was out botanizing, Dr. Webb had a seine drawn in the bay, bringing in an assortment of "exceedingly interesting specimens."

There still seemed time for another excursion before departure for El Paso, this time north to Mission San Luis Rey, architecturally the most beautiful of all the Spanish missions. Accompanying Bartlett on this junket were Dr. Webb, Pratt, and apparently Thurber as well. Since this mission, like others, had been abandoned and its future ownership was still being considered by the Board of Land Commissioners, a unit of soldiers was quartered there, protecting the mission against "plunderers and squatters."

Bartlett's party remained there several nights and was entertained by the sergeant in charge. The old orchard was thriving, but it was disappointing to find the famous mission garden in disarray, the only plant of interest being the old Peruvian Pepper Tree that Dr. Parry had seen earlier. Thurber collected a sample of the tree, including its berries, which tasted just like common (but unrelated) black pepper. A swamp invading portions of the big garden was now the hiding place of "the screaming heron and other water fowl." An Indian chief from a tribe he called Kechi provided Bartlett opportunity for an interview and to add another Indian vocabulary, with accompanying Spanish synonyms. The Indian came dressed for the occasion in the uniform of a Mexican officer, with a military cap and a red-faced blue coat trimmed with gold lace.

On May 26, 1852, Bartlett's command—organizing for the trip to El Paso, Texas, at quarters up valley near Mission San Diego—was ready to move. In addition to rations for seventy days, they packed a quantity of medical supplies including antiscorbutics: fruits, vegetables, and pickles. Bartlett himself, down with a touch of malaria, and Webb had remained in San Diego but came up later in their small, mule-drawn wagon to join the main company. The two mules were the very ones he had departed with from the coast of Texas in 1850. Unfortunately, the wagon fell into a gully, damaging the axle-tree, which became a source of annoyance throughout the rest of the journey.

The route led into the foothills, over Soledad Hill, where one of the loaded wagons had upset earlier, through a landscape covered with dense stands of wild oats, and passed through San Pasqual, where Kearny and the Mexicans had met in battle six years earlier. While the California summer was already turning the lower elevations brown, Thurber found that in the higher hills spring was still on the land. At San Pasqual he collected more than a dozen new plants, including three species of lupines and an attractive penstemon he himself would later describe as a new species, *Penstemon spectabilis*.

Bartlett and Webb finally caught up with the main group—the escort of fifteen soldiers commanded by Colonel Lewis Craig, and Whipple and his crew busy making astronomical observations—at the Indian village of Santa Isabel, set in its beautiful oak woodland. Bartlett's party included Webb, Pratt, assistant surveyor Malcolm Seaton, Leroux, and Thurber. Thurber was now wearing three hats: he was commissary, quartermaster (currently making arrangements to obtain an additional wagon from San Bernardino), and of course botanist. Several days were spent at Santa Isabel, and Thurber collected a further dozen plants, among them three different lotus pea species.

On June 1, 1852, after a hot day, the entire group of about fifty officers and men, six wagons, horses, and twenty-five pack mules was on its way by moonlight, with Bartlett and Craig in the lead. They reached a campsite at the miserable Indian village of San Felipe at two o'clock the next morning. During the ensuing day Thurber collected six more species of plants, including what was thought to be a new *Euphorbia*. Although the commission camp was thronged with Indians of what Bartlett called the "Diegeno" tribe, for once he did not solicit a vocabulary. That night they traveled over a mountain pass and into desert country, with cacti and many century plants. A daytime camp was made at Vallecito, where a small group of soldiers was stationed beside some sulfurous pools fringed by willow bushes, but there were no trees; a band of Indians dressed up for the visitors in their "holiday clothes." That night the trip was through increasingly barren countryside, past the bleached bones and desiccated carcasses of oxen, mules, and sheep from past migrant parties, and on to Carrizo Creek, incised into a deep arroyo. The curiosity here was the piles of thousands of sheep carcasses, said to have died from eating some poisonous plant of which Thurber could find no trace.

Now came the crossing at night from the hills into the desert plain toward the Colorado River, through land Bartlett described as "the most desolate spot we had seen," the deeply eroded slopes of clay without any vegetation. Having left Carrizo Creek on June 4, the party did not reach Fort Yuma until June 10 after the crossing of the *jornada,* where in some sites Burro-weed was the only common vegetation. Thurber did come across another common desert plant, the buckwheat Thurber's Spine-flower, eventually to be named after him by the British botanist George Bentham. Nothing remained of the extensive lake seen by Whipple and Parry when they passed through in 1849. But Bartlett was told about the lake, and about migrant parties hailing it as a miracle from heaven, "like the manna furnished to the Israelites of old."

June 6 proved to be an unfortunate day for the men in this desert land. The escort under Colonel Craig, in advance of the other men, encountered two army deserters from Fort Yuma, whom Craig tried to persuade to return with him. In the ensuing altercation, Craig and Sergeant Bale were shot, and the colonel died. When Craig's body was recovered, Bartlett wanted to take it for burial to Fort Yuma, but Dr. Webb said the corpse was already decomposing in the extreme heat. So just beyond the now-dry channel of New River, a deep grave was dug. At sunset Bartlett read the Church of England burial service, and at the head of the grave a small wooden cross was erected. Sometime later Indians apprehended the two deserters, and they were eventually tried and hanged at San Diego.

In the early morning of June 9 the Colorado River was reached, half a dozen miles below its junction with the Gila. Here were the remains of Fort Defiance, built in 1849 at the early ferry crossing. On a nearby ridge they found one of the Boundary Commission's iron monuments, installed just the year before and already knocked down by the Yuma Indians. The Colorado River was edged by cottonwoods, willows, some Mesquite, Seep-willow, and an abundance of Arrow Weed.

The next day everyone reached Fort Yuma, where Major Samuel

Heintzelman was in command. Tents were pitched beneath shelters constructed from pole frames covered with stems of Arrow Weed that provided moderate relief from the desert sun. After a nine-day rest—during which, among other things, Bartlett made some experiments to measure the velocity of the Colorado River, coming up with a speed of five and a quarter miles per hour—the Boundary Survey party was on its way east again, bound for El Paso.

On August 17, 1852, after eighty-one days and eleven hundred miles of travel, plus a few cases of scurvy, bowel complaints, rheumatism, fever, and ague, the Mexican Boundary Commission members finally reached the terminus of their journey on the eastern bank of the Rio Grande at Magoffinsville, Texas. Bartlett had been away about a year and a half. To Whipple and his survey party he paid a worthy tribute: "The country cannot award too much praise for its successful accomplishment under such privations, and hardships, together with so many and such serious difficulties."

Thurber was worthy of commendation, too, because he returned with a major collection of plants, more than one hundred species from California alone. He recognized that those from California "were mostly of well-known plants," but he did discover new species and genera. Part of Thurber's collection involved plants from the upper Napa Valley and beyond, where no botanist had collected before, and three dozen species gathered between San Diego and Fort Yuma. Parry had collected along the latter route in 1849, but not during the flowering season, and much of Parry's collection had been lost in a Panama fire.

In 1854 Thurber visited Asa Gray at Harvard and offered him the plants he had collected under Bartlett. From this collection and with some help from John Torrey, Asa Gray wrote a monograph memorializing Thurber and his plants, *Plantae Novae Thurberianae,* published by the American Academy of Arts and Sciences in 1854. Of all of Thurber's collections from the American Southwest, the most remarkable was undoubtedly the parasitic plant *Pilostyles thurberi,* named in honor of Thurber by Gray. Although it does occur in California, Thurber found his specimen across the Colorado River near the mouth of the Gila River, parasitic on a species of shrubby dalea, *Psorothamnus emoryi.* This diminutive brownish-purple flower, less than a sixteenth of an inch across, is the only North American relative of the famous parasitic *Rafflesia* from Southeast Asia, which produces the largest flower in the world, up to three feet across. Both Gray and Torrey were justifiably incredulous at Thurber's discovery.

Dr. Webb, amateur geologist and zoologist for Bartlett's Boundary Commission, had made collections of minerals, rocks, fishes, reptiles, and insects. Unfortunately, his collections from the trip out to San Diego from the Copper Mines had mostly been lost when wagons had to be abandoned. But several boxes of minerals collected later in California were sent safely home by ship.

Bartlett himself had accumulated observations on the Indians, made additions to his Indian language collection, and produced a beautiful suite of sketches. Most important from the standpoint of western literature and history, Bartlett amassed enough Californiana to fill almost two hundred pages of a rambling narrative he eventually commenced writing on his return to the East. But first came another excursion in the autumn of 1852, this time supposedly to join up with Major Emory well down the Rio Grande at Ringgold Barracks in Texas. As usual, it turned into a somewhat extended tour of southeastern Mexico, as the group circled through Chihuahua and Monterrey, and experienced an Apache attack during which the oblivious Thurber, busy collecting plants, barely escaped being lanced.

At Ringgold Barracks, Bartlett received word that the commission had temporarily been disbanded. He charged Thurber, in his role as quartermaster and commissary, to proceed with the entire train of animals, wagons, and all the public property except the instruments to San Antonio, Texas. There he was to reserve the best of the wagons, mules, and equipment, selling the rest to provide funds to pay off the remaining men of the commission. Thurber remained in Texas until June 1854, when he was relieved by members of the new commission and was finally able to return home.

Botanist Thurber went on to a distinguished professional career. He received a master's degree from Brown University, served on its faculty, and was later honored with a Doctor of Medicine title from University Medical College in New York. He became the first professor of horticulture in the United States, at Michigan State Agricultural College (now Michigan State University), and for years presided as editor of *The American Agriculturist*. On December 20, 1867, at a special dinner in New York City honoring John Torrey's fiftieth year of botanical publications, Thurber delivered the address, summarizing his longtime friend's contributions. When Torrey died in 1873, Thurber wrote his obituary for the Torrey Botanical Club *Bulletin*.

Thurber made a final contribution to the Golden State in 1880 by preparing the section on grasses for the flora of California by William

Brewer and Sereno Watson. Although as a naturalist Thurber paid little attention to animals, in 1890—the year of his death in his adopted home state of New Jersey—the bird collector A. W. Anthony named a subspecies of Oregon Junco from Mount Wilson in Southern California in Thurber's honor. During bachelor Thurber's final years, he suffered from painful rheumatism he attributed to his demanding experiences with the Mexican Boundary Survey.

On December 22, 1852, the Mexican Boundary Commission was officially disbanded, at least for the time being. Early the next year John Russell Bartlett was back in the civilized East with his family. His time of public service was not at an end. At home once again in Rhode Island, he was elected secretary of state for Rhode Island and Providence Plantations, an office he conducted with commendable efficiency for eighteen years. He died in 1886. His two-volume *Personal Narrative of Explorations and Incidents in Texas, New Mexico, California, Sonora, and Chihuahua: 1850–1853* became a classic.

THE GADSDEN PURCHASE: NEW BOUNDARIES, PLANTS, AND ANIMALS, 1853–1855

In August 1850 John Torrey had written to Asa Gray that "the [Mexican Boundary] Commission will certainly have a full staff of Botanists, when all three are on the ground." A year later, in another letter to Gray, Torrey wrote, "For a time, at least, there will be four botanists in the Boundary Commission." Then in January 1852 still another letter to Gray rightly warned that "the affairs of the Commission are in wretched plight . . . we may soon hear of Mr. B's recall."

Sure enough, early in May 1853, the Boundary Commission was reorganized, with "Mr. B" (John Bartlett) being replaced by a new boundary commissioner, Robert Blair Campbell, former congressman and U.S. Consul to Cuba, now living in San Antonio. And to this new commission William Emory, who had been at home with his family in Washington, was appointed as chief astronomer and surveyor. His old colleagues Lieutenant Nathaniel Michler and botanist Arthur Schott, the pair who had come out with him to Texas in 1851, were to serve with him. The new charge was to survey the Rio Grande between Laredo, Texas, and the Gulf of Mexico. Under Emory's efficient direction this survey was completed in short order, and Emory returned to Washington in early December, where adjustment of the Mexican boundary between El Paso and California was still a very major issue. The United States simply had to have land for a southern railroad route.

After considerable discussion and negotiation, the Gadsden Treaty was signed on December 30, 1853, which adjusted the Mexico–United States boundary southward to extend from the Rio Grande just north of Juárez somewhat irregularly westward to about twenty miles below the junction of the Gila and Colorado rivers, thus to include today's Sunbelt within the United States. Though not as large nor as cheap as the Louisiana Purchase, the Gadsden Purchase was still a bargain at fifteen million dollars. Meanwhile, nine months earlier Congress had authorized a number of surveys for proposed railroad routes to the West Coast. Secretary of War Jefferson Davis eventually chose his longtime friend Major Emory to head the railroad survey project, called the Bureau of Explorations and Surveys. In 1854 Emory already had the railway surveys under way and well staffed with a variety of naturalists.

In August 1854, however, a more appealing offer came Emory's way—namely, to become not only the new Mexican boundary commissioner, but also the surveyor and chief astronomer for the Gadsden Purchase survey, which would put him actively out in the field in familiar country, where he could also continue to pursue his scientific interests. John Torrey was delighted that Emory had finally been named commissioner. When Weller had left in 1850, Torrey had fully expected that the new commissioner would be Emory, who, as he wrote, "is well disposed towards naturalists." But now in 1854 some of the familiar botanical faces of the earlier surveys were absent. Parry was back in New York, working on his report for the early years of the boundary survey. Charles Wright was the botanist with the U.S. North Pacific Exploring Expedition under Captain Cadwallader Ringgold, and Wright would be in San Francisco with that expedition in the fall of 1855. George Thurber had been relieved of his commission duties in San Antonio in 1854 and had returned east. Dr. John Bigelow was now a botanist with the Pacific Railroad Survey on the 35th parallel west to California, headed up by Whipple. Only one familiar scientific face remained for Emory, Torrey's German disciple, Arthur Carl Victor Schott.

Two new Mexican boundary surveying parties were now organized. One was to proceed from the Rio Grande west, headed by Emory and assisted by a select staff of eleven. Commanding Emory's escort was Captain E. Kirby Smith, who was a good botanist himself and whose scientific observations would eventually be published by the Smithsonian Institution.

The other survey party would proceed from San Diego eastward to the Colorado River, and then on to the 111th meridian (near present-day Tucson), where the two parties were to meet in late spring of 1855. The San Diego survey group was under the command of Lieutenant Nathaniel Michler, the topographical engineer who had served as Emory's assistant under the old commission. With Michler, in addition to three other officers, was Schott, who had earlier served with both Emory and Michler, and would now be Michler's assistant. Schott proved to be an all-around accomplished naturalist, an "efficient artist, surveyor and collector" (Goetzmann 1979, 195). He not only made plant collections but also became the first major collector in the lower Colorado River country of mammals, birds, cold-blooded vertebrates, and fossils. His mentor John Torrey had given him a salutory personal recommendation as "an entertaining and agreeable man." Of the group of official and unofficial boundary survey naturalists associated with the Gadsden Purchase, the one who would leave his mark in California during 1854 and 1855 was unquestionably Schott.

Lieutenant Michler left the East Coast by ship on September 20, 1854, crossed Panama, and arrived in San Diego. By November 16 he was off with his surveying party including Schott, following the now-familiar route inland by way of Santa Isabel, San Felipe, Carrizo Creek, and across the desert to Fort Yuma, arriving on December 9. Lieutenant Robert S. Williamson's Pacific Railroad Survey party, with naturalists William Blake and Adolphus Heermann, had already been here the previous December, and Lieutenant John G. Parke's Pacific Railroad Survey, with naturalist Heermann, passed through on their way east in midwinter 1854.

Now Michler's survey camp was set up on the east side of the Colorado River near the Gila River, where already the new town of Colorado City had been surveyed, streets and blocks laid out, and lots sold. In the next century this southwestern section of Arizona would become prime rock-hound country, and around the survey camp in 1854–55 the men had a variety of souvenirs that they could pick up: agates, jasper, opals, chalcedony, quartz crystals, and even petrified wood.

Much of the local landscape seemed stark to Lieutenant Michler's eye, Fort Yuma on its volcanic hill having a "bleak, dreary appearance," and the area "destitute of every form of vegetation" save for a poisonous euphorbia, *Croton californicus,* which the local Indians reportedly used for rattlesnake bites. But off in the distance to the northeast could be seen an imposing desert dome called St. Peter (today's Castle Dome), and beyond Fort Yuma to the northwest the tall, jagged crag that Michler called

Castle Peak (now Picacho). Closer at hand, marking the course of the Colorado River, "we look with pleasure upon the bright green foliage" of the riparian vegetation, the large Fremont Cottonwoods, different willows, impenetrable thickets of Greasewood and Arrow Weed edging the water, with Mesquite and Screw Bean farther back.

For the survey party the trip inland from San Diego to Fort Yuma in midautumn had been fast, and the season was not a good one for botanical endeavors. But once by the Colorado River, Schott began dividing his attention during the many months of the stay between assisting Michler and pursuing natural history. To start off the new year of 1855, Schott on January 2 captured a subspecies of Botta's Pocket Gopher that zoologist Samuel Woodhouse had first collected near the San Francisco Peaks in northern Arizona four years earlier with Lorenzo Sitgreaves's Expedition. The original type specimen of the species had been collected at Monterey in 1827 by the eclectic Dr. Botta of Duhaut-Cilly's French expedition. During January 1855, Schott and Michler surveyed the meanders of the Gila and Colorado rivers above and below their junction and spent time repairing the instruments. Most of the plants Schott encountered, though new to him, had already been collected by others: the gray ghosts of Smoke Trees in the dry washes of the Colorado Desert, the Ironwood, two different palo verdes around Fort Yuma, and Bigleaf Mistletoe in the cottonwoods along the lower Colorado River.

On February 14 the survey camp was moved twenty-two miles downstream along the Colorado River to a beautiful Mesquite woodland near a freshwater lagoon and close to the new southwestern Gadsden boundary point. Flowers, appearing on the desert plants now, were collected by Schott, some of them proving to be new species. One was a blue-flowered shrubby pea from the Colorado River's floodplain that John Torrey would name after his protégé, *Psorothamnus schottii*. In March, out in the desert west of the river, Schott came across Crinkle Mats, rounded mounds of lavender-flowered borages in sandy patches, which Torrey later named. Torrey had already seen preserved specimens, collected in the same area by Emory during the winter of 1846, but these "were too imperfect for determination." Altogether, Schott collected about three dozen species of plants during his desert stay, which eventually made their way back to John Torrey.

By March 4 all triangulation was completed, an observatory erected, and the astronomical observations commenced. However, on that day a flash flood roared out of the Gila and down the Colorado, inundating the survey camp. By 2:00 A.M. a decision had been made to move into the sand hills. The warning long roll of the drum was sounded, the camp struck, and everything transported to higher ground. Michler, noting the new encampment's isolation, observed that off to the east there stretched a sandy desert for forty miles, while to the west there was now a sheet of water up to five miles across.

As the winter months passed into spring and on into summer and the flowers disappeared, Schott turned to the more demanding task of observing, collecting, and preserving animals. Especially evident up and down the Colorado River all the way to the salt marshes was the activity of Beavers, "miles and miles of river banks, which are inhabited, fortified, and covered over by the labors of this singular animal." Muskrats were also common. When Schott opened up a Muskrat "nest" he found it filled with Screw Beans, probably food for the winter. Schott collected a Pacific Kangaroo Rat, first found at Los Angeles by William Gambel; a Desert Pocket Mouse, another of Woodhouse's Arizona finds; and a Cactus Mouse, a new species of deer mouse with the type locality at Fort Yuma. Another mammal that Schott collected was the abundant Black-tailed Jackrabbit, a new subspecies not to be described for four decades.

The Indians around Fort Yuma searched the nests of the White-throated Wood Rat (a new species of *Neotoma* that Schott mistook for a Mexican species) for the rodents, which they considered a delicacy. One Indian visited Schott's campfire, killed several of the rats, and buried their bodies in the ashes; in fifteen minutes the dish, with a sprinkling of salt, was ready. The Indian offered to sell Schott one to eat, *"mucho bueno por comer"* (very good to eat), but Schott declined. Birds, of course, were easier to come upon than mammals, but to be collected they had to be shot. Schott had no binoculars with which to see the birds, nor were there bird field guides. Schott, as the area's first serious bird collector, obtained about twenty-five species in the countryside around Fort Yuma, ranging from Osprey, Great Horned Owl, and Sandhill Crane to Greater Roadrunner, Vermilion Flycatcher, and Lark Sparrow. He was particularly struck by the distinctive appearance of the Phainopepla, a slim black silky-flycatcher with white wing patches and a crest. This distinctive bird, of which Schott collected three, was originally obtained by Colonel George McCall, inspector general of the army, when he had been north of Vallecito in 1852, and the new genus *Phainopepla* would be designated by Spencer Baird in the Pacific Railroad Survey report.

As the desert warmed, the cold-blooded vertebrates began to attract Schott's attention. He collected eight species of lizards, including a chuckwalla, two horned lizards, and a Desert Iguana; four species of snakes, among them the Western Diamondback and the Sidewinder; and the Colorado River Toad. He also found unusual species of fish in the Colorado River, including back in the fall of 1854 the Bonytail Chub. It was declared extinct here by the California Fish and Game Commission in 1980. Likewise, the Colorado Squawfish that he caught, the largest "minnow" in North America, reaching a length of six feet, is today endangered if not extinct in the lower Colorado River because of changes in habitat as a result of impoundments upstream.

Two companies of artillery were now at Fort Yuma, and Schott was pleased to discover that among the officers were a number interested in natural history, including Major George Henry Thomas, who had already provided plants and animals for the Pacific Railroad Surveys. At Fort Yuma, Major Thomas collected three new composites: a plant named after Emory by Torrey, the white-and-yellow-flowered *Perityle emoryi* (Emory's Rock Daisy); a woody species Asa Gray would name *Baccharis emoryi*; and the descriptively named Spanish-needle, with only disk flowers that give rise to a globe of elongated, bristle-tipped seeds. Both Thomas and Schott collected an annual composite, the white-flowered Pebble Pincushion, named by Gray. Thomas also collected a California Leaf-nosed Bat, named as a new species by Baird, with Fort Yuma as the type locality. And both Thomas and Schott added specimens of the large Northern Chuckwalla lizard from around Fort Yuma.

In 1855 Thomas was transferred to Fort Mason in Texas, where he served until 1860. At Fort Mason he continued his natural history collecting, shipping a number of mammal skins and specimens preserved in alcohol to the Smithsonian. But his real fame came in 1863 during the Civil War. As major general commanding the Union Army of the Cumberland, his forces routed the entrenched Confederates from atop Lookout Mountain above Chattanooga, Tennessee, in the Battle above the Clouds, winning the indelible title of "the Rock of Chickamauga." After the war General Thomas continued to collect natural history specimens for the Smithsonian. In 1869 he returned to the Golden State as head of the military district of the Pacific in San Francisco, where he died on March 28, 1870.

While Michler was occupied along the Colorado River, Emory had moved west from Texas with his contingent into Apache Indian country, pushing his men to complete the eastern section of the new boundary survey and meet Michler. But by May 1855, still with no news of the California team, Emory had sent out an express rider from his encampment in Santa Cruz (northern Mexico) to find Michler. It turned out that the West Coast group had been considerably delayed by disillusioned emigrants from California who were stranded in the desert. Michler and Schott finally completed their survey line, and in August 1855, Major Emory notified Washington that the new Gadsden boundary survey was essentially completed.

The year 1856 would be devoted to preparing the final report on the entire Mexican Boundary Surveys. Already the experts were working on various aspects of natural history: Spencer Baird on the birds, mammals, reptiles, and amphibians; Torrey, Gray, and George Engelmann on the botany; Charles Girard of the Smithsonian on the fishes; and Charles Parry on the California geology and ecology. The Iowa Geological Survey's eminent paleontologist, James Hall, dealt with the overall geology and paleontology. Emory himself completed an invaluable map, drawn to a scale of 1:6,000,000, as well as dozens of other charts. The abundant artwork for the report was done by Schott, Augustus de Vaudricourt, and Austrian emigrant John Weyss, the latter providing the sketches of the boundary marker sites. Emory rented office space in Washington, D.C., optimistically for only seven months, for his select staff to work on the publication. Altogether three volumes were published. The first narrative volume, which included geology, was ordered printed in ten thousand copies on August 15, 1856. So much expense was associated with the two remaining volumes (botany and zoology)— with their plethora of excellent illustrations including twenty-two bird engravings that alone cost $365.50—that only three thousand copies were authorized. Finally in 1859 the publication of the set was completed. Meanwhile, numerous specialized articles on natural history aspects of the Mexican Boundary Surveys had appeared in scientific periodical literature.

Professor Gray wrote of the botany volume, "It must be ranked as the most important publication of the kind that has ever appeared." Certainly the three volumes represented at the time the most comprehensive southwestern United States natural history publications ever attempted by the young government of the United States.

PART FOUR

Iron Horses and River Steamers

The 1850s Surveys

SHORTLY AFTER THE CALIFORNIA GOLD RUSH and statehood, the United States government organized expeditions charged with finding transportation corridors for trains, and later for steamer travel on the lower Colorado River. All but one of the Pacific Railroad Surveys involved California, where there would be three east-west and three north-south expeditions. For the first time since the U.S. Exploring Expedition, surveys included at least one botanist, zoologist, geologist, and artist. Later even more natural scientists were involved with the collections and reports. Competition for positions with the surveys was high, with the result that many experienced and able civilians were hired. The surveys in aggregate blanketed the map of California, and the results were published in more than a dozen massive government tomes with excellent illustrations and detailed maps, as well as in a plethora of additional articles. Ives's Colorado River expedition to study that watercourse's potential for steamer travel was equally productive.

Williamson's Railroad Survey for Southern Mountain Passes, 1853

A vast mine of new information of great value for the study
of the history, ethnology, zoology, palaeontology, botany,
and geology of the West . . .
—George Albright, *Official Explorations for
Pacific Railroads* (1921)

Anyone who attempted to reach California by land or water prior to
1850 waxed enthusiastic about the idea of a railway from sea to shining
sea. Even before gold made the West Coast especially attractive, New
York entrepreneur Asa Whitney, interested in trading with China, proposed in 1845 that Congress grant a right-of-way from the Great Lakes
to the end of the Oregon Trail for train tracks. About the same time Missouri senator Benton, vocal exponent of "manifest destiny," pushed for
a railroad to the West Coast.

With the coming of the California gold rush and statehood, as well as
the growing popularity of the Northwest, a transcontinental rail became
a national obsession, understandably fraught with regional and political
implications and controversies. Whitney favored a northern route while
Mississippian Jefferson Davis—who became secretary of war in March
1853 under the new president, Franklin Pierce—naturally looked with
favor on a southern route, as had the Mexican Boundary surveyors earlier.

The Pacific Railroad Survey bill, passed by Congress in early March
1853, provided an immense challenge to the new secretary of war.
Within ten months Davis had to submit a comprehensive report on all
practical railroad routes to the West Coast, based on intensive field surveys. These surveys, to be carried out by the Army Corps of Topographical Engineers, were to consider possible routes and, equally important, evaluate each region's natural resources, utilizing "parties of
persons capable of making collections and observations in natural his-

tory." The Pacific Railroad Surveys represented a major—and for the time expensive—governmental exploration, which for the natural sciences would systematically consider things botanical, geological, zoological, paleontological, and ethnological.

Of the four east-west surveys finally approved, one involved the Northwest and was headed by Washington's territorial governor, Isaac Stevens. The other three concerned California: one crossing from central Colorado into Northern California approximately along the 38th parallel; a 35th parallel survey from New Mexico west to the Colorado River and on to the coast; and a southernmost survey between San Diego and Texas approximately along the 32nd parallel. Several additional surveys within California were to hunt for feasible mountain passes, as well as for rail routes southward through the state and northward into Oregon.

On July 10, 1853, Army Corps of Topographical Engineers lieutenants Robert S. Williamson and John G. Parke—with their professional staff, a spring wagon equipped with an odometer, four six-mule teams, five teamsters, eight additional personnel for general duty as cooks and field men, and a military escort—departed from the U.S. military post at Benicia. The expedition had been staging there for nearly a month, since the engineers' arrival in nearby San Francisco by sea from New York and Panama on June 20. The survey orders were to examine, between the 32nd and 35th parallels, Sierra Nevada passes leading east from the San Joaquin and Tulare valleys for the most feasible route across to the Colorado River, and then to head to San Diego.

Williamson and Parke were no strangers to exploration in the Far West. Two years earlier Parke had served with naturalist Samuel Woodhouse and artist Richard Kern on Brevet Captain Lorenzo Sitgreaves's Expedition down the Zuni and Colorado rivers, Parke and Kern completing a monumental five-by-seven-foot map of the Territory of New Mexico. Earlier Parke, after graduating second in his West Point class in 1845, worked as a topographical engineer with the Iowa-Minnesota boundary survey. In 1849 Lieutenant William Warner, with Williamson as his assistant, was assigned by General Persifor Smith, Division of the Pacific, to explore for a possible railroad route out of the Sacramento Valley and over California's northern mountains. During this survey Warner had been killed by Indians.

Now in 1853 Williamson and Parke were not only to survey for railroad routes, but also "the topography, meteorology, geology, natural history, the character of the Indian tribes of the country, &c, will be studied as closely as circumstances will permit." The pair had an impressive group of professionals with them. Charles Preuss, the survey's draftsman, was Frémont's cartographer for three of his western expeditions, including those through the southern Sierra Nevada and the Mohave Desert. Prior to the trips with Frémont, Preuss had worked back east for Ferdinand Hassler of the U.S. Coastal Survey and for Old Northwest surveyor Joseph Nicollet.

The survey's physician and naturalist was Dr. Adolphus L. Heermann, whose German father had served for years as a surgeon in the U.S. navy. Heermann's western adventures began in 1843 in his late teens, when he accompanied one of Sir William Drummond Stewart's plush expeditions to the Rocky Mountains. They ran into aging John James Audubon, who, unbeknownst to Adolphus, in 1821 had flirted so enthusiastically with Heermann's mother, one of his art students—"well of figure and graced with acquirements"—that Audubon had been fired and left unpaid by Heermann's father. Also encountered by Stewart's expedition was Frémont, ready for his first expedition to California. Heermann quickly became Stewart's promising young naturalist, affectionately called "Dolly the Great."

Back in Philadelphia, Heermann met naturalist William Gambel at the Philadelphia Academy of Natural Sciences, who excited him about California. The two even discussed a joint junket to the West Coast. In 1846 Heermann received his M.D. degree from the University of Maryland. He met Spencer Baird of the Smithsonian Institution and Sir Charles Lyell, the great British geologist, and birded in Florida with John Krider in early spring 1848. The next year he was off for the West Coast, crossing Central America and collecting in western Mexico. During his voyage north to California by ship, Heermann baited a hook with refuse scraps, cast the line off the stern, and hauled in about ten Black-footed Albatrosses.

By 1851 naturalist Heermann was headquartered at Sacramento, visiting gold camps and birding at the nearby American and Cosumnes rivers. He scoured the Sacramento Valley for birds during his several years of residence, collecting at least two dozen species, including the state's first Black-chinned Hummingbird in Sacramento. The California Gnatcatcher, today seriously endangered, was first encountered by Heermann in San Diego in 1851. He visited Baja's Coronado Islands off San Diego, where he collected eggs and nests of some red-billed gulls. This gull is "one of the most handsome of the numerous family to which it belongs," wrote John Cassin, who named it *Larus heermanni*, Heermann's Gull.

Heermann visited San Francisco, where he collected two dozen species of fish in the bay. Also, he boated out to investigate the sea birds of the Farallon Islands. There the frenzied harvesting of Common Murre eggs shocked him. The Farallon Egg Company monopoly was in full swing providing eggs for the San Francisco market. Heermann wrote that the annual income from the eggs ranged up to two hundred thousand dollars, and in 1854 more than half a million eggs were sold in a two-month period. Heermann did some hunting and collecting there himself, including the egg of a Tufted Puffin, his collecting hand wrapped in a thick handkerchief to protect it from the parents' massive red-and-yellow beaks.

In 1852 Heermann returned east with about twelve hundred bird skins, as well as eggs, nests, and some mammals for the Philadelphia Academy and the "National Collection" in Washington (the Smithsonian). Heermann's lengthy "Notes on the Birds of California, Observed during a Residence of Three Years in That Country" appeared in the January 1853 academy journal. Baird supported Heermann's application to join an expedition to Japan, while John Cassin of the academy wanted him to make a collecting trip to Texas. When neither opportunity materialized, Heermann returned to California and by good fortune joined Williamson's staff.

Williamson, not having filled the position of geologist-mineralogist, asked Spencer Baird to locate a candidate. A good choice was made in William P. Blake, who had graduated from Yale's Sheffield Scientific School in 1852, where he ranked as a "distinguished pupil" and assistant. Because of his last-minute appointment, Blake was two weeks delayed in reaching California, but he proved worth waiting for. He provided invaluable geological observations and made a good collection of plants for John Torrey. Blake appreciated the geological sites in California, as compared to those in the East, because they were not completely covered by trees, "permitting the outline and surface of every hill and mountain to be seen." In later years Blake became a mineralogy professor at Yale and assisted Baird in the preparation of Smithsonian exhibits for the 1876 Centennial Exposition in Philadelphia. The second-choice candidate, who missed out when Blake accepted the position, was Fielding B. Meek, working at the time with Ferdinand Hayden and destined to become one of the great western American geologists and paleontologists.

Williamson's staff also included Isaac Williams Smith as civil engineer; Charles Koppel for assistant civil engineer, who as a talented artist made many fine sketches for the final report; and Lieutenant George Stoneman of the First Dragoons, who commanded the escort. Stoneman, an important Union general during the Civil War, was elected governor of California in 1883.

On the morning of departure, July 10, 1853, the expedition crossed the Carquinez Strait on a little ferryboat to Martinez, unfortunately not taking the time to have the newly acquired mules shod there. With Mount Diablo on the east, they headed south down San Ramon Valley to Livermore Valley, which proved gravelly and destitute of trees except along streambeds. The fields of sere wild oats made the rounded hills of sandstone and conglomerate look barren. Heermann spotted a Golden Eagle soaring above one of the Livermore foothills. On July 17 they discovered that the San Joaquin Valley was a vast sheet of water, the river having overflowed its banks. The next day a ferry established by Andrew Grayson, later to become famous for his bird paintings, took the survey party across the main San Joaquin River, where Heermann collected his first Swainson's Hawk of the survey. That evening they camped among the verdant Valley Oaks and shrubs beside the Tuolumne River six miles above its mouth.

The party delayed a week at Fort Miller, an army post established in 1851 on the San Joaquin River in the foothills for the protection of miners against rambunctious Indians. The mules were finally shod, the expedition wagons repaired, and more provisions obtained, while Heermann collected his first Western Whiptail Lizard. Blake studied the interesting local geology, where old flows of basalt capped the older granitic bedrock, and Koppel made sketches of what the survey report called the "table-hills." Miners, including Chinese, still successfully placer-mined here, washing out sand and gravel by pan or wooden cradles, watching for a "show of color." In what remained of the ravaged stream the local Indians were catching salmon, which they sold to the expedition. The major excitement was the arrival of the head of Joaquin Murieta, "the notorious robber-chief," in a large jar filled with whiskey, brought by Captain Harry Love and his posse after a fight at Murieta's Cantun Creek stronghold across the San Joaquin Valley. The site of old Fort Miller is now under the waters of Millerton Lake.

The party left Fort Miller on July 31. Crossing desolate plains toward the Kings River, they encountered California's first full-term senator, Dr. William M. Gwin, who had initiated the bill in Congress about constructing a railroad on the 35th parallel. In January 1853, Gwin had prevailed upon Richard Kern to provide a detailed, illustrated endorsement of the route before the Senate, recommending a crossing of the southern

Sierra Nevada over Walker Pass. Now Gwin was on his way back from previewing the old Tejon Pass (to the east of the current Tejon Pass) as a possible railroad route.

Near Woodville on a branch of the present-day Tule River the survey party ran into Frémont's old guide Alexis Godey, "a formidable rival to Carson," as Frémont put it. Godey informed Williamson that he had been at Walker Pass, hoping to meet Edward Beale, his Mexican War comrade, who was returning from a Washington, D.C., trip. Having missed Beale, Godey was on his way south to await Beale at his ranch in the Tejon. Godey agreed to join Williamson's expedition temporarily as guide, provided he could leave as soon as Beale showed up.

On August 8 Williamson established a base camp, Poso Depot, up Poso ("O-co-ya") Creek in the lower foothills of the Sierra northeast of present-day Bakersfield. From here, with Godey as guide and taking along Preuss, Parke, Smith, five more of the survey party, and ten soldiers from the military escort, Lieutenant Williamson on August 10 started his extended side trip east up Poso Creek into the southern end of the Sierra to reconnoiter possible railroad passes. Remaining at the depot were naturalists Blake and Heermann, artist Koppel, and Lieutenant Stoneman with the remainder of the escort.

Williamson was particularly interested in Walker Pass, named after Joseph Walker, who is presumed to have led his trapping party east over this route in early May 1834. Walker's trappers had zoological interests of a commercial sort—Beaver pelts—and one member of the party, Zenas Leonard, made a significant botanical find. In Leonard's published journal (1839) appeared his account of the first discovery the previous autumn of the Giant Sequoias in a grove—Tuolumne or Merced groves—north of Yosemite. As Leonard exclaimed, some of the trees "would measure from 16 to 18 fathoms round the trunk at the height of a man's head from the ground." After surveying Walker Pass, Williamson proclaimed it "badly situated and impracticable" for a railroad.

Williamson's party continued south to the end of the Sierra Nevada, on August 17 entering a beautiful, high prairie called "Teh-ee-chay-pah" by the Indians (Tehachapi). After scouting the region for several days, the men left Tehachapi and proceeded west to the Tejon plain, part of the Tulare Valley adjacent to the southern mountains. These mountains, now considered part of the Transverse Ranges, were called the "Bernardino Sierra" by geologist Blake, who felt they were an extension of the Sierra Nevada to the Coast Ranges.

The sloping Tejon plain, with the entrance to the historic Tejon Pass on Tejon Creek near its eastern extremity, was a beautiful place, with several small streams, fine oak groves, and an abundance of grass. Williamson's group headed west across the Tejon plain, then went south on a good pack trail over the Cañada de las Uvas Pass (today's Tejon Pass), which seemed excellent. Next they circled eastward, and returned north over old Tejon Pass with its wagon road to Los Angeles, "one of the worst roads" Williamson had ever been on. Then they headed back toward the Poso Depot. Williamson ranked Tehachapi Pass best for a railroad route across the southern Sierra and Cañada de las Uvas best into the Los Angeles basin.

Williamson's party reached Poso Depot on August 29, finding everybody well, "suffering only from a desire for a more active life." The naturalists, however, had kept busy during the twenty days. Geologist Blake "had a fine opportunity to examine the geology of the vicinity." He found fossilized shark teeth and ten species of fossilized shells around Poso Creek, as well as a new species of freshwater snail, Jupiter Ramshorn. Heermann collected plants and vertebrates in the area, including a new euphorbia and three new buckwheats, one of which, *Eriogonum heermannii*, was named after him. He also added a yellow monkeyflower, a bright red Indian paintbrush, the greenish-white Narrowleaf Milkweed, the Blue Elderberry, in an alkaline site the saltbush Bractscale, and Common Unicorn Plant, its bizarre "devil's claw" fruit to be pictured in a survey plate (figure 22). Heermann captured a Common Garter Snake, a Western Garter Snake, and three Gambel Deer Mice, as well as a very cocky Ash-throated Flycatcher, a Northern Mockingbird, and the "Louisiana" (Western) Tanager, first collected by Lewis and Clark. And he collected a new species of turtle, the Pacific Terrapin (now *Clemmys mormorata*), which would become the famed mud turtle of the San Francisco market.

For three days while Williamson was away, Heermann, Koppel, Blake, and Stoneman headed southwest for Kern Lake. The second day they reached a swampy sheet of water, beyond which rose the foothills of the Coast Ranges. Heermann observed Bald Eagles and "counted three nests within sight of each other." The adventurers "scouped up enough of the warm, muddy water to make coffee," then headed back on a hard fifty-mile ride to Poso Depot.

The day Williamson's detachment returned to Poso Depot, the wagon train that had been sent north to Fort Miller for provisions returned. Three days later the entire party, with its ten heavy wagons, proceeded into the lower San Joaquin Valley (Tulare Valley), rafted across the Kern

Figure 22. Sketch of Common Unicorn Plant fruit, collected west of Poso Creek in 1853 by W. P. Blake on Williamson's Pacific railroad survey. (From the Pacific Railroad Surveys, vol. 5.)

River, where there were tules (reeds) more than eight feet high, and headed south. Heermann continued adding to his bird collection, including another Lewis and Clark species, Lewis's Woodpecker. The basin south from Kern River, "a country without water, not sandy, but arid, bearing artemisia & c.," proved good habitat for reptiles: Heermann collected a Desert Crested Lizard, Collared Lizard, Foothill Alligator Lizard, Sagebrush Lizard, and a Long-nosed Snake, as well as an Ord's Kangaroo-rat.

THE RAILROAD SURVEY'S TEJON DEPOT

The expedition returned to the Tejon on September 3, 1853. A beautiful woodland-meadow site at the edge of the foothills was selected as the Tejon Depot camp, from which to commence "our operations for surveying with the spirit-level." The depot, surrounded by hills except to the west, was about three miles northwest of the mouth of Tejon Canyon, near Tejon Creek, which comes out of the southern Tehachapi Mountains into the Central Valley. There was a scattering of Indian rancherias, crudely cultivated and irrigated, and nearby two squatters, Ridley and

Brooks, anxiously waited to obtain title to land they were half-heartedly cultivating.

Here the expedition members met Lieutenant Edward Beale, just back from the East after what Williamson's report called "a long and arduous journey across the plains." Edward "Ned" Beale had been appointed California's first superintendent of Indian affairs in 1852. Concerned about the mistreatment of generally peaceable California Indians by white settlers, he considered establishing large reservations on vacant land where the Indians could become self-sufficient by raising crops and herding livestock. The purpose of Beale's recent trip to Washington had been in part to solicit government support for such reservations. Returning, he joined his cousin Gwin Heap's western surveying expedition, with the famous Antoine Leroux as guide, following a route over the Rocky Mountains used later in the summer by Lieutenant Gunnison and then by Frémont on his Fifth Expedition.

After discussing his California Indian reservation idea with the survey men, Beale decided that the Tejon would be an excellent location. On September 12, 1853, Beale supposedly held a council meeting at old Tejon Pass to discuss such a reservation with the local Indians. The council is said to have lasted for two days, with over a thousand Indians showing up, and the Tejon was agreed upon as the site. Blake revisited the Tejon during the late summer of 1854 and found that the Indians were harvesting wheat, barley, corn, beans, turnips, melons, cabbages, and many acres of pumpkins. But a few years later the reservation failed, finally closing in 1863, with the Indians moved elsewhere.

To begin the surveying from Tejon Depot, Lieutenant Williamson dispatched Lieutenant Parke to make a hasty reconnaissance of the country to the southwest in the direction of Los Angeles. Meanwhile, on September 5 Williamson's main survey party, including Smith, Preuss, and Blake, with the necessary rod men and chain men, commenced the survey of old Tejon Pass off to the southeast. Preuss busied himself making sketches of the hills and ravines, while Blake carried out geological examinations. He perceptively observed that the range they climbed was almost at right angles to the Sierra Nevada to the north, forming a "Transverse Chain" (this region is now called the Transverse Ranges). The main party returned to the depot on September 18.

The valleys of Tejon Creek and Grapevine Creek (Cañada de las Uvas Pass, the modern Tejon Pass and the route of Interstate 5) to the southwest proved great places for the naturalists. In the foothills Grapevine Creek coursed through a delightful site where later Fort Tejon was established, in an open woodland of gigantic oaks and interspersed meadows; some of the oaks, as geologist Blake noted, were eight feet in diameter. Beyond the surrounding grassy slopes rose wooded summits several thousand feet higher. Here Heermann shot a Western Gray Squirrel. Near Tejon Depot he obtained two Botta's Pocket Gophers, and at old Tejon Pass he collected four California Ground Squirrels. During evenings at the Tejon, Heermann occasionally heard the throaty hooting of the Great Horned Owl. On the way to old Tejon Pass he procured more than a dozen species of birds, most of them new to him. Among the more interesting finds were Ruby-crowned Kinglets with a small flock of Mountain Chickadees, the latter species first collected by Heermann's Philadelphia friend William Gambel in New Mexico in 1841, and an Oak Titmouse, originally collected by Gambel near Monterey. The choice find at Tejon Canyon, in the small creek near the pass, was the West's unique stream bird, the American Dipper, a first for the region. Here, also, Heermann caught a Canyon Tree-frog and a Pacific Treefrog. One dusk he shot several Violet-green Swallows flying south above the pass, seemingly in migration. Another migrant, seen atop the pass on October 1, was a Black-throated Gray Warbler. During hunting excursions around the depot, the California Condors were an awesome sight. Hours went by without seeing a single one, "but if we shot a deer, before its body could grow cold the vultures [condors] were seen rising above the horizon and slowly sweeping towards us, intent upon their share of the prey." Heermann observed one pull a dead deer from where it had been concealed and totally consume it within an hour. On another occasion he observed four condors drag the body of a young, hundred-pound Grizzly Bear more than two hundred yards.

Despite his preoccupation with animals, Heermann did find a new species of parasitic orange dodder, *Custata subinclusa*, on a willow at the top of old Tejon Pass, and collected a specimen of Rubber Rabbitbrush, as well as one of *Chrysothamnus teretifolius*, which was "all over the mountains around the Tejon Valley."

THE MOHAVE DESERT EXCURSIONS

On October 5, 1853, Williamson completed the survey of Cañada de las Uvas Pass. Parke had returned earlier from the Los Angeles basin, having explored a good pass from San Bernardino east into the desert. In 1875 San Gorgonio Pass would become the Southern Pacific Railroad's route through the Peninsular Ranges to Los Angeles. Parke then helped

Williamson complete the Cañada survey, the two camping on the future parade ground of Fort Tejon. Next Williamson had to determine a railroad route across the Mohave Desert to the Colorado River. Williamson and Parke, with a large party including zoologist Heermann and geologist Blake, on October 10 traveled up over old Tejon Pass and headed southeast toward what Blake called the "Great Basin" (Antelope Valley) because it resembled a miniature version of the actual Great Basin. Blake's Great Basin was a landscape of desert shrubs, including the strange Joshua Trees, commonly called bayonet trees because of their strong sharp leaves, looking at a distance, in the words of the survey report, "like bands of Indians."

Their southeastward route eventually brought them to a large playa extending for miles and looking like the frozen surface of a lake (the dry Lake Rosamond, fifteen miles southwest of present-day Edwards). In this desert brushland Heermann startled a Short-eared Owl out of its bushy hiding place, and he felt certain that the Cactus Wren he shot was the first ever collected north of Mexico. There were great numbers of Sage Sparrows, and Heermann commented that "where the Larrea [Creosote Bush] flourishes," the Sage Sparrow, Cactus Wren, Ladder-backed Woodpecker, and Common Raven "are the only birds inhabiting these large and desolate plains."

Instead of continuing eastward, the Williamson party turned southwest toward San Francisquito Pass (northwest of present-day Palmdale), passing varicolored volcanic rock outcroppings—reds, vermilions, lilacs—with thin layers of obsidian. On October 12 they camped in the foothills at the edge of a real and beautiful sheet of water, Lake Elizabeth. The next morning the men proceeded to the summit of San Francisquito Pass. After several days of surveying, the party returned by Lake Elizabeth and back to the valley floor. On the seventeenth they discovered a new pass to the southeast affording access west to the Santa Clara Valley and named it Williamson Pass. On the way Heermann made an unusual spotting of a pair of White-tailed Kites and collected a new species of poisonous prickly poppy, Chicalote, in full yellow bloom, as well as a fish, the Threespine Stickleback, originally named as a new species after Williamson but now recognized as an endangered subspecies.

By the eighteenth they had come upon "a bold stream." They named it Johnson River, and the men tried a bit of unsuccessful gold panning. The next day they struck the Spanish Trail (Mormon Trail) and went north over Cajon Pass and on to the Mohave River, a broad but shallow stream in a sandy bed lined with willows and cottonwoods. The Mohave Depot was established there on October 20, near present-day Victorville.

At Mohave Depot the expedition divided into three survey groups, leaving the escort under Stoneman at the depot. Isaac Smith took naturalists Heermann and Blake, artist Koppel, and a few others, plus the spring wagon with instruments and the chronometer, and returned to resurvey Williamson Pass, heading on into the Los Angeles basin before turning back. Lieutenant Parke with Preuss went upstream along the Mohave River, looking for "a practicable pass," while Williamson's party proceeded eastward downstream.

After Williamson left the depot on October 23 with four men, the streambed was soon dry. Consequently a mule was loaded with two ten-gallon kegs of water for the trip farther into the desert. Altogether they traveled about seventy-five miles, sometimes by the Spanish Trail, sometimes along the river when it reemerged, and at other times cross-country. Finally concluding that there was no feasible route across to the Colorado River, they turned back near an imposing black volcanic cinder cone three hundred feet high (Pisgah Crater), set in a three-mile circle of lava flow.

Williamson's party reached Mohave Depot on November 1, 1853, the same day Parke returned from his upstream survey. Parke reported that his group had encountered high, rugged mountains with precipitous ravines and sideslopes, no place for a railroad. Smith's party arrived back on November 7. His party had run into Parke and Preuss on October 23, then the next day crossed San Francisquito Pass, reaching Mission San Fernando Rey de España on October 30. A day was spent at Los Angeles, where Heermann saw small flocks of "quiet and gentle" Mountain Plovers "gleaning their sustenance" on the plains near the town. Heading back to Mohave Depot on November 2, the party resurveyed Cajon Pass.

The whole team was to proceed now to Jonathan Warner's ranch in the valley of Agua Caliente. Parke, with the wagon train and most of the men, including Blake, departed on November 9, crossed over Cajon Pass to San Bernardino, and returned to the desert through San Gorgonio Pass. Here Blake, observing the deep sand drifts at the eastern side of the pass, conjectured that the warmer, rising desert air created a "vacuum" that sucked the cooler oceanside air through the wind gap, where it then dropped its load of airborne sand. Blake realized that this phenomenon also accounted for prevailing incoming gales at the Golden Gate. Parke's company traveled southward through the Colorado Desert, passing across the vast salt flats of ancient Lake Cahuilla (Salton Basin, which became in 1906 the site of an accidentally rejuvenated Salton Sea). Blake rightly suspected that the area was below sea level, the first such observation for the United States. After examining Warner's Pass, Parke's survey proceeded to Warner's Ranch, arriving on November 28.

The day before Parke's departure from Mohave Depot, Williamson, Heermann, the pack mules, and Lieutenant Stoneman—with the entire escort because of possible unfriendly Indian encounters—once again headed east down the Mohave River, still hoping to reach the Colorado River. After passing through the granitic Lower Narrows, on occasion they traveled between clay bluffs with delicate tints of purple, pink, blue, and yellow. Where the river lost itself in the desert, Heermann saw "upon the floating sand hills" his first Sidewinder rattlesnakes. There were Antelope Ground Squirrels and several kinds of lizards, including a Western Skink. By November 16 the Mohave River disappeared into a huge salt lake, Soda Lake, with an adjoining playa beyond. Hereabouts Heermann first encountered the beautiful Gambel's Quail and pursued small gray-and-yellow Verdins among the mesquite trees, but the birds were too quick to be caught. Soda Lake appeared to be "the true sink of the Mohave River." Williamson estimated that it was still a hundred miles to the Colorado River, "with a mountainous country between, and neither wood, water nor grass that we knew of. To attempt to reach that river would have been madness. Our only alternative was to turn back." Less than a year later Whipple's 35th parallel party crossed from the Colorado River west to the "Soda Lakes" (lake and playa) and on to the coast. In the future the Santa Fe Railroad would follow virtually the same route.

Williamson's group returned by wagon road toward the western edge of the Mohave Desert, traveling for fifty-five miles without water, then southward along the foothills on a crude road through the Colorado Desert. On November 29 Williamson's command reached Warner's Ranch to join Parke and his men, "and once more all the survey party was together." Both groups had come northeast from the desert along Carrizo and Vallecito creeks, on to San Felipe Creek and over Warner's Pass (at 3,557 feet) to Warner's Ranch, concluding that the route was no good for a railroad.

On December 1 Parke and Stoneman departed with one detachment down Warner Creek to the coast and on to San Diego, while Williamson traveled with Blake and Heermann southeast to Fort Yuma, taking the wagon with the odometer to measure distances. Heermann's birding in the Colorado Desert and around Fort Yuma proved profitable. Near Little Lagoon south of the Salton basin Heermann saw a "Black Crested Fly-catcher" (Phainopepla), and at the Colorado River there were flocks of twenty to thirty "on the wing at one time," flicking their tails incessantly when perched on Mesquite. Also at Little Lagoon Heermann collected a wintering Ruby-crowned Kinglet that was foraging in riparian shrubs. Around Fort Yuma the Dark-eyed Juncos were winter residents, as were Brown-headed Cowbirds, and he procured a "brilliantly *plumaged* but small" Vermilion Flycatcher. Army surgeon Hilhaus informed him that in spring these striking birds were quite common around the fort. Less than a century later they would be listed in California as a species of special concern because of riparian habitat loss. Heermann added the Gila Woodpecker, numerous along the river, to California's roster of bird species, as well as what was undoubtedly a Red-naped Sapsucker. Heermann opened up abandoned ball nests of Cactus Wrens, still remaining in bushes, and collected a variety of feathers which "often indicated that certain species of bird were to be found in their neighborhood," sagely noting that "the naturalist, thus put on the alert, will more readily obtain such of those species as may have escaped his eye." His report of the Northern Caracara, a raptor, along the river near Fort Yuma with some Turkey Vultures was only the second record for Alta California. The first was seen at Monterey in the fall of 1847 by French naturalists with the *Vénus*. There have been only about a dozen California records since Heermann's time.

Williamson's party came back to Warner's Ranch along their former route and continued to San Diego. They arrived at the mission on December 19, joined the next day by Parke, who was returning from a side excursion to Jacumba Pass on the Mexican border, judged another poor pass. Williamson's trip to San Diego, with a dip into Mexico, had been frustrating. With a broken barometer and a chronometer jarred out of adjustment, he had been unable to complete his mapping. In 1855 civil engineer Charles H. Poole was employed to complete the survey to Fort Yuma. Blake, on the other hand, enjoyed the trip, collecting petrified wood, garnets and other gems, fossilized oyster and pecten shells from the Carrizo Creek beds, and several freshwater snails from lagoons, including a Corkscrew Physa and Desert Tryonia. He also got to see the extensive Algodones Dunes. At the Alamo Mocho campsite in Mexico he actually reclined on a dune to examine the sand grains—perfect spheres of quartz, agate, garnet, and chrysolite.

At San Diego the Williamson-Parke Pacific Railroad Survey was essentially disbanded. More than thirteen hundred miles had been covered, more thoroughly and in "much more detail than that of any other Pacific railroad survey" (Albright 1921, 143–44).

Before leaving San Diego, Blake did some seashell collecting and discovered a thick layer of embedded shells twenty feet above the present shoreline, conjecturing that it "looked like the remains of an old beach," although the shells could have been taken there by birds or people. Blake and Heermann returned to San Francisco on the steamer *Southerner*, beachcombing at the various ports. Heermann's find at Santa Barbara was a fossilized "giant boat shell" (Onyx Slipper Shell). Charles Preuss, Williamson's draftsman, returned home in very poor health, partly as a result of sunstroke. On September 1, 1854, he wandered away from his residence in Washington, D.C. A search party finally found him at a farm a few miles out of town, where he had hanged himself from a tree. On expeditions Preuss was not always the most pleasant companion, yet Frémont appreciated that he had "a cheerful philosophy of his own which often brightened dark situations."

Blake prepared an exhaustive geological report for the Williamson-Parke survey, some 310 pages, with illustrations and an appendix including Blake's catalogue of geological collections, plus six articles on fossils and soils by various authorities, including Louis Agassiz. Blake provided a colored geologic map of California, with a good depiction of the greenish serpentine (now California's state rock) and his "California Sandstone" in the Coast Ranges, along with a series of geological cross-sections, including the Coast Ranges, Sierra Nevada, and a series from old Tejon Pass south to San Diego and Fort Yuma. Blake eventually become director of the School of Mines at the University of Arizona, and it has been said that his geological reconnaissance report on Alaska's Stikine River influenced Secretary of State William Seward's recommendation to purchase Alaska in 1867.

John Torrey reported on the eighty-seven plants collected by Blake, some from the San Joaquin Valley but mostly from the desert, and by army men at Fort Yuma, with some fine illustrations by a young New York artist, E. Dwight Church. A number of Fort Yuma's officers were interested in natural history and contributed specimens for the survey, including lieutenants DuBarry and R. E. Patterson, surgeons Hilhaus and R. O. Abbott, and especially Major George Henry Thomas, who was with one of the two artillery companies. Thomas also provided specimens for the Gadsden Purchase survey during 1854 and 1855. Among the Fort Yuma plant specimens was a bristly annual borage that proved to be a new species of *Cryptantha.* Torrey's comment about the water-leaf *Phacelia ciliata*—"this species is to us one of the rarest of the genus"—may have been true then, but it is widespread now. There was a new southwestern species of spineflower, *Chorizanthe corrugata.* Two plants were named after Thomas: a new buckwheat, *Eriogonum thomasii,* and what was originally thought to be a new species of jimsonweed, a hallucinogenic nightshade, *Datura thomasii* (today *D. discolor*). Thomas's mammal contributions included three Round-tailed Ground Squirrels, described by Baird as a new species, a new deer mouse (the Cactus Mouse), and the Pacific Kangaroo Rat.

Botanists Elias Durand and Theodore Hilgard reported on Heermann's ninety-two plants, mainly from around Poso and Tejon depots. The two men were later criticized by Torrey for making "a bad business of their Report," including some "very objectionable drawings by Hilgard." There were reports on birds by Heermann and mammals by Baird. Dr. Edward Hallowell's report on reptiles and amphibians noted that with this California collection added to previous collections, "it may be said that its [California's] herpetology . . . is almost as well known as those of our older States." Fishes, mostly marine, collected on Williamson's expedition were included in the four-hundred-page illustrated report by Charles Girard, Baird's assistant at the Smithsonian. There was also a section on meteorological and astronomical observations. Williamson wrote the forty-three-page overview report on the expedition.

FORT TEJON: XÁNTUS AND THE CAMELS, 1857–1859

This fellow is from the Smith. Inst. and going to catch
poisonous insects, and all kinds of wild beast.

> —Alexander Forbes (quoted in a letter written
> by Xántus to Spencer Baird)

Several years after the 1853 railroad survey, Spencer Baird indirectly made one of Williamson's sites zoologically famous. This was the spot along Grapevine Creek below Cañada de las Uvas Pass where Williamson's survey parties had camped and collected. In 1850 Lieutenant George Derby had recommended it as a choice location for an army fort. In June 1854 construction commenced, and the following year Fort Tejon was completed. Meanwhile, Baird had received specimens collected at the site by Williamson's naturalists and had perused the survey reports. Excited, he wrote: "There is no locality from which [specimens] would be more acceptable. We have a few things brought by Lt. Williamson from that vicinity, just enough for us to wish for more."

Baird soon thereafter located a collector, a Hungarian named János Xántus. Immigrant Xántus, at loose ends, had enlisted in the U.S. army in 1855 but detested his first assignment at Fort Riley, Kansas. There he met Surgeon William Hammond, however, who collected birds for Baird. Hammond tutored Private Xántus and introduced him to Baird by letter, initiating a seven-year correspondence. Xántus became one of Baird's protégés and was soon sending specimens to him and to the Philadelphia Academy of Natural Sciences. In 1857 Baird managed to get Xántus transferred into the medical corps and sent west to Fort Tejon.

Xántus, newly promoted to "hospital steward" (sergeant), was supposed to run the dispensary, but since no one was ill, he busied himself collecting for Baird. Then Major Alexander Blake arrived as the new commanding officer and warned the soldiers they could not shoot within the garrison or leave the fort to hunt. When Xántus objected, Blake promised him one day a month for collecting, or perhaps two "if I behaved myself good." Upon learning of this, Baird had his father-in-law, General Sylvester Churchill, put pressure on Blake, who then let Xántus collect freely if he made mounts of the best specimens, especially birds, for Blake. When a special hunting gun arrived from Baird, and Xántus tried it out in the compound, he was nearly thrown into the guardhouse. And it did not smooth the Hungarian's cause when his pet Grizzly Bear cub ate the commandant's dog. Meanwhile, Xántus's fellow soldiers were busily drinking the hospital steward's good medicinal alcohol, which, of course, he was using to preserve specimens. Two memorable visitors came to the fort while Xántus was there. Heinrich Möllhausen, the artist and naturalist, on his way to join Lieutenant Joseph Ives's Colorado River expedition, encouraged the Hungarian to sell his specimens to the king of Prussia. Xántus was irked when officers, soldiers, and even Indians, none of whom would help him, enthusiastically assisted Möllhausen with his collecting. Lieutenant Edward Beale also arrived with a train of twenty-eight camels. He hired local citizens for his escort and armed them by raiding the fort armory of all its weapons and ammunition, leaving Xántus, as Xántus wrote, "entirely naked (scientifically speaking)!"

Beale's camels represent a story in themselves. The last time camels, ungulates that originated in the New World, had frequented Southern California was during the heyday of the prehistoric La Brea tar pit camelids during the Pleistocene. Now, with increasing travel across southwestern deserts, the United States Army, in particular, began contemplating introduction of modern Old World camels, the classical desert beasts of burden. Through promotion by Secretary of War Jefferson Davis, Congress passed an appropriation bill on March 3, 1855, to purchase and import Bactrian (two-hump) and Arabian camels for military purposes. Navy lieutenant David Dixon Porter, who would later gain fame for his Civil War accomplishments, was responsible for obtaining the "ships of the desert" in the Near East and initially having thirty-three of them shipped aboard the *Supply* to Indianola, Texas, by mid-1856. To accommodate one gigantic upstanding camel's hump, a hole had to be cut in the *Supply*'s deck.

Now, during the summer of 1857, Edward Fitzgerald Beale—a former navy officer, superintendent of Indian affairs for California, and fittingly dubbed a "pioneer in the path of empire" by *New York Tribune* reporter Bayard Taylor—was charged by Secretary Davis with leading the camels from their Texas depot west cross-country to California, essentially following Lieutenant Whipple's route. By October 18, 1857, Beale's Camel Corps was encamped on the eastern shore of Colorado River in the Mohave Desert. Beale had been warned that camels could not swim, but tied saddle-to-saddle the beasts "swam boldly" across the broad river and clambered onto the shore of the Golden State at what became known as Beale's Crossing. From here the wagons and the main train proceeded to Fort Tejon, while Beale himself, riding the white dromedary Seid and accompanied by Turkish camel driver Hi Jolly on Tuili, continued to Los Angeles at eight miles an hour.

The next year, when Beale with a small camel caravan returned from Fort Tejon to the Colorado River, he encountered at Beale's Crossing Captain George Johnson's steamboat, the *General Jessup,* returning downstream. The camels were sent back to their quarters at Fort Tejon, while Beale and his men were ferried across the river and continued east to investigate winter conditions on the newly surveyed wagon road to Fort Defiance.

After the outbreak of the Civil War, the remaining camels that had irritated Xántus were transferred to Los Angeles and sold at auction or otherwise dispersed. In 1864 the camel project was finally abandoned. The last known of Beale's "ships of the desert," Topsy, died at the old Los Angeles Griffith Park Zoo in April 1934, not all that far from La Brea.

Xántus lasted more than sixteen rough months at Fort Tejon, finally leaving the army. But during his stay he had, as Baird wrote, "exhausted the natural history of the vicinity." His specimens filled thirty-five boxes, with among other items two thousand birds of 144 species and 119 plant specimens, including eight new species. Probably his most memorable find represented a new family and genus of lizards closely associated with Joshua Trees that Baird named in his honor, *Xantusia vigilis,* the Desert Night Lizard. Today the site in Grapevine Canyon beside Interstate 5 has been developed as Fort Tejon State Historic Park.

The 35th Parallel

Whipple and the Naturalists, 1853–1854

In the early autumn of 1853 the survey for the 35th parallel had already started west from Albuquerque (New Mexico) for California under the command of Lieutenant Amiel Whipple, who was by now very familiar with parts of the Golden State. He was also well respected by all who served under him. As one of his natural scientists, H. B. Möllhausen, later wrote, Whipple possessed "special professional qualifications united with particularly pleasing manners which inspired confidence in all who approached him."

There was a tremendous number of applications for this expedition, and many qualified individuals had to be turned down. Numerous government expeditions were under way at the time, including oceanic ones, so Whipple experienced delays in procuring necessary scientific equipment. Two sections of this expedition were to depart from the East, one under Whipple and the other under Lieutenant Joseph Christmas Ives. The sections, each with its own support staff, met in central New Mexico and continued west to California. For part of the southwestern trek Whipple was fortunate in having as guide the experienced frontiersman Antoine Leroux, who had most recently served briefly with the Gunnison survey in western Colorado.

In mid-February 1854, Whipple left the Bill Williams Fork valley where it reached the east bank of the Colorado River, accompanied by a talented quartet of naturalists: Dr. John Milton Bigelow, a surgeon and botanist; Dr. Caleb B. Kennerly, a surgeon and naturalist; Jules Marcou, a geologist; and Heinrich Balduin Möllhausen, an artist as well as a zoologist.

Bigelow—not related to the older New England botanist Jacob Bigelow—was born in Peru, Vermont, on June 23, 1804. He had been imbued with enthusiasm for botany by Professor John Riddell at the Medical College of Ohio, where he received his M.D. degree in 1832. He
married the same year and settled down to medical practice in Lancaster,
Ohio. Bigelow became a close friend of fellow Ohioan and moss authority William Sullivant, and Sullivant had recommended Bigelow for the
Mexican Boundary Survey in 1850. There was some controversy involving the plethora of botanists recommended for that survey. Dr. Charles
Parry had already been promised. Then botanist George Thurber of
Rhode Island, friend of the new commissioner, John Bartlett, was added
as "computer," and botanist Bigelow as "Surgeon." Torrey's observation
about Bigelow was "I think he is not a botanist." Meanwhile, Bigelow
had written a letter naming Thurber as his "assistant." Thurber's response
was that he would not "play second fiddle to such a poor stick." The
Boundary Commission finally "left it to themselves to settle the question
of botanical rank." With the Mexican Boundary Survey Bigelow got off
on the wrong foot, unable to appreciate that plants new to him were not
necessarily new to science and unaware that botanist Charles Wright,
with whom he would serve, and Charles Parry had much previous experience in the Southwest. Furthermore, Bigelow never enjoyed the demands
of the boundary survey and developed little devotion to the West, but he
did later make valuable suggestions to Torrey relative to publication of
the Mexican Boundary Survey plants.

In mid-1853 Dr. Bigelow, surprised at being invited to be surgeon and
botanist for the 35th Parallel Pacific Railroad Survey, resigned from the
Mexican Boundary Survey before its completion and joined Whipple's
expedition. En route to join Whipple, he stopped for a few days in St.
Louis to discuss cacti with Dr. George Engelmann. Bigelow, in pursuit
of his "delightful Science," finally came into his own with Whipple.
Though the oldest member of Whipple's staff, he proved to be "a general favorite," always patient, ready with a quick laugh, kind and atten-
tive when doctoring, a zealous botanist, and an enthusiastic sportsman
in the field. His talent in the latter area was questionable, since he shot
only two things throughout the survey: a rattlesnake that took seven bullets to kill, and a cap someone hung on Bigelow's pistol, the gun accidentally discharging and drilling a hole in the hat. When artist Möllhausen first met Bigelow, he characterized the botanist as "a graybeard"
who had served with the old Mexican Boundary Surveys.

Zoologist Caleb Kennerly, born in Virginia in 1829, had the good fortune to be a student at small Dickinson College in Pennsylvania with a young professor of biology named Spencer Baird. Baird was impressed by this young man, who became his protégé. In one letter of recommendation Baird described Kennerly as "the most notorious snake, salamander, bug, cave-bone, wolf, panther, and tadpole catcher in the community." The Bairds treated the young man as a son, and, as Baird wrote, "Our house has been his home." In 1852 Kennerly received his M.D. degree from the University of Pennsylvania and soon became acting assistant surgeon with Lieutenant Ives's contingent of Whipple's railroad survey. His closest companion during the expedition was artist Möllhausen, with whom he often made collecting junkets.

Geologist Marcou was born in a small town near the Auvergne Mountains of central France in 1824. Because of poor health he dropped out of the Collège de St. Louis, under a doctor's advice to spend some time in the out-of-doors. Although Marcou was originally interested in mathematics, the outdoor life shifted his attention to plants and then to fossils. One of his early publications on the latter so impressed Louis Agassiz, the great Swiss natural scientist, that he recommended the young man for the Sorbonne chair of mineralogy in 1846. Two years later the French geologist Adrien Jussieu at the Jardin des plantes in Paris endorsed Marcou as the botanic garden's three-year "Traveling Geologist," with North America as his chosen destination. There Marcou became reacquainted with Agassiz, newly at Harvard, and took his first field excursion with Agassiz to Lake Superior in 1848. This summer expedition—which included a group of nine Harvard students and six volunteer naturalists of whom the tall, elegantly mannered Marcou was one—was a real north-country adventure, complete with birch-bark canoes and Ojibwa Indians; and every evening Agassiz lectured about the day's adventures. A coincidental encounter here was with a twenty-nine-year-old field geologist named Josiah Whitney, destined to become California's premier state geologist. In 1852 Marcou moved from the North Woods to the American Southwest, examining the geology of the semiarid Texas panhandle. Then in 1853–54 he joined Whipple's railroad survey to California, becoming the first internationally important geologist to have crossed North America.

Artist Heinrich Möllhausen, in his late twenties, arrived in the United States from Germany in 1849, encouraged by the great world geographer Alexander von Humboldt. His first venture into the West was with Duke Paul William of Württemberg, whose excursion got only as far as Fort Laramie, just over the border from Nebraska. Möllhausen returned to

Berlin in January 1853 but was back in the United States in the spring and heard about the railroad surveys. The expeditions of Stevens and Gunnison already had a naturalist staff, so following the advice of friends Möllhausen applied to Whipple's survey and on May 10 was appointed "topographer or draughtsman." Möllhausen was instructed to keep a journal, "making such drawings and memoranda as may appear to you of value or interest." He not only produced some marvelous artwork for the survey but also maintained a wordy journal and helped Kennerly with the zoological collecting, especially birds.

Early on the morning of February 20, 1854, with a chorus of ducks and Coyotes in the background, the survey group, scaling a gravelly ridge above the valley, looked west onto a beautiful vista, the waters of the Bill Williams Fork flowing into the Colorado River. The Colorado meandered down a magnificent valley from the north, while to the south the joined rivers disappeared into a chasm edged by black crags. The astronomical and magnetic instruments were set up, observations made, and then the men were on their way upriver on a well-beaten Indian trail to their first Colorado River campsite. As another member of the survey team, John Sherburne, recalled, at eventide the good Dr. Bigelow, to lighten the luggage load, "with the advice of those assembled," broke out eight bottles of Brandy, which "clear & in Toddies *mit sugar*" were drunk to toast the mighty Colorado.

The next few days, moving north, the men sometimes followed the river, on other occasions detoured away when the gorge became too rocky and narrow. The two heavy wagons and unnecessary baggage had to be abandoned, but scientific collections and instruments in Whipple's small, two-wheeled *carretella* cart, with its odometer, and his personal *carretella* were retained. They encountered numerous Indians, whose friendliness was encouraged by gifts of tobacco. The weather was pleasantly warm and flowers were abundant. Kennerly added a number of new lizards to the collection, as well as the unique Gila "trout" (Bonytail Chub). On February 23 the party left the desert mountains and traveled into "the great valley of the Mojaves," where today Needles is located. Here the Whipple party started its crossing of the five-hundred-yard-wide Colorado River, with a sand island in midstream.

The crossing proved a challenge. Among items brought from New Mexico were three much-worn canvas pontoons that could be lashed together and contained waterproof sacks lined with gutta-percha, a rubbery latex from Malaysian trees. Air was pumped by attached bellows

Figure 23. Whipple's party crossing the Colorado River near Needles in 1854 on the 35th parallel survey, by Heinrich Balduin Möllhausen. (Courtesy Oklahoma Historical Society.)

into the pontoons, which were set under the body of the cart to make a raft. When the hybrid craft was launched, said Whipple's report, "it sat upon the water like a swan" but resembled a bloated Venetian gondola (figure 23). The Indians watching the activity departed in disgust, having hoped to ferry the party across in their own boats.

By the next morning two of the air bags had collapsed, and the wagon bed was filled with water. Lieutenant Ives, who later in the decade became famous as a river "mariner," set to mending the many holes. Air was put back in, and "the pontoons again danced lightly upon the water." Ropes were attached to each end of the flatboat. With one rope in hand, a man on an air mattress paddled across to the island in midstream and towed the loaded craft over; then the boat was pulled back and reloaded. On the second trip the packed raft flipped over. At that moment word came that an important Indian chief was arriving. The "gondola" was forgotten, "bottoms upward, men underneath it, entangled among boxes and struggling for life." Meanwhile, rafts of driftwood had been constructed, and a few Indian craft commandeered. Bigelow was seated in one boat when a soldier carelessly threw in his musket, causing it to discharge and sending a bullet through Bigelow's stocking, but it only grazed his ankle. Unfazed, the botanist coolly remarked, "It's just as well as if the ball had gone into the air, a miss is as good as a mile."

Everything was unloaded on the island and then transported across the Colorado River to the west side. Here the current was stronger, and numerous times a loaded raft flipped. On Möllhausen's swamped raft, the artist was the only one of seven who could swim. Assistant Surveyor William White and one of the Mexicans nearly drowned. All the mules were able to swim across, and the cart carrying some of the survey instruments, field notes, and scientific collections managed the crossing safely. But when one of the rafts capsized, the notebooks of many members of the expedition, including Bigelow, Whipple, and Marcou, were drenched. Indians dived into the river, recovering some instruments and John Sherburne's meteorological journal.

By eventide of February 27, with assistance from the Mohave Indians, the 35th Parallel Survey was safely within California. Zoologist Kennerly celebrated that night by collecting the expedition's initial California mammal, a packrat. The first full day in the Golden State was spent assessing losses and cleaning and readjusting instruments. Wet clothing, papers, journals, books, and rolls of bedding were spread out to dry. There seemed to be hundreds of Indians frolicking about, jumping, racing, shouting, laughing, "having a hell of a time" as Assistant Astronomer George Garner observed. Some were still diving into the river and retrieving equipment, including two broken barometers, a powder flask, a rasp, a tin cup, and a good prismatic compass. Among valuables too soaked to save were some of the naturalists' reference books. The company was cheered, however, when Lieutenant Joseph Tidball turned up that evening with two bottles of brandy to celebrate arrival in California, with "Brandy Clear" and "Toddies" for all.

The expedition, now having traveled about a hundred miles across the desert, was just north of today's Providence Mountains. Two Mohave Indians, Chief Cairook and warrior Iretéba, led the Whipple party through an arid landscape dominated by expanses of what botanist Bigelow called "the celebrated but totally useless" Creosote Bush, "one of the most repulsive that can well be imagined." Bigelow collected a number of prickly pear and barrel cacti, to be sent to Engelmann in St. Louis for identification. Geologist Marcou reveled in the desert geological setting of black basalt flows, cinder cones, stark block ranges of granite and ancient metamorphics, and glaring white playas. He even analyzed water from some of the alkaline seeps. Möllhausen was disappointed that except for a few horned lizards no wildlife was to be seen, until he came upon a desiccated hummingbird with outspread wings lying on the sand, looking as if it had been "suddenly struck by death while on its flight." He tucked the bird in an envelope and later sent it to Europe.

Water was sparse, and there was little wood, twigs of Greasewood and yucca stalks the only fuel for campfires. Although the desert itself had been very warm and summerlike, on March 6, 1854, encamped in the Providence Mountains, the explorers felt a chill wind, snowflakes floated in the air, and snow patches still lay on the ground. Whipple divided his expedition into three parties: a small advance one, then Whipple's main team with the pack train carrying the reports and specimens, and finally the herders with the animals. Whipple sent the parties on one at a time so that springs could recover before the next contingent arrived. On March 8 the lower extent of Mohave Lake (Soda Lake) was skirted, and the next morning after a severe twelve-mile march across sand the travelers "were gladdened by the sight of the Mojave River," a beautiful stream that unfortunately disappeared a short distance to westward. The men quickly stripped and endured a very chilly bath. Soda Lake and its adjacent playa had been reached from the west the previous year by Williamson's railroad survey team, which had concluded that there was no feasible route across to the Colorado River.

The expedition's only food now was some unsalted mutton, beans, cornmeal, and coffee without sugar. "Smoking tobacco" consisted of ground corn, tea, and willow leaves. About March 11 the survey party encountered the Spanish (and Mormon) Trail, which followed the Mohave River bed, swinging south beyond present-day Barstow and eventually beginning its ascent from the desert through the Upper Narrows granite gorge. In the foothills the countryside became less desertlike. Water sparkled in the river channel. There were Joshua Trees, Singleleaf Pinyons, and California Junipers. Along the stream zoologist Kennerly collected a wintering Cinnamon Teal, a migrating Greater Yellowlegs, and in the brushland a Western Scrub-Jay and California Quail, with a Red-tailed Hawk shot in flight.

On March 14, along the Mormon Trail northeast of Cajon Pass, Whipple's survey encountered a party of Mormons heading from the Mormon settlement in San Bernardino to Salt Lake City. The spokesman bore sad and unsettling tidings about one of the other railroad surveys. Although details were sketchy, it was related that Captain Gunnison and several of his men had been ambushed and killed by

Paiute Indians near Lake Sevier in southern Utah the previous October. Two days later, having left the Mohave River and following what many years later would approximate the Santa Fe railroad route, the party reached the summit of Cajon Pass, with a nip in the spring air. Jules Marcou contemplated the "quite interesting" geology of the range of mountains here (figure 24), with "white marble and red porphyry being found, and indications also of gold and silver mines." He conjectured that the apparent upheaval seemed much more recent than had been supposed.

Marcou could not have been at a more intriguing geological site than where he found himself on March 16 (and where geologist Blake with Williamson's survey had been in 1853). This was the meeting of the east-west San Gabriel and San Bernardino mountains of the Transverse Ranges, and to the south the beginning of the north-south Peninsular Ranges. The great San Andreas Fault slashes through on a diagonal, with a variety of associated faults and a striking mishmash of geological formations. In 1857 the great Fort Tejon earthquake occurred in this area, almost as severe as the San Francisco earthquake of 1906. About seventy years after Marcou's visit, geologist Levi Noble, recognizing that here there was a multimile horizontal displacement between adjacent portions of the San Francisquito sedimentary formation, as well as tilted fault scarps, helped set the stage for plate tectonics theory.

Botanist Bigelow also enjoyed Cajon Pass. He collected a diminutive Red Maids, a Toyon in bud, a white gooseberry and a red currant, and a goodly assortment of ferns: Coffee Fern, Cliff-brake, California Polypody, Goldback Fern, Great Chain Fern, and Wood Fern. David Stanley, the expedition's quartermaster, exclaimed about the scenery: "I have seen nothing so wild, majestic, and awful as the view." The woodland of this mountain country was composed of pinyons, oaks, and Western Sycamores on the lower slopes and ravines, while "towards the tops of the mountains are pines and firs, excellent for timber." Kennerly collected his first Steller's Jay. Immediately over the pass to the southwest "the vegetation changes like magic," with chaparral and oaks. Bigelow encountered Chamise, yerba-santa, and Birch-leaf Mountain-mahogany. A white-flowered and a bright cerulean Ceanothus (California-lilac) reminded him that back home in Ohio there was only a single species, *Ceanothus americanus* (New Jersey Tea). He collected specimens of "the new remarkable genus *Fremontia*" he had seen figured in John Torrey's *Plantae Frémontianae* (now *Fremontodendron californicum,* Flannelbush). Unfortunately the beautiful evergreen oak with very large acorns,

Section of Cajon Pass

Figure 24. Cajon Pass in the Transverse Ranges, 1854: sketch by Charles Koppel (from the Pacific Railroad Surveys, vol. 5) and geological cross-section by Thomas Antisell (from the Pacific Railroad Surveys, vol. 7).

favored by the Indians, could not be identified, because the "proper books of reference" had been lost in the Colorado River.

Whipple's expedition proceeded into the San Bernardino Valley, passing an itinerant Mormon vendor trying to sell a smuggled keg of whiskey. Whipple was sure the liquor would remain untouched because his men had no money, but one of the teamsters fooled him and finally had to be suspended from a tree, with only the tips of his toes touching the ground, until he sobered and quieted down. The track continued through fields of grass and wild oats to Leon Prudhomme's Rancho Cucamonga, where they arrived on March 18. A number of American emigrants here had come across the Great Basin just after the ill-fated Gunnison expedition. Prudhomme had his own vineyard and produced many barrels of wine, but none pleasant to the taste of the Whipple party savants. His gardens were protected by fifteen-foot hedges of prickly pear cacti, which Bigelow was told had been imported from Spain by the Jesuits—most unlikely since cacti are essentially New World plants.

In early evening of March 19, 1854, at Cucamonga a good omen appeared, "a rainbow at sunset promised fair weather for the morrow." The next morning was beautiful, and the Whipple corps traveled past

Mission San Gabriel to camp overlooking Los Angeles. The collection of animals preserved in containers of alcohol, including fish, reptiles, mammals, and amphibians, was carefully repacked in a large keg of liquor for convenience of transportation. Several days later the steamer for San Francisco arrived at San Pedro, and the officers of the survey party—except for William White and John Sherburne, who had both been smitten by the allure of Southern California—were on their way north and then en route to Washington, D.C. Before leaving San Francisco Bay, however, Kennerly put a line in the water and added some marine fish to the survey's collection, including a weird-looking Plainfin Midshipman, studded with phosphores resembling a midshipman's brass buttons (figure 25). Unfortunately, in Panama, thanks to the express company's negligence, the alcoholic keg containing about three dozen specimens—including horned lizards, rattlesnakes, three new species of fish from the Mojave River, the now-rare Bonytail Chub out of the Colorado River, and Kennerly's pack rat collected that first night in California—disappeared. In his formal zoology report to Whipple, Kennerly wrote that "not only the expedition has suffered very seriously, but . . . science itself has met with an almost irreparable loss."

Dr. Bigelow was also smitten by California. After his arrival in San Francisco by steamer, he decided to remain in the Golden State until late spring, to enjoy the flowering season. Bigelow was around San Francisco throughout April 1854, and he took his first field trip up to Point Reyes on April 1, where he collected a mariposa lily. During April 4 and 5 he made a sortie across the bay to Oakland, probably being the first to botanize seriously in "the mountains above Oakland." Bigelow came away with forty-seven species, including the first specimen of Western Leatherwood ever found west of the Mississippi River (now on California's endangered species list).

The second week of April Bigelow headquartered at Corte Madera on the northwestern San Francisco Bay coastline. He explored the local Redwood country and then became the second botanist (after Thurber in 1852) to visit the upper Napa Valley, returning to the coast at Point Reyes, Tomales and Bolinas bays. The last week of April Bigelow collected around Benicia and Martinez, in the Vaca Valley, and especially up Napa Valley again. On May 1 he was elected a corresponding member of the California Academy of Natural Sciences, his initiation fee and monthly dues kindly paid by zoologist Dr. Williams Ayers. During May Bigelow visited Sonoma, Santa Rosa, and once more Napa Valley before

Figure 25. Painting of Plainfin Midshipman, 1854, collected by Caleb Kennerly in San Francisco Bay. (From the Pacific Railroad Surveys, vol. 10.)

heading off to the Central Valley and into the Sierra Nevada, first south of Sacramento in the Stanislaus and Mokelumne river areas and around Ione. In mid-May he spent at least a day among the Giant Sequoias in what is now Calaveras Grove State Park; there he collected his second California orchid, Rattlesnake Plantain. There were also a currant and a gooseberry, both growing atop the trunk of a fallen Giant Sequoia, the currant bush twenty feet above the ground. Then Bigelow proceeded north to the Yuba River, up into the mountains, and back into the Sacramento Valley to Sutter Buttes on May 25, where he obtained a beautiful lavender Foothill Penstemon and a Diffuse Daisy. Bigelow's botanical farewell to the Sierra Nevada was on the South Yuba River the next day, where he collected an Oregon Boxwood on a slope up-stream. On the return trip to San Francisco, he garnered a Wild Hyacinth, both in flower and in fruit, signaling the end of spring for John Bigelow in California.

During the weeks in the Golden State in 1854, the energetic Bigelow collected over eight hundred specimens, including a number of grasses, sedges, and ferns. These ample collections, as Asa Gray noted, "were brought home in perfect order, among them a number of new genera and more than sixty new species, as well as much information on "heretofore imperfectly known plants." In addition, Bigelow made a large collection of California mosses for his friend Sullivant, who later jokingly called the survey's published chapter on the mosses *Musci Bigeloviani.*

Bigelow collected a prostrate shrub in Redwood country on April 12 that Torrey named after "the accomplished commander of the expedition," *Whipplea modesta* (Yerba-de-selva). Asa Gray named a number of the new species in honor of their indefatigable collector, among them *Coreopsis bigelovii* from Mohave Creek and *Helenium bigelovii*, a "handsome and well-marked species" from a swamp near Santa Rosa. Gray also named some of Bigelow's new species after other plant collectors, including *Hemizonia fitchii* for Reverend Augustus Fitch, who had been making collections in the Sacramento Valley. Bigelow discovered the plant in the valley where it had been detected earlier by Fitch. The lily *Odontostomum hartwegii* had been found by Bigelow in Ione Valley and was named by John Torrey after Hartweg, who first discovered it. *Clintonia andrewsiana* was collected first by Bigelow at Tamul Pass, but better specimens were found later by Dr. Andrews, a "gentleman who assiduously examined the botany" of California.

Some of Bigelow's finds were worthy of special note. The Yellow Bush Lupine he encountered on sand hills at Point Reyes had supposedly not been reported since Menzies had collected it in the 1790s. At Duffield's Ranch in the central Sierra he collected "fine specimens, in full flower," of the rare Snow Plant Frémont had first discovered. Bigelow found Scarlet Pimpernel, introduced from Europe, already common in the Golden State. Mountain Mint, *Pycnanthemum californicum,* obtained by Bigelow along the Sierra Nevada's Mokelumne River, proved to be the only species of the genus west of the Great Plains. He correctly recognized that a goodly number of California plants were undoubtedly introductions; for instance he suggested that the common wild oat, "now spread over the whole country, many miles from the coast, may have been introduced by the Spaniards," as was also undoubtedly true for Annual Bluegrass and Dwarf Nettle.

Most of Bigelow's collecting sites are known today, but Duffield's Ranch, somewhere in the Sierra near Sonora, took decades to track down. From May 10 to May 15, 1854, Bigelow collected around this obscure ranch, amassing many specimens, including some new to science. The ranch was the type locality for the Three-bracted Onion, later of particular interest to University of California professor Willis Linn Jepson, at work on his comprehensive *Flora of California* and his *Manual of the Flowering Plants of California.* During the summer of 1918 Jepson in-

vited fellow botanist Professor George Peirce of Stanford to join him on
a two-week junket to search for Duffield's Ranch. As Jepson wrote, every
Sonoran old-timer was "willing and anxious to tell us about Duffield
ranch," but vague on location. The old-timers were finally whittled
down to taciturn seventy-eight-year-old Ben Soulsby, who did recollect
a Duffield mine but no ranch. Jepson was suddenly struck with an idea.
He had brought with him volume 4 of the Pacific Railroad Surveys,
which included Bigelow's plant collections. Opening the large book, he
pointed to the Duffield Ranch name. Soulsby contemplated, thumbed
through the volume, reached the title page with the publication date of
1856. Finally he closed the book, ran one hand through his iron-gray
hair, and slowly drawled, "Now I bethink myself. There was a Duffield
ranch." Shortly Jepson and Peirce found the site up in the mountains
northeast of Sonora where Bigelow so long ago had discovered six new
plants, including the Three-bracted Onion.

Dr. John Bigelow returned to New York from San Francisco by
steamer in June 1854. He turned his collected plants over to John Torrey, made a brief visit home to Ohio, and by mid-July was in Washington, D.C., together with Möllhausen, to work on the final report with
Whipple, who was eager to have a plant section ready for publication as
soon as possible. The plants were to be examined and described by Torrey and especially by Gray, who would get the bulk of the specimens for
the Harvard herbarium. Torrey got a set of duplicates, and some were
sent to Joseph Hooker of the Royal Botanic Gardens, Kew. Bigelow's
hopes for duplicates that he could sell proved a bit optimistic. By 1855
work on the report was completed, and Bigelow returned to his comfortable home, family, and medical practice in Lancaster, Ohio.

In 1856 the Whipple survey reports were published in volumes 3 and
4 of the Pacific Railroad Surveys. In volume 3 appeared Whipple's itinerary and his discussion of the route's topographical features, his report
on the Indian tribes, and the geology and fossils. Geologist Marcou had,
without permission, taken his field books back to France with him.
When they were returned, William P. Blake, geologist for Williamson's
survey, translated them and wrote up the survey's geology section, with
eastern paleontologist James Hall doing the fossils. Volume 4 included
the zoology overview by Kennerly, then specific write-ups on the mammals, birds, reptiles (by Baird), and fishes (by Girard). In the lengthy
botany section, Bigelow provided an admirable ecological description of
the route, with Dr. Engelmann describing the cacti and Bigelow's old
friend Sullivant preparing the material on the mosses and liverworts.

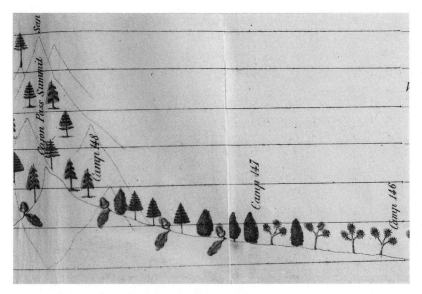

Figure 26. Western portion of botanist John Bigelow's panoramic tree transect for Whipple's 1854 35th parallel survey. (From the Pacific Railroad Surveys, vol. 4.)

Professor Torrey wrote up the botanical collections. A unique inclusion was Bigelow's lengthy fold-out representing in profile the entire route west to the coast with colored symbols for the various tree types where they occurred across mountains, deserts, and plains (figure 26).

At the completion of work associated with the 35th Parallel Pacific Railroad Survey, Amiel Whipple was assigned by the Army Corps of Topographical Engineers from 1856 to 1861 to supervise operations aimed at opening up the Great Lakes for large shipping vessels. When the Civil War commenced, the topographical corps was made responsible for aeronautics, specifically observation balloons. Whipple was assigned to decide which "aeronauts" would be invited to work for the army. His first choice was Thaddeus Lowe, who had built his first balloon in 1858 and anticipated making a balloon trip across the Atlantic. Lowe had already begun filling his balloon to observe the Battle of Bull Run when Whipple interviewed another candidate, John Wise, who promised to deliver a balloon for only two hundred dollars. Whipple had Lowe's balloon removed from the gas main and requested that Wise rush his balloon to the front. But on the way to the battlefield Wise's balloon was wrecked, and Lowe went on to become chief of army aeronautics. Years later Lowe became prominent in Southern California, establishing the state's second astronomical observatory.

When the balloon operations were transferred to the Quartermaster Department in 1862, Whipple was appointed a brigadier general of the U.S. Volunteers. He served with the Army of the Potomac in a corps headed by Pacific railroad survey colleague George Stoneman, especially at the battles of Fredericksburg and Chancellorsville in 1862 and 1863. Whipple was with his Third Division at Hazel Grove the morning after the fierce two-day battle when he was critically wounded in the abdomen by a sharpshooter. He died in Washington on May 7, at the home of Major John Sherburne, another of his railroad survey colleagues. President Lincoln posthumously promoted Amiel Whipple, that indomitable western army officer and naturalist, to major general and rode in an open carriage with Whipple's funeral cortege to the Capitol.

During 1859 Bigelow's botanical collections from the Mexican Boundary Survey were finally published in a government report. Despite being easygoing, Bigelow was irritated all over again because Asa Gray back in 1850 had published Charles Wright's Mexican Boundary Survey collections without permission of the survey or of the other boundary-survey botanists, whose findings were only just now appearing. At that time Bigelow had written to John Torrey, "The humble collector who undergoes much fatigue & privation as well as danger should not be forgotten or neglected in the roll; for if we cannot make the music we are necessary in raising the wind so essential in successfully playing the organ of fame."

In 1860 Bigelow moved with his family to Detroit, where Whipple was busy with the topographical engineers' shipping survey for the Great Lakes. Bigelow was appointed again to Whipple's staff, this time not as a botanist but as a meteorologist. He was deeply saddened by Whipple's death during the Civil War. Remaining in Detroit after the war, Bigelow became a professor of medical botany and materia medica at Detroit Medical College and later served as surgeon at the Marine Hospital in that city. He died on July 18, 1878.

Parke Heads East while Beckwith Heads West, 1854

PARKE'S 32ND PARALLEL PARTY

On January 24, 1854, Lieutenant John G. Parke departed from San Diego with fifty-five men to survey eastward along the 32nd parallel. His staff included as topographer Henry Custer, a Swiss, who would serve again with Parke at the end of 1854 on the railroad survey along the Southern California coast and in 1857 with him on the Northwest Boundary Survey. In 1867 Custer became one of Clarence King's 40th parallel topographers. Also with Parke from Williamson's survey were naturalist Adolphus Heermann and Lieutenant George Stoneman with twenty-eight dragoons. The only piece of new equipment Parke was unable to obtain anywhere was a mountain barometer. The party was bound for Fort Yuma by way of Jonathan Warner's ranch in the Agua Caliente valley, then east to retrace Parke's earlier route between the Pima villages and Tucson, and on to the Rio Grande at Doña Aña, north of El Paso. Parke's trips in the West and John Pope's surveys from the east to the Rio Grande would complete the Pacific Railroad Survey along the 32nd parallel.

Because of the winter season, there was little for Heermann to record or collect during the trip to Fort Yuma. At the Colorado River near the fort he found Gambel's Quail "more or less abundant"; Abert's Towhees numerous in sheltered thickets, scratching for insects and seeds; and two common woodpeckers, Gila and Ladder-backed, in Mesquite, especially after the expedition had crossed the Colorado and come to the mouth of the Gila River. Parke's party headed east along the south bank on an old wagon road into the desert, bound for the Pima and Maricopa villages. Soon there were no reptiles, amphibians, mammals, or flowering plants. Collecting improved as spring slowly replaced winter, but Heermann despaired when the cans in which cold-blooded vertebrates had been packed became leaky and neither could be repaired nor the alcohol replaced. All these specimens had to be discarded, leaving only a few preserved in bottles.

Dr. Adolphus Heermann, who had truly proved himself a "compleat naturalist," suffered a premature end. After preparing his reports in Philadelphia for the survey, he spent two winters collecting birds in Texas, eventually settling in San Antonio during the Civil War. On September 2, 1865, Heermann—out alone to do some collecting—apparently stumbled and fell with his gun, which discharged and killed him.

LIEUTENANT BECKWITH SURVEYS
NORTHEASTERN CALIFORNIA

The highly interesting scenery . . . bears . . . most
strikingly the character of a volcanic country.
                                        —Dr. Jacob Schiel

In the late winter of 1854, when the 38th Parallel Pacific Railroad Survey left Salt Lake City and continued across the Great Basin toward Northern California, it was a very different expedition from the one that had departed from eastern Kansas under topographical engineer John Williams Gunnison in mid-June 1853. On October 26, 1853, north of Lake Sevier in southern Utah, a reconnoitering field party led by Captain Gunnison was ambushed by Paiute Indians. Eight of the twelve men were massacred, including Gunnison and naturalists Frederick Creutzfeldt and Richard Kern. This was the nation's worst government survey disaster and the first deaths of survey naturalists in the line of duty. It was particularly poignant since both Kern and Creutzfeldt had nearly died during Frémont's private expedition five years earlier.

Replacing Gunnison was First Lieutenant Edward G. Beckwith, Gunnison's former second in command, of the Third Artillery rather than the topographical engineers. Once before Beckwith had assumed a dead leader's command, that of Captain Herman Thorne in 1849 at Fort Yuma. In February 1854, Beckwith received permission from Secretary of War Jefferson Davis to pursue a route from Salt Lake City across to the Sierra Nevada along approximately the 41st parallel, rather than the 38th parallel. He was allotted $23,000 for expenses, all that remained from Gunnison's budget.

Beckwith's survey party spent the winter of 1853–54 in the Salt Lake City area. Remaining with Beckwith from Gunnison's staff were Brevet Captain Robert M. Morris and Lieutenant Lawrence S. Baker of the Mounted Riflemen, assistant topographer James A. Snyder, and astronomer Sheppard Homans, who was charged with tracking the expedition's westward progress. Continuing as geologist was Jacob "James" Schiel, who held a Ph.D. in chemistry from the University of Heidelberg, where he had lectured in chemistry and geology. Mysteriously, "M.D." appeared after Schiel's name in his geology report for the survey. He served as the survey's surgeon but never had formal medical training. Schiel described Creutzfeldt, who was killed in the Gunnison massacre, as his "assistant," but this young naturalist had actually been Gunnison's official botanist.

Because Beckwith found no replacement for Creutzfeldt, he assigned James Snyder to the botanist post. Since the eighteen-year-old youth had been a personal friend of Richard Kern as well as his assistant, Snyder was particularly devastated by the massacre. He remembered that Kern had been apprehensive about the side trip with Gunnison, telling Snyder the night before his departure how to dispose of his possessions if he did not return. The next morning as Kern bade him good-bye, Kern told Snyder that "he should not see me again. A tear fell from his eye at the time." Snyder, three days after Kern's death, wrote to his own father, "I am all alone, and have lost one whom I loved dearly, and who had been as a father to me." As Beckwith's naturalist, the youthful Snyder possessed little scientific expertise and seemingly lacked enthusiasm for the assignment, since he collected plants only haphazardly and paid little attention to animals.

By the end of 1853 three prior expeditions had made the trip over the Continental Divide in southern Colorado. The first was the Gwin Heap–Edward Beale party, shortly followed by Gunnison's. The third, four months later, was Frémont's second privately funded railroad survey across the Colorado mountains and westward. Frémont, like Gunnison, had encountered disaster, this time in the snow-covered southern Utah mountains. When Frémont's party finally struggled into Parawan on February 8, 1854, food was gone, the horses and mules eaten, and one man dead. Except for the ten Delaware Indian guides, the expedition members were physically and mentally exhausted.

Frémont continued on to California several days later, but two of his scientific staff abandoned him and headed north in a wagon for Salt Lake City, arriving on March 1, 1854. These two were Baron F. W. von Egloffstein, a talented Prussian immigrant who had been serving as Frémont's topographical engineer, and Solomon Nunez Carvalho, the artist and daguerreotypist. At Parawan Carvalho was in considerably worse condition than Egloffstein, with matted, unkempt hair and weighing 101 pounds, so weak he had to be lifted into the wagon in which the two men departed. Upon reaching Salt Lake City, the pair obtained lodging at Blair's Hotel. Beckwith, hearing of the two Frémont men in town, invited them to dine with him at Mormon Apostle E. T. Benson's home. Beckwith offered each a position with the expedition to California. Carvalho, who with Frémont's expedition had been making "beautifully clear" daguerreotypes (according to Jessie, Frémont's wife), declined the offer and after regaining strength continued to California on his own. Fortunately for Beckwith, Egloffstein accepted the invitation, replacing Richard Kern as artist and topographer.

Beckwith, like other Pacific Railroad Survey leaders, was searching western mountain barriers for routes and passes suitable for railway lines. He continually divided his survey expedition into smaller exploring parties, keeping track of length and degree of ascents and descents, mainly traveling westward but on occasion returning eastward to review an area.

Despite northeastern California's isolation, Modoc Indians, wild river canyons, snow-encrusted peaks, black volcanic fields, dense coniferous forests, and semiarid brushland expanses, the countryside by 1854 was not entirely unexplored or unsettled. In 1841 the inland contingent from Wilkes's United States Exploring Expedition with geologist James Dana had come south along the western side of Mount Shasta and into the Sacramento Valley. In that expedition's report Wilkes, based on Hudson's Bay Company sources, had designated the main headwater branch of the Sacramento River as the "Pitts River" (correctly Pit River, after Indian pit traps for wild game dug along its banks). Frémont in 1846 had left frontiersman Peter Lassen's ranch in the northern Sacramento Valley and explored northeastward up the Pit River, crossing and naming Round Valley (now Big Valley) and venturing as far as the Modoc Plateau. Two years later Peter Lassen, leading an emigrant wagon train, had blazed a crude track from the established Nevada-Oregon Applegate Trail cutoff near the Oregon border southwest into the Pit River country and eventually through intervening mountainous and canyon terrain into the northern end of the Sacramento Valley. Lassen's route was a long and difficult one for wagons.

In June 1849, during the early days of the Mexican Boundary Surveys, Captain William Warner and Lieutenant Robert S. Williamson had been sent up the Sacramento Valley to search for a potential railroad route over the eastern range and north into Oregon. They tried out Lassen's Road, which seemed too rugged for trains. Near Goose Lake on the Oregon border Indians attacked and killed Warner. In 1851 prospector William H. Nobles had plotted a wagon route out of the Great Basin that passed directly westward from the big bend of the Humboldt River in Nevada, across the Smoke Creek Desert into the mountain front, on to the Sacramento Valley over Nobles's Pass (in the northwest corner of today's Lassen Volcanic National Park), and thence to Shasta City. Nobles's Road, completed in 1852, was definitely shorter and easier than Lassen's. By 1854, answering the lure of the gold rush, many prospectors and numerous scattered mine camps dotted the region, some along river courses with their own crude ferries. Also there were resident lumbermen, lumber mills, and even a few serious settlers.

On June 15, 1854, Beckwith's surveying expedition came northwest from Nevada's Pyramid and Mud lakes to camp beside Buffalo Creek at what Beckwith called the "east base of Sierra Nevada." The party then entered California across Painters Flat and headed toward the headwaters of Smoke Creek. As they traveled westward through a shrubland of Big Sagebrush and Greasewood, they confronted a vista of eroded volcanoes, domes, cinder cones, upright volcanic plugs, igneous dikes, reddish-white rhyolitic bands, and black lava swaths. There were more than two dozen volcanoes of various ages and forms, interspersed with marshes and dry meadows, steep ravines, and rocky terraces of outcropping basaltic strata. The previous year when Gunnison's expedition had crossed Colorado's Continental Divide at Cochetopa Pass, Schiel had sketched volcanic formations there—columnar basalts and "red trapp-porphyry"—but nothing compared to the dramatic landscape here. By the end of the next century dozens of geological publications would deal with this volcanic region.

At the head of Smoke Creek, so-called East Madeline Pass, rose a high, somewhat isolated landmark Beckwith decided to call President's Peak (later changed to Observation Peak), an andesitic volcano. Members of the survey party, after an exhausting climb, reached the 7,964-foot crag-topped summit and enjoyed an impressive view of the surrounding countryside, with snowy Mount Shasta far to the northwest beyond the broad Madeline Plains. Egloffstein made a sketch of the vast valley across to West Madeline Pass. Schiel, the first professional "geochemist" to explore the region, was especially excited by this land where volcanic rocks "became masters of the ground." The fine-grained phonolite (extrusive andesite), he found, "would ring beautifully to the hammer" and clinked when one walked across the numerous platy flows.

Over the moderate eastern pass between volcanoes, the survey proceeded northwest across the old lake sediments and extrusive volcanics of the Madeline Plains and into the mountain escarpment. Greater Sagegrouse were encountered in the sagebrush and alkaline Greasewood flats, and Common Snipe in the few wet meadows, but, as Beckwith's report notes, there was "no evidence of larger game." They met a few Indian families, the women quickly disappearing while the men, lured by presents, temporarily joined the travelers.

Beckwith and his men continued over West Madeline Pass, while Egloffstein at 8:00 A.M. on June 26 finished a sketch from a peak overlooking Madeline (Willow) Creek. The mountains ahead, as Beckwith wrote, were "beautifully dark with forests of timber—the first we have seen in twelve months really worthy of the name." The surveying party dropped into Frémont's Round Valley, reaching the Pit River, the surveyors' arrival greeted by spiraling smoke from Indian signal fires.

The party followed the Pit River southwest across to the end of Round Valley, where the stream narrowed into the so-called First Canyon. Lassen's old emigrant road was encountered, heading southward. It was evident that the expedition's wagons and animals could not make the trip farther down the river, in its eighty-foot-high, rock-walled gorge. So Beckwith with a small crew proceeded downstream to the junction with the Fall River, coming back to rejoin the main party the next evening.

On June 30, Beckwith decided to resurvey the Mud Lakes area along the Nevada border. The expedition proceeded southwest on Lassen's crude road, entering a majestic coniferous forest, some trees more than five feet in diameter and up to 140 feet tall. Egloffstein scrambled atop a ridge, which "afforded an extensive view of the mountain country around us," and made a panoramic four-panel sketch of the awe-inspiring vista to the northwest: a wilderness of ridges, buttes, valleys, and chasms, dominated in the background by the snow-white Mount Shasta, with a horseman in the foreground on Lassen's Road.

On July 2, 1854, the men struck Nobles's Road in upper Pine Creek Valley, making their evening camp east of Mount Lassen near Summit

(Bridge) Creek. Early the next morning, crossing a high rocky butte above the Susan River, the men got a fine view northeast down onto a large lake they named Eagle Lake. The freshwater lake was a remnant of the ancient Lake Lahontan that once covered the Great Basin lowlands in this region. This trip from Pit River toward the lower portion of the Susan River impressed geologist Schiel because of "its most magnificent pine forests," with tall, dense stands of Jeffrey, Pacific Ponderosa, Lodgepole, and Sugar pines, White Firs, and Incense Cedars.

At four in the afternoon of July 3, the party, dropping steeply down from the plateau, came into the broad Susan River valley, where two men were erecting a log cabin and planting a small vegetable garden. Isaac Roop and his brother, Ephraim, had just arrived in this verdant, well-watered valley in June, becoming the first settlers of the region. On Independence Day the expedition traveled to the north shore of the vast, shallow Honey Lake lying beyond a long ridge—a twenty-three-mile march—and camped near today's Wendel Hot Springs. When they measured the hot springs' temperature, the water was 205 degrees Fahrenheit, while 150 feet downstream it was still 170 degrees. Meanwhile, snow remained atop some of the surrounding mountains to the west.

The next morning Beckwith assigned a detachment including Snyder and Egloffstein, escorted by Lieutenant Baker and a detachment of riflemen, to continue around Honey Lake toward Nevada and connect with Beckwith's earlier survey line at Smoke Creek. Honey Lake is an ephemeral body of disagreeable alkaline water, sometimes a shallow, muddy lake but more often a dry playa, its tan dust swirling eastward with the prevailing wind. It had been, like Eagle Lake, an arm of ancient Lake Lahontan, and the surveyors could discern an old shoreline. If Beckwith's survey had come by during the 1868 wet spell, they would have found Honey Lake twenty-five feet deep. The detachment returned the same day, and the next morning the entire company headed west, traversing the lakeshore, where dozens of ducks and some American White Pelicans were milling above the men's heads. After camping on Willow Creek, the men were back at the Roops' by the evening of July 7.

On July 8, as Beckwith's party headed west up the Susan River, Snyder collected his last plant of the expedition recorded with a specific date: a mint, *Monardella odoratissima*. The men commenced the ascent to Nobles's Pass, traveling pretty much along Nobles's Road, which dipped in and out of the northern boundary of today's Lassen Volcanic National Park. Shortly before reaching the pass on the tenth, where the road circled around Prospect Peak, geologist Schiel, looking off to the south, was impressed by "a mountain 800 to 1000 feet high of conical shape, and formed of black lava, apparently a monument of the latest disturbing forces in these regions." Down from the rounded summit of this isolated butte were volcanic sand and small pieces of lava, one of which he added to the expedition's collection. The expedition report named it Black Butte, but it is unquestionably today's Cinder Cone (6,907 feet). Present-day Black Butte lies five miles farther east.

West of Nobles's Pass the expedition continued down the gradual western slope into the Sacramento Valley. As they descended, the forest panorama changed from Jeffrey Pine and White Fir to Pacific Ponderosa Pine and Incense Cedar, then open oak woodland and grassland dotted with black boulders of basalt, with gray-foliaged Foothill Pine on layered volcanic ridges. At the foothills camp near McCumber's Mill on July 11, Schiel was elated to find a new extrusive rock unlike any others he had seen, "reddish gray rough rock, interspersed with crystals of mica, pyroxene." With little time to examine the rock, he "did use the blow-pipe but to no effect." For several days the expedition had been traveling on red soil the local miners claimed was gold-bearing because it resembled the color of gold-bearing sediments in the Sacramento Valley. Schiel knew better, recognizing it as merely decomposed volcanic rock.

The destination of Beckwith's Pacific railroad survey was Fort Reading on Cow Creek, a beautifully wooded riparian site a few miles northeast of a bend in the upper Sacramento River. This military post, the first and largest in Northern California, had been established two years earlier on Nobles's Road and named after frontiersman and early settler Major Pierson B. Reading, who had served with Frémont's battalion. The survey party was about three miles from Fort Reading at three in the afternoon on July 12, 1854, when Beckwith recorded a temperature of 106 degrees in the shade. They reached the fort by sundown and received a warm welcome from Lieutenant Colonel George W. Wright and his officers of the Northern Military District of California. Over the next several days, as horses and mules were shod and supplies replenished thanks to Quartermaster Captain M. S. Miller, the visitors were hospitably entertained.

On July 15 the survey crew returned to the northwest toward the Pit River, intending to follow it east to its junction with the Fall River (Frémont's 1848 route), thus completing the Madeline Pass survey. Some twenty-one miles north of the fort they encountered a small mining village, Churntown, composed of a dozen "miserable log-huts," the temperature so oppressively hot that not a miner was to be seen at work.

The Pit River excursion took seven arduous days. The river canyon, dammed in 1945, is now inundated by the extensive Shasta Lake (covering fifty square miles), but when Beckwith made his trip the river gorge was so steep and narrow that the survey party "was forced to keep to the hills south of the river." Where Fall River meandered south and met the Pit, Beckwith turned south, following an Indian trail, to explore the Canoe (now Hat) River valley as a possible route to Nobles's Pass. Beckwith noted that the region to the east was "infested throughout with hostile savages." Many Indians were indeed seen in Canoe Valley, but they fled on spotting the expedition. Although Hat Creek's headwaters eventually do reach Nobles's Road east of the pass, the route, at first through an open valley, finally proved too mountainous below Nobles's Road and unfit for wagons (or trains). The party returned to Fort Reading by July 26.

During Beckwith's entire survey, from Salt Lake City on April 4 to Fort Reading on July 26, the party had been blessed by good weather and a lack of mishaps. The exploration party was duly disbanded, public property was sold at auction, and the men bade farewell to Fort Reading. The Washington, D.C.–bound company took along Snyder's plant collection, Schiel's black shale from the Pit River and his chunk of lava from "Black Butte," violet and white feldspar specimens, and a chunk of gneiss from the base of the Sierra Nevada. Quartermaster Miller of the fort donated a chunk of gold-bearing quartz and "also a petrified oak block" from Cow Creek. Schiel brought back a rock from the headwaters of the Pit River that he later had a man in his laboratory analyze; it was about 58 percent silica. With his microscope Schiel subsequently examined soil gathered from near the Pit River, finding fossilized infusoria (microscopic plants and animals). He sent soil samples from Honey Lake and the Pit River to Professor Jacob Bailey at West Point. Bailey, an authority on microscopic algae, reported that both samples contained myriads of minute skeletons of freshwater siliceous diatoms, known and unknown.

Egloffstein made numerous exquisite landscape sketches during the Beckwith expedition that were included in volume 2 of the Pacific Railroad Surveys. A full plate shows the western end of Madeline Pass from a summit overlooking Madeline Creek, and another the mountain passage of the Pit River. One folding panoramic view encompassed from Observation Peak (East Madeline Pass) west across the Madeline Plains to West Madeline Pass and beyond (figure 27), the other the magnificent "Northern Slopes of the Sierra Nevada," looking northwest from near Lassen's Road.

Figure 27. View northwest toward Mount Shasta in the last panel of the
Madeline passes panorama, by Baron F. W. von Egloffstein on Beckwith's
1854 survey. (From the Pacific Railroad Surveys, vol. 11.)

When John Torrey and Asa Gray began working with the plants from
botanist Snyder, they found that the collection contained only twenty-
nine specimens from California. Since the region was mostly unmapped
initially, the collecting sites were often not specific—for example, "in a
valley of the Sierra Nevada"—and in some cases lacked specific collection dates. Snyder had begun his limited California plant collecting east
of the mountain front in desert scrub country, finding Hop-sage (which
the hunters called greasewood), Winter-fat, Checker Mallow, Cushion
Buckwheat, purple-flowered Nodding Buckwheat, and a new species of
horsebrush, to be named *Tetradymia glabrata* by John Torrey and Asa
Gray. One of the yellow evening-primroses, *Camissonia claviformis,* that
Snyder found is of historic interest because the species was apparently
also one of the few plants Frémont collected on his last expedition to
California.

Snyder procured the remainder of his species in the mountains, including a red columbine in a canyon, the California Poppy, Bitterroot, a
Common Camas in a high wet meadow, a larkspur in a mountain valley, and from Round Valley near the Pit River a species of *Lomatium*. At
Nobles's Pass on the morning of July 3 he collected the first two plant

species from what would become, in 1916, Lassen Volcanic National Park: Common Pussypaws and a mariposa lily, *Calochortus leichtlinii.* Snyder's notable California floral specimen was a crimson clover he found while crossing East Madeline Pass about June 22. Appropriately, it was named *Trifolium beckwithii* by Sereno Watson of the Gray Herbarium. The only drawing of a Snyder specimen in the published report was of a lavender annual, *Phacelia humilis,* that Snyder collected "near the summit of the Sierra Nevada."

Beckwith, back in Washington, put together the report for both his and Gunnison's expeditions, with the assistance of astronomer Homans, Schiel, Snyder (now employed as Beckwith's assistant) with his plants and meteorological and barometric observations, and Egloffstein, Beckwith's "topographer, to whom I am indebted for superior topographical sketches" and "who is still engaged in making the elaborate maps of the survey." Especially valuable was Egloffstein's "Map No. 4—from the valley of the Mud Lake to the Pacific Ocean," showing in detail the expedition's complicated treks across northeastern California, including the campsite dates. The artist became famous for his almost three-dimensional cartography.

At the end of October 1854, Beckwith sent Snyder's plants to John Torrey, and shortly Torrey divided the specimens—two hundred species for the 1854 expedition—into two sets, one for himself and one for Asa Gray, but unevenly, so that Gray got those "belonging to your own favorite orders." Torrey wrote that the few Snyder specimens were in nice condition and that Gray should "take pleasure in studying them." For the report Torrey and Gray prepared, they shared two hundred dollars. Torrey could not bear to charge any more.

Because Lieutenant Beckwith was an artilleryman rather than a topographical engineer, his report was essentially disregarded by topographical engineer Andrew Atkinson Humphreys, who had taken over from William Emory as commander of the Bureau of Explorations and Surveys, and likewise by Secretary of War Jefferson Davis, who as a southerner understandably favored a much more southerly train route. With respect to the northeastern California region, Beckwith recognized two significant hazards, the steepness of the divides and the attendant problems with snow. It was a tribute to Beckwith that after the Civil War, when the first transcontinental rail line was completed, some of the sections of the Central Pacific Railroad across the Great Basin did follow Beckwith's 1854 survey route. But neither of the main passes in California—Nobles's and Madeline—was ever used for a major railroad line.

From the California mountains Beckwith brought back as a "souvenir" the dried roots of *Lomatium triternatum*. Commonly known as Lewis's Lomatium, the type specimen had been collected by Meriwether Lewis near the Clearwater River in Idaho on May 6, 1806. Lewis described in his journal "a root 5 or 6 inches long eaten raw or boiled by the natives." Lieutenant Beckwith's report noted that the roots were "collected very largely by the Indians. When dried they are hard but brittle, and have a mild sweet taste. They afford a good proportion of the food of some tribes." Beckwith was obviously of the same mettle as those earlier army officers Lewis and Clark. In 1898 University of California botanist Willis Jepson, as a tribute to Beckwith, placed a pinkish-rose buttercup in a new genus, *Beckwithia* (now back to *Ranunculus andersonii*). Though not collected in California by this railroad survey, this buttercup does occur in the Golden State's sagebrush country through which Lieutenant Beckwith's expedition traveled.

Expeditions Up and Down California

DOWN THE COAST WITH LIEUTENANT PARKE
AND ANTISELL, 1854–1855

In November 1854 Lieutenant John Parke was undoubtedly still recovering. He and Lieutenant Williamson had surveyed from the Sierra Nevada down to San Diego in 1853, and in January 1854 Parke had led his own railroad survey eastward along the 32nd parallel to New Mexico. Now, back by ship from New York with his party, he was at California's Benicia army depot ready for yet another survey: to "determine the practicability" of a railroad line running west of the Coast Ranges from San Francisco to Los Angeles.

Parke was assisted by Albert H. Campbell, Brown University Class of 1847, a civil engineer with artistic talent and a sharp eye for geology. Campbell had already served as engineer and surveyor with Whipple's 35th parallel survey in 1853. Also on Parke's staff was Henry Custer, his topographer on the survey from San Diego east the past winter. Dr. Thomas Antisell, a U.S. Army Medical Corps assistant surgeon, was Parke's surgeon and geologist. Antisell arrived in New York City from Ireland in 1848 with his M.D. degree. He became acquainted with botanist-geologist John Torrey, practiced medicine for several years, and then taught briefly in New England. Torrey recommended him for the geology position, expecting him to collect some plants as well—a "supererogation," as Torrey put it, beyond the call of geological duty.

283

Parke's escort would serve under Lieutenant William A. Winder, Third Artillery.

By November 20, 1854, the expedition had obtained stores and camp equipage and arranged transportation. They went across the straits to Martinez and to San Jose, the official starting point. Antisell collected five species of plants en route for Torrey: Blue Nightshade, Nit Grass (introduced), *Rorippa curvisiliqua* (an uncommon yellow cress), Coyote-brush, and *Hemizonia parryi* ssp. *rudis* (a tarweed).

South from San Jose the party searched for potential railroad routes, following the little Pajaro River to Monterey Bay. Then they continued up the Salinas River, with stops at missions including a side trip to Mission San Antonio, where in December an annual composite, a species of *Lessingia,* was added. The survey continued to the Salinas River headwaters in the La Panza Range. Parke anticipated that a usable route might lead from here to the Cuyama Valley in the southwestern San Joaquin. Despite the time of year Antisell collected thirty-nine plant species in the upper Salinas Valley, including Biennial Sagewort. In preparing the botanical report, Professor Torrey noted that this species had not previously been found west of the Rocky Mountains. It later proved to be European in origin.

In February 1855 the survey party, continuing south, encountered the Santa Ynez Mountains north of Santa Barbara, the beginning of the east-west Transverse Ranges that extend across California from the ocean to the Mohave Desert. For the expedition the Santa Ynez Mountains presented "a decided obstacle." The only well-known passage was over Gaviota Pass, and during a detailed survey there the expedition supplies ran low. A hasty trip was made down to Santa Barbara for more supplies, plus pack mules for scouting a possible rail route down the shoreline from San Luis Obispo. Meanwhile, the naturalists collected more than two dozen kinds of fossils, dominated by oyster shells, from the steeply dipping brown and yellow sedimentary outcrops in the vicinity of the pass.

Campbell took a small party to explore an existing pack route that crossed the Santa Ynez Mountains over San Marcos Pass (today State Highway 154) toward the source of the Santa Ynez River. Frémont had taken this path on Christmas Day 1846 to avoid a rumored ambush by the Californios at Gaviota Pass. Campbell's survey included the base of the San Rafael Mountains and the valley of the Santa Maria River, crossing many faults including the major Santa Ynez Fault, which parallels on the north the east-west coastline. Antisell and

Campbell were aware that Southern California was "remarkably subject to earthquakes"; fifty-nine had been recorded during the previous six years.

At Santa Barbara Parke decided to go back and determine whether indeed the Salinas Valley route could be joined with the inland road to Los Angeles. Starting from San Buenaventura (present-day Ventura), the surveys ran north into the mountains, hoping to reach the Cuyama Valley (San Joaquin Valley around State Highway 33) and connect with the inland road to Los Angeles. Before leaving, Antisell collected a shrubby canyon-sunflower, *Venegasia carpesioides*, which only Douglas and Nuttall had collected earlier. During the late winter of 1855 each of Parke's frustrating surveys into the Transverse Ranges encountered precipitious canyon walls and summits buried deep in snow. Finally the party moved with the wagon up the Santa Clara Valley, eventually reaching the road to the San Joaquin Valley. Traveling over Cañada de las Uvas Pass (today's Tejon Pass) and into Cuyama Valley, they attempted to connect with the survey's Salinas River route. Working cross-country to the headwaters of the Santa Maria River, the men scrambled over a "most forbidding" divide to a tributary of the Santa Clara River, where they wisely concluded that the rugged country was unsuited for a rail route. At last the survey party was back on the road to Cañada de las Uvas, having entirely circled a mass of ranges and canyons that even today constitutes in large measure a roadless wilderness, including the Dick Smith Wilderness Area and the isolated Sespe Condor Sanctuary.

Campbell, when in the vicinity of the San Andreas Fault's "big bend" west of Cajon Pass, surveyed without recognizing the conjunction of four major faults or the significance of the east-west Transverse Ranges. But from his field observations, he realized that the mountainous country to the north and east consisted of numerous parallel mountain ridges with intervening valleys. Parke optimistically wrote that Campbell was able "thereby [to] unlock this whole system," and he considered Campbell's examination of "the geological dynamics of this entire region" as "one of the most important ones made." Dr. Antisell, in his geological report for the survey, noted that there were similar granitic outcrops from Monterey Bay's Point Pinos south and then east toward the Mohave Desert, and that there were serpentine rock and "volcanic protrusions" along the coast. He correctly disagreed with geologist Blake, who claimed that the rise of the easterly San Bernardino Range had caused the uplift of the Santa Ynez Mountains. The limited extent of geological knowledge in the 1850s made it difficult for the survey scientists to interpret California's geology accurately, and even today its interpretation remains a challenge.

During the fourteen days of this particular section of Parke's explorations, the surveyors experienced "more hardships and difficulties, from ruggedness of country, snow, and rain, than we had before met with." Also, the chaparral, or as Parke named it, the *chemizal*—the dense mass of bushes covering the slopes and summits, "chiefly dwarfish oak, manzanita, and a shrub called redwood"—made "very trying the travel." "It seldom exceeds ten feet in height and from the toughness of the wood and density of the growth it is often impracticable to penetrate these thickets without free use of axe."

Now following the coast road south and inland through the Simi Valley, the men headed to Los Angeles, completing the first division of the coastal survey's work, four months after leaving Benicia. On April 3, 1855, near San Bernardino, the escort commanded by Lieutenant Winder joined Parke. Leaving most of the company, including Antisell and Campbell, with Winder, Parke, now in familiar country, proceeded with a small party to the Mohave River and followed its course to its final disappearance into Soda Lake. He was charged with determining if "the valley of the Mohave River joins that of the Colorado, and will afford a practicable route for a railroad." Parke ascertained, as had Whipple and Williamson before him, that no stream valley continued from Soda Lake toward the Colorado River.

Aware that Whipple the previous year had come west across the Mohave Desert and located the only water holes, Parke decided to continue his survey no farther in that direction. His remaining orders were to follow up on Whipple's suggestion for a survey extending from Soda Lake diagonally southwest across the Chemehuevi Valley to the Colorado River. But having heard from local trappers and Indians, as well as receiving reports from the 1853 Williamson expedition, that there was "not a drop of water for over a stretch of one hundred and ten miles," Parke deemed such an excursion "too hazardous" and returned to Los Angeles.

While Parke sailed from San Pedro to San Francisco to replace broken instruments and obtain additional funding, the main party was sent on to San Diego. It being prime springtime, Antisell collected some sixty-eight species of plants in the Los Angeles–San Bernardino area and more around San Diego. Two especially beautiful species were mariposa lilies, first collected by David Douglas. One had large white petals splotched blood-red (Butterfly Mariposa Lily), the other lilac petals marked with purple spots (Splendid Mariposa Lily). Near San Diego the Tree Poppy, with its large, satiny yellow petals, caught Antisell's fancy, "a low and rather ornamental shrub" in flower and fruit in May, as well as the unique Bladderpod, a shrub with inflated fruit that exuded a strong tobacco aroma. Antisell also continued his geological work, including the collection of "bitumen" from tar seeps around Los Angeles. Although he found about three dozen fossils, he missed the prehistoric mammals trapped in the tar, as did others until University of California paleontologist John C. Merriam discovered them in the early twentieth century.

Parke arrived in San Diego from San Francisco on May 23, 1855. He found his expedition reorganized and refitted for the next jaunt, the trip to Fort Yuma, up the Gila River to the Pima Indian villages, and on to the Rio Grande, making "additional examinations and surveys" recommended after his previous year's trip on the same route. Three days later the expedition was off, Lieutenant G. T. Andrews having replaced Lieutenant Winder with the escort, on what was now becoming a familiar route to Fort Yuma by way of Warner's Pass. Campbell was along, and Antisell collected nearly two dozen plant specimens, including typical southwestern species such as Screw Bean at San Felipe, Smoke Tree at Carrizo Creek, agave and Desert Willow at Vallecito, Arrow Weed along the Colorado River, and Ironwood at Fort Yuma.

Antisell made a nice collection of rocks and minerals during the Parke expedition, especially from the San Diego–Colorado River section. His fossil collection went to Timothy A. Conrad of the New York State Survey, while initially he turned his plants over to Lieutenant Parke. John Torrey was not sure whether Parke would deliver the plants to himself or Asa Gray, but Torrey finally received them in mid-February 1856. Antisell had warned that "a good many of the specimens got injured in crossing the Isthmus! This was owing to Carelessness somewhere." True, some had apparently been damaged by water.

The final report on this coastal survey included Parke's overview and a botanical section by John Torrey, but nothing on zoology. Antisell provided a detailed geological report on California, focusing primarily on the counties of Santa Cruz, Monterey, San Luis Obispo, Santa Barbara, and Los Angeles, 129 pages comprising thirty chapters. Included were chemical analyses, a list of the minerals collected, the order of the strata, and a paleontology report prepared by Conrad. In addition, accompanying Antisell's report was the "Geological Plan of the Coast Range of California from San Francisco Bay to Los Angeles," with colored features and a number of cross-section sketches of the geological formations. Antisell's map was sufficiently detailed to show the dispersed location of serpentine rocks and even the exposure of the white Salinian granite at the tip of the Monterey Peninsula. He is recognized for having realized how recently these Southern California coastal mountains were uplifted. After the survey was completed, Antisell had a long career ahead of him as a professor of chemistry at Georgetown University and chief chemist for the U.S. Department of Agriculture. He died on June 17, 1893, in Washington, D.C.

Albert Campbell provided three lithographic views of California for the published report: San Luis Obispo harbor, Warner's Pass from San Felipe, and the southern end of the Santa Ynez Mountains and San Buenaventura Valley. From 1857 to 1860 Campbell served as superintendent of the Pacific Wagon Road Office, for which Parke's colleague Henry Custer would also work. Campbell was a Virginian, and like many other Virginians he sided with the South during the Civil War, becoming a major and chief of the Topographic Bureau for the Confederacy. Fittingly, after the Civil War Campbell spent his professional life as a chief engineer for a number of railroads.

LIEUTENANT WILLIAMSON AND NEWBERRY
JOURNEY TO THE OREGON LINE, 1855

A pile of stones . . . identified the boundary between
Oregon and California.
                                        —Henry L. Abbot

Lieutenant Robert S. Williamson of the Army Corps of Topographical Engineers was no stranger to exploration in the Golden State's northern mountain wilderness. He and Lieutenant William Warner had explored for a possible railroad route out of the northern Sacramento Valley and over California's northern mountains in 1849. In May 1855 Williamson arrived back in San Francisco by ship from New York with his civilian assistants and his aide, Lieutenant Henry L. Abbot. His orders from Captain Andrew Atkinson Humphreys, in charge of the Pacific Railroad Surveys Office, were to organize a surveying party to explore for a railroad line up the great inland valley into eastern Oregon, the "California and Oregon Line," as well as a possible route across the Sierra toward the Carson River in Nevada. The orders stated that "the geological information is considered especially valuable," and "your attention is directed to the botany and natural history of the country."

Making up Williamson's staff were Abbot, charged with preparing the report; H. C. Fillebrown, assistant engineer; Dr. E. Sterling, physician and zoologist; C. D. Anderson as computer and amateur naturalist; John Young, draftsman; and as geologist and botanist John S. Newberry. Dr. Newberry, a Connecticut Yankee, received his M.D. degree from Cleveland Medical School in 1846, followed by two years of study in Paris. He was thirty-three when he joined the Williamson survey. Later, he served under topographical engineer Lieutenant Joseph Ives on the Colorado River Survey in 1857–58, and in 1859 with Captain John Macomb's expedition in the Southwest. Williamson also had nineteen men with the pack train, commanded by Pack Master Charles Coleman. Unlike the other surveys, Williamson's expedition took no wagons. The only vehicle was a lightweight two-wheeled cart to transport the diverse instruments: chronometers, sextants, two compasses, surveyor's chain and pins, odometer, thermometers, an aneroid and four cistern barometers, and two telescopes.

Because of the high possibility of encountering hostile Indians, a large escort commanded by Lieutenants Philip Sheridan, John B. Hood, George Crook, and Horatio Gates Gibson was added at Fort Reading. All of these officers later became famous. Sheridan was involved in important battles of the Civil War, rising to major general and in 1883 succeeding Sherman as general in chief of the U.S. Army. Hood, a Kentuckian, commanded the Texas Brigade for the South in the Civil War and was responsible for evacuating Atlanta ahead of Sherman's Union forces. Crook also served in the Civil War, commanding the federal army of West Virginia, but his future fame lay in the West, where he fought in the Indian wars for two decades, capturing the Apache chief Cochise and later Geronimo. Gibson was wounded in October 1855 during the Indian battle of Grave Creek in Oregon while detached from Parke, and he served the Union during the Civil War. When Gibson died in 1924, just short of one hundred, he was the oldest living graduate of the U.S. Military Academy (Class of 1847).

About noon on July 10, 1855, the California-Oregon survey party left the U.S. Army supply depot at Benicia beside the Carquinez Strait. The road extended through low hills to the marshy edge of Suisun Bay, then inland through a landscape dotted with oaks, much of it under cultivation. Among the many birds observed were Cliff Swallows, building their mud nests on San Pablo Bay's perpendicular cliffs, and Barn Owls inhabiting cliff holes. The first mammals encountered were Norway Rats, swarming about the wharves and cellars of Benicia, and in the surrounding countryside was the faint waft of Striped Skunk aroma. It was locally suggested that skunks could be caught in a box trap and plunged into water to drown, or be suddenly killed "very dead" by rifle ball or shot. Better yet, have a small dog attack the skunk, and while the skunk was engaged, pick him up by the tail and dispatch him with a definitive blow, "his system of defense in such circumstances being inoperative."

The third camp was at a little settlement of three houses called Vacaville, where an elusive Virginia Rail was collected. The next day it was on past Rio de los Putos, a region already planted with peach and fig trees, as well as vineyards. The Sacramento River, reached at Knight's Ranch, was too muddy to ford, so the expedition dropped south to the settlement of Fremont and crossed by ferry to the mouth of the Feather River before continuing north. On July 18, near the expedition's camp on Deer Creek about twenty miles south of Red Bluff, there were excited reports of a Grizzly Bear in the bushes, "but the monster proved to be only a burnt log."

Traveling up the Sacramento Valley, the survey party watched the graceful flights of the California Condors. Wrote Lieutenant Abbot: "As I sometimes recall the characteristic scenery of California . . . here and there, between the rounded hills, orchardlike clumps of oak, a scene so solitary and yet so home-like, over these oat-covered plains and slopes . . . always floats the shadow of the vulture." Abbot and Anderson engaged in a steeplechase after Coyotes. The two horsemen each chased a Coyote at breakneck speed until the Coyotes flagged. Anderson was proving to be an enthusiastic naturalist. He had hunted Wolves on the prairies of Texas and was now becoming adept at collecting specimens. Needing a Badger for the zoological collection, Anderson tried to trample one under his horse's feet, running over the mammal several times but not injuring it. Then he jumped off and, with sheath knife out, ran to kill it. Undaunted, the Badger bared its teeth and attacked. While Anderson frantically scouted for a club, the Badger escaped into its den.

Birds abounded on the trip up to Fort Reading, including Yellow-headed Blackbirds, Yellow-billed Magpies, Bullock's Orioles, and Burrowing Owls. Near Sutter Buttes was a Clark's Nutcracker, and a Lewis's Woodpecker in flight mimicked a small crow. There were Greater Roadrunners all along the route, while abundant everywhere were Horned Larks, "which rise before the traveller at every step . . . and, by their numbers and ceaseless twitter, give life to scenes as monotonous as the prairies of the west."

At Fort Reading on July 21, Williamson's corps, like Beckwith's a year earlier, was well received. A preoccupation now at the fort was hunting the plentiful Black-tailed Jackrabbits. The survey party discovered that the fort's recently assigned surgeon, Dr. John Fox Hammond, had already "very successfully investigated the natural history of the vicinity." A Mexican War veteran, Hammond had been serving at Fort Barrancas in Pensacola, Florida, and collecting on the side for Spencer Baird. Requesting duty in California, he arrived in San Francisco in February 1854 and became surgeon at Fort Tejon and eventually at Reading. Among specimens Hammond provided from around Fort Reading were a Great Horned Owl, a Shadow Chipmunk, two Beechey's Ground Squirrels, two Ord's Kangaroo Rats, a House Mouse, and a Western Spadefoot Toad that Baird would name *Scaphiopus hammondii* after the surgeon. Also, Hammond donated two new aquatic species from the Pit River: the Pit Sculpin and the distinctive Pit-Klamath Brook Lamprey, which occurs only in that river, the adults reproducing but never feeding.

On July 28 the party, accompanied by Hammond, headed northeast toward the mountains, with Bartee, a hunter better known as "Old Red," as scout. Initially the route was through a dense pine-oak forest with a manzanita understory, and for a time Lieutenant Abbot and his group were lost. On the track of Lieutenant Beckwith, Williamson crossed over Nobles's Pass, with impressive "Lassen's Butte" off to the south. The sight and smell of smoke in the air intimated that Indians had set fire to the forest. On the last day of July the company bade good-bye to Hammond, who returned to the fort, promising to take readings with the barometer loaned to him.

Leaving Nobles's Road and heading north on August 1, the expedition encountered an expanse of lava, a "rocky pedregal of scoriaceosus trap, which taxed our patience to the utmost." The mules suffered, and the instrument cart tipped over, rendering the chronometers useless for the rest of the survey. Then the Pit River Indians began showing up, first a single naked one, who returned the next day with more naked companions. By reputation these Indians were treacherous, skilled with bows and arrows, the latter tipped with sharp arrowheads that separated from the shaft when shot into flesh.

The expedition now divided into two groups, one heading down the Pit (Sacramento) River, the other proceeding up to the mouth of the Canoe (Hat) River. The instrument cart had completely broken apart, but the two wheels were saved to carry the odometer. Newberry accompanied the Pit River party and unsuccessfully attempted to shoot a Mountain Lion. On another occasion at sundown he went fishing alone.

As the stars appeared, a solitary Coyote atop a ledge across the stream commenced howling, "so mournful and sinister that I was fain to look upon it as an evil omen, and gathering up my fish and groping my way back to camp was quite disposed to congratulate myself on arriving there without adventure." On one day's march several hundred Northern Harriers (Marsh Hawks) were observed, together with some Short-eared Owls. A Blue Grosbeak, "this pretty and musical bird," was seen at the Pit River. Throughout the mountain country the Mountain Quail was common, a favorite pet of the gold miners, often kept in confinement and "much admired for their trim figures, elegant plumage, and chivalrous bearing." Numerous birds of various kinds inhabited the prairies of the upper Pit, including enough Sharp-tailed Grouse to afford "fine sport and an abundance of excellent food." Newberry was attracted by a chalky bluff two miles from the trail, with a warm spring watering a verdant meadow. Returning to the little oasis early the next morning, he filled his "plant case with flowers, . . . obtained frogs and snakes and chalky, infusorial earth enough to load down the boy who accompanied me," shot a number of the grouse, and flushed a new bird the size of a turkey, the Greater Sage-grouse.

In mid-August Williamson decided to proceed north by the western side of Lower Klamath Lake in California and then explore the Oregon Cascades, while Abbot, with the main party, would head up the Lost River to Upper Klamath Lake in Oregon, explore the Deschutes River valley, cross the Cascades, and meet Williamson at Oregon City. The last zoological adventure in California was the discovery on the morning of August 14 of a huge Northern Pacific Rattlesnake coiled up under a blanket at camp. That snake was killed, as well as several more in the evening camp. As the men headed north they remained apprehensive, remembering that they slept on the ground without tents.

After their Oregon surveys, the expedition's return trip to Fort Reading met with difficulties. Because of a major Indian uprising in Oregon, with killings and pillaging, their escort had been appropriated by the commandant of Fort Dalles. Both Williamson and Sterling became ill and Williamson returned early to California, leaving Abbot in charge with twenty-eight men, including about ten Mexican packers, to return on their own. Abbot's group had worn-out mounts and only five rifles for the entire company, but the Indian threat failed to materialize. On this return trip, in the Siskiyou Mountains on November 6, 1856, "a pile of stones by the roadside identified the boundary between Oregon and California." The next day the party reached Yreka, an 1851 mining town where gold was still mined in the streets. Two days later there was an evening snowstorm at the Fort Jones camp, sixteen miles southwest of Yreka. On November 12 the men scrambled up Scott's Mountain, the highest summit on the return trip (6,829 feet), by a steep and rocky ascent, with four inches of fresh snow on the ground. Three days later the Oregon and California survey, back at Fort Reading, was officially disbanded.

Up the Colorado River with Ives and the Naturalists, 1857–1858

Natural features whose strange sublimity is perhaps
unsurpassed . . .

— Joseph Christmas Ives

Many were the naturalists, and others, who had visited, crossed, tramped, and collected around California's side of the Colorado River since botanist Thomas Coulter's brief visit in May 1832. In 1850 Lieutenant George Derby of the U.S. Army Corps of Topographical Engineers, a flippant author who hid behind the pseudonyms "Squibob" and "John Phoenix," was assigned to survey the lower Colorado River to see if supplies could be carried by boat up to Fort Yuma. In the Gulf of California he was provided with an oversize 120-ton ship that barely managed to get upstream using flood tides. Derby pressed on, this time by rowboat, and carried out a monthlong river survey, concluding that it was possible to get as far as Fort Yuma by water, provided one used a small stern-wheel boat with a thick, flat bottom and a powerful engine. Within two years there were indeed commercial freighters steaming between the Delta and Fort Yuma.

In mid-October 1857 a party of surveyors and natural scientists assembled in San Francisco, proposing to head for the Colorado River at Fort Yuma and from there to travel upstream by steamboat as far as they could go. This San Francisco detachment, directed by geologist John Newberry, was part of Lieutenant Joseph C. Ives's enterprising expedition "to ascertain the navigability of the Colorado." Interest in navigating the Colorado River beyond Fort Yuma had been growing. The Colorado Steam Navigation Company was already active downriver with its vessels. In 1856 the company's president, George Alonzo Johnson, unsuccessfully approached Secretary of War Jefferson Davis about underwriting a private expedition on public money up the Colorado River. The project was backed by Senator John Weller, the first Mexican Boundary commissioner, and in fact, Congress appropriated $75,000 for such a venture. With a change in administration in Washington, the new federal Bureau of Western Exploration and Surveys now administered the appropriation, under a new secretary of war, John Floyd, and the expedition became another topographical corps survey, under Lieutenant Ives's command. Ives was a perfect choice. From Yale he had gone on to West Point, graduating in 1852. In 1853 he was assigned to lead one contingent of Whipple's Pacific railroad survey along the 35th parallel, subsequently returning to Washington for three years at the Pacific Railroad Survey Office. His wife, Sara, was Floyd's niece. Further encouragement now for such a survey was the "Mormon War" brewing in Utah, with the possible necessity of getting military supplies by water readily up to the Mormon land of Deseret.

Ives had a special shallow-draft iron steamboat, the *Explorer,* built in sections in Philadelphia, transported to Panama, and then, during November 1857, carried by a small schooner up the Gulf of California to Robinson's Landing, just within the mouth of the Colorado River. The fifty-four-foot ship was reassembled there by Ives; A. J. Carroll, the vessel's constructor and engineer; and a team of eight men. Not surprising after the arduous journey, some of the ship's hull sections were in sad condition and had to be hammered back into shape. The ship, a sternwheeler, had a small cabin and a large observation deck for the helm, pilot, flagstaff, and survey observers, built above the paddle wheel. The gigantic three-ton boiler sat amidships. While the *Explorer* was being fitted together, Ives made topographical sketches and notations in his field book of Robinson's Landing and tried out his "little photographic apparatus." Naturally there were problems. The desert light was glaring, the processing chemicals deteriorated rapidly, and finally the light-tight indiarubber photography tent blew away, "apparatus and all." Although none of Ives's photographs survived, his lithographed photo of the *Explorer* at Robinson's Landing appears in the published account of the expedition.

The new steamer was ready to back off her ways at full tide on the moonlit night of December 30. On New Year's Eve, with two heavily loaded skiffs lashed alongside, towing a sloop, and riding only half a foot above water level, the *Explorer* was on her way north to Fort Yuma, with river pilot Captain David Robinson in command, to meet Newberry's overland group from San Diego.

Newberry had been naturalist and surgeon with Williamson's railroad survey north to Oregon in 1855. For Ives, Newberry served as physician and geologist, and was generally in charge of natural history work. Baron Friedrich von Egloffstein of the Frémont and Beckwith expeditions was along to make topographical sketches and zoological collections, and Heinrich Möllhausen, who had been with Whipple and Ives on Whipple's railroad survey, was appointed "artist and natural history collector."

Möllhausen had returned to Germany after the survey, but maintained an enthusiastic desire to serve again with a western expedition. While Möllhausen was a house guest of Alexander von Humboldt, on March 24, 1857, Humboldt wrote Secretary of War Davis about Möllhausen's continued interest. As a result, Ives extended an invitation for his former colleague to join the Colorado River Expedition. The actual surveying of the river would be carried out by Casimir Bielawski, a Polish American hydrographer from San Francisco. At Fort Yuma Ives picked up a military escort of twenty-five men under the command of Lieutenant John Tipton, and a pack mule team coming by way of Fort Tejon would eventually proceed up the riverbank, to supply any inland junkets from the ship. Someone with the pack train added a Mountain Plover from the Tejon to the expedition's bird collection.

Newberry's overland party shipped south from San Francisco at the end of October, putting in at Monterey, San Luis Obispo, Santa Barbara, and San Pedro before arriving at San Diego about November 1. The sea voyage south past the coastal ranges gave geologist Newberry opportunity to compare them with the Sierra Nevada, which he had seen during the Pacific railroad survey. South of San Luis Obispo the Sierra seemed to have a subdivision that tended westward to the coast and "usurps the place of the ranges forming the coast line northward." Newberry quibbled with Pacific railroad survey geologist Marcou's idea on this "question of interest" needing to be settled. Later this subdivision was recognized as the Transverse Ranges.

At each ocean landing the naturalists went ashore to collect. Redwoods were seen south of San Francisco, and perhaps Foothill and Sugar pines. Monterey Pines and Coast Live Oaks were abundant around Monterey, but probably not the Coulter Pines reported on the mountains immediately back of town. During the brief pause at San Luis Obispo the California Fuchsia was collected, as were the Pan-American *Aster chilensis* and two other composites, the common tarweed *Hemizonia fasciculata*, and *Stephanomeria paniculata*. At an even briefer stop in San Pedro they picked up at an alkaline site the unusual Yerba-mansa, a plant first described by Thomas Nuttall. The party lingered in San Diego about a fortnight, and Newberry, despite his other responsibilities, obtained two dozen more species of plants at the seashore, along the San Diego River, and on the tablelands.

Newberry's detachment left San Diego in mid-November, using the regular road to the desert. En route Newberry collected plants, especially composites, at San Pasqual, Santa Ysabel, and Jonathan Warner's ranch. They stopped at "The Oaks" near San Felipe, and upslope in the mixed evergreen forest Newberry made the first collection of Bigcone Douglas-fir, the only other member of the genus *Pseudotsuga* in North America. Along Carrizo Creek he added a Chuparosa bush, and across the Colorado Desert past Alamo Mucho a Rush Milkweed "in the most dry and sterile soil." Passing New River, the party arrived at Fort Yuma on December 20, 1857.

During the three-week wait for Ives and the steamer from the Gulf, the expedition members camped about a mile below Fort Yuma. Until he became indisposed, Newberry, assisted by Möllhausen, made further "valuable natural history collections." Meanwhile, Egloffstein nearly completed a series of topographical sketches of the surrounding desert country. This being wintertime, few reptiles, amphibians, or invertebrates were encountered, and minor attention was paid to mammals. A variety of residential and wintering birds were seen around the fort. No new species were discovered, but Gila and Ladder-backed woodpeckers, Ash-throated and Vermilion flycatchers, Say's Phoebe, Black-tailed Gnatcatcher and Verdin, Phainopepla, and Gambel's Quail were all collected. Winter visitors included Mountain and Western bluebirds, Ruby-crowned Kinglet, Yellow-rumped Warbler, Dark-eyed Junco, White-crowned Sparrow, Abert's Towhee, flocks of Sandhill Cranes, and a few White-faced Ibis. The Crissal Thrasher was probably the second of the species to be collected in the country, the first having been from New Mexico.

On the last day of December, one of George Johnson's ships, the *General Jesup,* left Fort Yuma under government orders with army troops, military supplies, and a howitzer, responding to rumors that Mormon agents upstream were actively recruiting Mohave Indians for the so-called Mormon War. Before Johnson had left Robinson's Landing he had gloated to Ives that the *Explorer* would still be on the ways when the *General Jesup* was reaping publicity for its navigation of the upper Colorado River.

During the evening of January 8, 1858, some Indians near Fort Yuma reported the approach of a steamer. Early the next morning the *Explorer*'s whistle was heard, and soon the boat came into view. Now all members of Ives's expedition were together and eager to head upriver. Dr. Newberry was still ill, but influenced by the prevalent feeling that Fort Yuma, because of its isolation and loneliness, was considered "the Botany Bay of military stations" and thinking a change of air would be beneficial, he decided, sick or not, to go along.

The night before departure a memorable party was thrown at the fort, with single-sex dancing, much consumption of liquor including Rhine wine, and raucous singing accompanied by geologist Newberry on violin, engineer Carroll and topographer Egloffstein on flutes, and Lieutenant Ives and artist Möllhausen on guitars. The toasts were timely: from the garrison men, "Think of us when the Mohave Indians lift your scalps," and from the *Explorer* crew, "Remember us when the Mormons burn you and your fort to the ground."

Early the next morning, January 11, with many a hangover, survey members collected supplies for six weeks—provisions, arms, ammunition, and luggage—at the wharf. By noon everything and everybody— some two dozen men—were loaded, steam gotten up, and the *Explorer*'s adventure began. The pack train under G. H. Peacock left sometime later, traveling up the California shoreline to "near the head of navigation, wherever that may be." Celebrating the departure, garrison people and Indian men, women, and children crowded the bluffs on both sides of the Colorado River. As the *Explorer* began steaming upstream, to the delight of onlookers it ran aground within sight of the fort. The river was low, and once under way again the crew often had to jump into the water to wrestle the steamer off a sandbar.

Twelve miles above Fort Yuma the *Explorer* left the desert flats and entered the Purple Hills, the rocky entrance quickly named Explorer's Pass by the surveyors. The expedition camped here, and while Möllhausen made his sketch of the pass, Newberry noted that the hills were "composed of granite and mica slates, associated with which are purple porphyries and trachytes, in sufficient quantity to impart to them their prevailing color." Despite the feeling of wilderness, a Yuma Indian runner surprised them early the next morning, bringing mail for the survey party that had arrived at Fort Yuma the previous night.

Seven miles above Explorer's Pass a range of the Purple Hills crossed the river, creating what was named Canebrake Canyon because of the dense river-edge reeds of Phragmites. "The scenery became wilder, and

Figure 28. "Steamboat 'Explorer' (Chimney Peak)," watercolor and gouache, from Ives's Colorado River Expedition, 1858, painted by Heinrich Balduin Möllhausen. (1988.1.1, courtesy Amon Carter Museum, Fort Worth, Texas.)

the variety of colors assumed by the rocks adds to its beauty." Periodically, stretches of smooth water tempted the boatmen to relax, only to break the rudder on some hidden rock. Just after leaving the camp at Canebrake Canyon, with "an open looking stretch of water ahead that gave encouragement of a good day's run," the *Explorer* grounded on a sandbar. It took waders two hours to work the steamwheeler free. Philadelphian Carroll considered "the Colorado the queerest river to run a steamboat upon that he has ever met with in his experience as an engineer."

Typically during daylight the naturalists watched from the wheelhouse and bench atop the paddle wheel, with Captain Robinson at the tiller and one of the crew standing at the bow with a sounding pole (figure 28). Cinders from the tall stack singed holes in the naturalists' journals. At day's end, when the *Explorer* pulled to shore, Egloffstein often did some zoological collecting while Möllhausen busied himself with his sketch pad, recording the desert landscape and geological features or, when opportunity afforded, making numerous pictures of the Indians. Newberry collected a variety of plants that favored either the sandy shoreline (such as a common and an uncommon species of spurge, "Euphorbias") or the rocky slopes (Sweetbush and the shrubby Pygmy-cedar, both composites). One of his most interesting plants, found in a sandy spot, was Bugseed, an uncommon, weedy, annual goosefoot from Eurasia whose branches form into a ball that catches the wind like a tumbleweed, scattering its seeds as it rolls along. In this same stretch of the Colorado River Newberry collected Barnyard Grass, from Eurasia; Finger Grass, native to the Great Plains; and several weedy California native grasses: Sand Dropseed, Bearded Sprangletop, and Satintail. Surrounded by this interesting flora and spectacular geological landscape, Newberry became "quite restored to health."

At a break in the Purple Hills, a broad vista opened to the northeast, with a distant view of the Dome Rock Range, in future years to become a major mining district. Off to the northwest came periodic glimpses of the jagged spires surrounding Chimney Peak, the modern Picacho Peak complex, where another mining district would develop. At campsites Newberry found evidence of miners having worked veins of copper and galena. Watching Newberry come aboard loaded with rocks, Ives recognized that the steamboat "affords facilities for transportation not ordinarily enjoyed by exploring parties nor scientific collectors, and the doctor has already laid in a large assortment of specimens." Egloffstein's zoological ventures were less successful. He caught only one fish, and "that a poor variety." The only noteworthy mammals encountered were a dozen Bighorn Sheep scampering up a gravel hill near Lighthouse Rock, upriver beyond Picacho Peak. The naturalist had to be satisfied with an occasional flock of ducks and geese flying over, and a glimpse of a wading Great Blue Heron.

In mid-January the *Explorer* safely navigated the rocky pass through the Chocolate Mountains and steamed into the expansive "Great Colorado Valley" (today's Palo Verde Valley, with Parker Valley to the north). Ives noted a fair amount of seemingly fertile bottomland, but Newberry astutely commented that the soil was "so charged with alkali as to be unproductive."

Although there were no challenging canyons here, the Colorado River braided into many channels, which meandered around sandy shoals, making navigation "more difficult than any yet experienced." But the scantily clad Chemehuevi Indian sirens provided a pleasant distraction, the "pretty daughters of the wilderness," as Möllhausen called them, running through the water after the ship, laughing and mischievously throwing handfuls of sand at the explorers.

As the expedition approached the northern boundary of the Great

Colorado Valley, a chain of mountains spiked with towers became visible about ten miles upstream (the Monument Range, now the Whipple Mountains). During their several days here, a number of plants were collected near the close-by Riverside Mountains, including an evening-primrose, blazing-star, wild cucumber, and ground-cherry. On January 25 a "rare and interesting" Ghost Flower was spotted, the yellowish blooms streaked with red. Its generic name, *Mohavea,* refers to the type locality where John Charles Frémont first collected this desert annual. Other annuals were also gathered, especially several species of *Phacelia,* and a diversity of cacti including two species of fish-hook cacti, a barrel cactus, Silver Cholla, and Beavertail.

Dr. Newberry made a geological trek into the Riverside Mountains, whose colorful barren rocks, with "patches of purple, brown, blue, ash, cream, red, . . . would scarcely be exaggerated if represented by the colored diagrams of the geological lecture-room." On his hike he found in the rugged outcroppings indications of silver, lead, iron, and copper, with veins reminding him of the gold-bearing rocks of the Sierra Nevada. Ives regretted that "the nature of our duties does not permit any lengthy examinations. A careful search might develop ample stores of treasure, which the close proximity of water transportation would greatly enhance in value." Later the Riverside Mountains became a beehive of prospecting activity.

On January 30, just before entering Monument Canyon, the *Explorer* met the *General Jesup* on its way downstream. Ives learned that the *Jesup* had gone north to near the present-day site of Davis Dam, where rapids prevented further navigation. The men had erected a rock cairn on shore to designate the spot. On the way back downstream, just above where the river reached the California border, the *Jesup* had encountered a company of men, mules, wagons, and camels on the western shoreline. It was Edward Beale's camel expedition surveying a road from Los Angeles and Fort Tejon to Fort Defiance in the New Mexico Territory. Captain Johnson ferried Beale's party east across the river before the *Jesup* continued south to the meeting with Ives.

No longer feeling a need for Lieutenant Tipton's escort, Ives sent the troops on with the *Jesup,* asking Tipton to find and take charge of the supply mule-train coming north. The *Jesup* continued downstream to the monolith of Lighthouse Rock, projecting into the middle of the river, where the steamer struck an underwater outcrop and sank. No one was drowned, but everyone, including Tipton's escort, had to hike about twenty miles down to Fort Yuma.

The Ives expedition camped on February 1 at the mouth of Bill Williams Fork, down which Ives and Möllhausen had traveled with Whipple just a few years earlier. But the fork this winter was a mere trickle and was nearly missed amid the dense thickets of willows. The next evening the men camped a short distance up the Colorado, while Newberry and Möllhausen returned to Bill Williams Fork to go fishing. On a nearby gravelly hilltop Newberry collected Slim Tridens, a tufted grass common on desert surfaces all along the river, and he got his second specimen of the green-trunked Blue Palo Verde, first encountered at Fort Yuma. Here on the volcanic cliffs of Monument Canyon were scattered Saguaro, a giant cactus to be pictured in the lithograph of Möllhausen's landscape titled *Shore on Lower Colorado*.

On up the river the Monument Mountains closed in; the gradual slopes changed into "rough and confused masses of rock and the scenery at every instance became wilder and more romantic." Suddenly the narrow canyon opened out into the broad Chemehuevi Valley. Many Chemehuevi Indians lined the sandy shoreline bluffs, excitedly watching the passage of the steamer. One group attracted the vessel's attention, so Ives had Captain Robinson head for the bank, where a diminutive chief waited. It turned out he had spent much time with Whipple, Ives, Bigelow, and Möllhausen during the railroad survey trip along the Colorado River in 1854. The pilot of the *Explorer* was not enthusiastic about landing and would not stop the steamer, so the chief jumped aboard and soon saw his Indian companions disappearing from sight as the steamer continued north. Later, he and his tribe members admitted that they considered the temporary shanghaiing a "high compliment."

The expedition camped at the headquarters of the Chemehuevi Nation, put the "captured" chief ashore, and hosted interested Indians of all ages. Zoologist Möllhausen seized the opportunity to obtain assistants, sending Indian children out on collection forays for lizards, snakes, pocket mice, and kangaroo rats and rewarding them with strings of beads. The youngsters, thinking that Möllhausen planned to eat the animals, were, Ives reported, "delighted that his eccentric appetite can be gratified with so much ease and profit to themselves."

When the steamer had first come into the valley, a gale had been blowing, picking up drifts of beach sand that periodically darkened the sky. Upon its leaving the valley, the gale picked up again. As the *Explorer* entered Mohave Canyon, to the east could be seen "a cluster of slender and prominent pinnacles earlier named by Lieutenant Whipple 'The Needles.'" Ahead was a low purple gateway into a massive red-walled corridor hundreds of feet high. The channel became tortuous, and ahead they heard a roaring of rapids. The stream, dividing around a gravelly island, rushed and churned up sheets of foam, while the ship kept grating over rocks. With a flowery flourish Ives wrote in his journal that "a scene of such imposing grandeur as that which now presented itself I have never before witnessed. . . . Brilliant tints of purple, green, brown, red, and white illuminated the stupendous surfaces and relieved their sombre monotony. Far above, clear and distinct upon the narrow strip of sky, turrets, spires, jagged statue-like peaks and grotesque pinnacles overlooked the deep abyss."

The approach of twilight "enhanced the wild romance of the scenery." As light faded, the rock outcrops became somber. "A solemn stillness reigned in the darkening avenue, broken only by the plash of the paddles or the cry of a solitary heron, startled by our approach from his perch on the brink of some overhanging cliff." At a sudden turn of the river, the expedition left Mohave Canyon and came into the broad Mohave Valley, camping here on February 10 (near present-day Needles). Away from Fort Yuma about a month now, they had rations for only two more weeks. The Mohave Indians signaled the ship's approach with a smoke pyre flaming atop a small peak. Since women and children as well as braves clustered on the sandy banks of the Colorado River to welcome the river explorers, it seemed evident that this new tribe of desert Indians "had no immediate hostile intentions."

The next morning Lieutenant Ives gave a speech to the assembled Mohave Indians, a somewhat unique presentation to Indians since he never referred to the "Great Father at Washington." Ives's address in English was translated into "indifferent Spanish" by navigation surveyor Bielawski. Spanish-speaking Diegueño guide Mariano translated the crude Spanish into Yuma for Yuma Indian guide Maruatscha (who preferred to be called "Captain"), who then translated what was left of the speech into Mohave. Among other things, Ives inquired about the two Mohaves, Chief Cairook and Iretéba, who had led Whipple and himself across the desert to the Mohave River in 1854. No one seemed familiar with Iretéba, but Ives was assured that Cairook was still alive and "retained his authority."

Leaving the Mohave villages, the expedition actually encountered Iretéba at one of its stops. Ives invited him to join them on shipboard until the arrival of the mule train, which was carrying gifts for the Indians. Iretéba told Ives that Cairook was in the vicinity, hoping to meet the ship. Before leaving California waters both former Indian guides and Ives were reunited, and the trio enjoyed reminiscing about the railroad survey, Chief Cairook particularly asking after Lieutenant Whipple, "for whom he had conceived an exalted opinion." The two Indians also wanted to know more about that survey man (Lieutenant Tidball) "who could carry his teeth in his hand," taking them in and out of his mouth. Later, Cairook's attractive young wife joined the group, the only woman to travel aboard the *Explorer*. Reflecting on her experience, Ives later surmised, "I am afraid that the trip turned her head, and that she must have been quite unbearable to her friends after she left us."

After Camp 47 about February 16, 1858, at the head of Mohave Valley, the expedition left California's river shoreline and soon passed where Beale had been ferried across. Möllhausen sketched both Boundary Hill and Beale's Crossing, later adding the camels. In the Arizona-Nevada borderland desert country, Ives noted, the "winter has given place to spring," with the days warming but the nights refreshingly cool. The *Explorer* steamed north past the present site of Davis Dam, with today's Newberry Mountains to the west, and up to the "*Jesup* Rapids" that had daunted Captain Johnson. Captain Robinson lightened the vessel's load and built up a head of steam, and over the rapids went the *Explorer*. In celebration stream boulders were piled up into a pyramid, atop which a stake marked "*Jesup*'s Halt" was driven. Möllhausen made a sketch.

The *Explorer* continued upstream with difficulty, fighting more rapids and being buffeted by a twenty-four-hour sandstorm. Provisions were running low, and some men showed signs of scurvy. Finally, near the entrance to Black Canyon, the "head of navigation" was reached with a crash. The *Explorer* hit a hidden rock, men at the bow were thrown overboard, and Ives, Newberry, and Möllhausen were catapulted headfirst into the hold. The group spent the week of March 6–13 at this spot, repairing the *Explorer* and awaiting the arrival of the pack train. By skiff Ives and Robinson fought their way on up the Colorado River through the perpendicular walls of Black Canyon to what they concluded was the mouth of the Virgin River (actually Las Vegas Wash, beyond present-day Hoover Dam) before returning to their encampment. Then the repaired *Explorer* proceeded downstream to the Mohave Valley, where Lieutenant Tipton's pack train had finally arrived.

Most of the expedition—including Ives, Newberry, Egloffstein, and Möllhausen, together with packers, laborers, Tipton with his military escort, and Iretéba as guide—soon proceeded eastward overland, visited the Grand Canyon, and eventually reached Fort Defiance in the New

Mexico Territory. Meanwhile, Captain Robinson piloted the *Explorer* south to Fort Yuma at a leisurely paddle, "in order not to run any risk of losing the collections and the field-notes." About April 1, twenty-five days out of the Mohave Valley, the *Explorer* arrived at Fort Yuma.

Ives eventually got back to California by stage from Texas to San Diego and then headed east again to Fort Yuma, where he sold the *Explorer* to Captain Johnson. Then he proceeded to San Francisco and on by ship to New York and Washington. While working up the material from the Colorado River Expedition, during 1859 and 1860 Ives became the superintendent and construction engineer for the unfinished Washington Monument. He returned to California in 1860 to survey the eastern state boundary at Lake Tahoe. The next year the Government Printing Office published his *Report upon the Colorado River of the West*. Ives had written the narrative, Newberry the 154-page geology section, the quartet of Gray, Torrey, Thurber (from the Mexican Boundary Surveys), and Engelmann the detailed floral inventory, and Spencer Baird the zoology section. Many of the lithographs were based on sketches and watercolors by Möllhausen, while the accompanying maps were produced by Egloffstein, who also included some of his spectacular, fold-out panel landscape panoramas. The year of publication saw the start of the Civil War. To the regret of many northerners, Ives went with the Confederacy, eventually becoming aide-de-camp to his friend Jefferson Davis. Three years after the war, back in his home state of New York, he died a premature death at forty.

Of Ives's team of naturalists, Newberry went on to a distinguished career in geology. He accompanied Captain John Macomb's exploring expedition in 1859 from Santa Fe to the junction of the Green and Colorado rivers, providing the geological report as well as his own sketches. He observed about the eroded slickrock Colorado Plateau, as did later geologists, that "nowhere on the earth's surface, so far as we know, are the secrets of its structure so fully revealed as here." When Newberry's former tutored student Ferdinand Hayden in 1862 published his summary of the upper Missouri geological investigations, Newberry was one of the contributors. In 1866 Newberry became professor of geology and paleontology at Columbia University, establishing the School of Mines there, as well as directing the Ohio Geological Survey from 1869 to 1874. Although he gained the most fame for his work with fossilized plants, especially from the coal measures, Newberry's work on fossilized fishes was classic, and as an American paleontologist he came to rank with Joseph Leidy of Philadelphia.

Despite being a German citizen, at the onset of the Civil War Baron Egloffstein turned his talents and money to battle. He personally planned and organized the 103rd Regiment of New York Volunteers as an elite corps for German Americans. Egloffstein fought under General Burnside near New Bern in North Carolina in 1862, where he was seriously wounded; in 1863 he was breveted out of the service as brigadier general. Back in New York, he helped found the Geographical Institute and edited *Contributions to the Geology and the Physical Geography of Mexico*. In 1865 he patented a halftone process and opened the Helio Engraving Institute in New York. In 1878 he and his family moved back to Germany, residing near Dresden, where in lingering poor health he died in 1885.

Artist Möllhausen, upon completing his work in the East for Ives, sailed for Berlin on September 1, 1858, never returning to the New World. He became custodian of royal libraries at Potsdam, but not forgetting the American frontier, he turned his hand to writing. His diary was published and translated, as well as a lengthy two-volume work, and he wrote dozens of novels, articles, and short stories, many about western adventure in the tradition of James Fenimore Cooper, gaining him the moniker of "the German Cooper." Möllhausen died in 1905, but one year before his death he reminisced that "one would like to push back the inevitable onward march of civilization, before which the shaggy buffalo and the brown hunter disappeared, and, with them, the last of the romance of the 'Far West.' "

PART FIVE

The California
Geological Survey

If you get good men, your work will go well, and you will
be relieved.

—James Hall (Brewster 1909, 189)

AGAINST THE BACKDROP OF THE United States Civil War, California initiated its own scientific survey, using civilian scientists under the leadership of geologist Josiah Whitney. The project emphasized the state's geological resources, but also included the broad spectrum of natural
history. Over the years of the survey a number of young naturalists became involved, many of whom later went on to distinguished scientific
careers. Despite the challenges of frontier country and lackluster political and financial support from the state government, the naturalists per-
severed, and the resulting fieldwork was prodigious. The book *Up and
Down California* by Whitney's assistant, William Brewer, understates
the geographical coverage of the survey. But in the eyes of politicians and
vested interests, there was too much emphasis on fossils and flowers, and
too little on mining and the economic potential of petroleum.

The California Geological Survey's Formative Years, 1849–1862

Dr. John Boardman Trask of Massachusetts first set foot in California on October 15, 1849, crossing the Colorado River to Fort Yuma on a scow fashioned out of a wagon body. Also on the "ferry" was John Woodhouse Audubon, John James Audubon's younger son, leader of a large party of gold seekers and emigrants that included Trask. Audubon had finished working with his father on a monumental book of American mammals with paintings by father and son, and on this trip to California, in addition to mining, he hoped to continue collecting and painting objects of natural history. But so far on the trip, his paints and canvases had been abandoned in the desert, arsenic powder for preserving specimens carried off by the wind, his paper for drying plants used for gun wadding, and his few specimens "lost or thrown away."

Johnny Audubon, like his father, was a good naturalist, and Trask was equally talented. On the excursion Trask demonstrated his areas of expertise: as a medical doctor he administered to cholera victims; as a mineralogist he identified silver, gold, and quantities of lead in a rock Audubon showed him; as a botanist he recognized Creosote Bush dominating the arid desert near Arizona's Pima villages; and as an organizer and leader he captained the party's mess number 12.

Arriving in San Diego on November 4, Audubon's company reorganized for the overland journey to San Francisco, while Trask and seven others went by ship. Audubon continued his bird observations and collecting, watching Brown Pelicans diving into Mission Bay for fish, Long-billed Curlews and Black-bellied Plovers feeding along the bayshore, and in the water great numbers of Horned Grebes (he shot two of the latter). The countryside seemed alive with hundreds of what Audubon called "California Marmots" (Beechey's Ground Squirrel). At a small lake near Mission San Luis Rey de Francia were myriads of Gadwalls and American Wigeons, which provided a welcome supper. A curious California Condor glided down to within fifty yards before soaring aloft and circling away over the hills. The mission, now in American possession, was in disrepair, but its garden still had a bounty of olive and fig trees and grapevines.

At Los Angeles, the "City of the Angels," Audubon, viewing the dilapidated community, concluded that the angels must be "fallen ones." The party divided, all but eleven going by ship to San Francisco from San Pedro. Audubon, with ten men and forty-six mules, continued north into the San Joaquin Valley, where in the foothills he was excited to see, for the first time, both Lewis's Woodpecker and Steller's Jay, as well as numerous California Condors. With the midautumn wet season the valley was awash with ponds, overflowing streams, and boggy ground. Near the San Joaquin River hunters shot a Tule Elk, a Pronghorn, two Sandhill Cranes, and three Canada Geese. Later they encountered hundreds of geese flying overhead and a huge herd of wapiti, north of today's Tule Elk State Reserve. In this lonely country with "not the track of a white man to be seen," the mourning howls of the Coyotes, the haunting honks of the geese, and the "discordant note of the night heron" increased the sense of solitude.

At Stockton Audubon's party boarded a side-wheeler steamer that despite gale-force winds reached San Francisco on December 21, to rejoin Trask and the rest of the company. On December 29, 1849, most of the men shipped up the bay on the steamer, bound for gold rush country. Dr. Trask was soon immersed in prospecting, while Audubon and the other gold seekers traveled up and down through the gold belt south of Sutter's Fort into spring. The winter weather was miserable, with continuous rains and flooding, and no one in the company was getting rich. However, for Audubon there were new birds and mammals to be collected and drawn, but except for a Yellow-billed Magpie he began to dissect, he had "no time to skin and preserve specimens." Later, when he did collect, his specimens were eaten by his hungry companions.

Little enthusiasm for mining remained, but naturalist Audubon began reveling in a newfound Sierran Eden. Formerly drab hillsides were "now fresh and beautiful with every shade of green and brilliant flowers of all colors." The Mountain Dogwood, which Audubon's father had painted as a perch for Band-tailed Pigeons in *The Birds of America,* was in full bloom. The younger Audubon penned in his journal, "The ultramarine jay [Western Scrub-Jay] is here by dozens, robins, fly catchers, chats, finches by hundreds. I see daily new birds and plants that a year's steady work could not draw, but if our government would send good men, what a work of national pride could be brought out! Geology, botany, entomology, zoology, etc. . . ."

Audubon shortly thereafter departed from California disillusioned, commenting that "California will for the present lower the moral tone of all who come here." He had done some oil painting and made sketches, about two hundred altogether, mostly of gold diggings and scenery, but had time only to blot in "what will answer hereafter." One painting he finally finished, which later appeared in *Quadrupeds of North America,* authored by his father and Reverend John Bachman, was of a dead Sea Otter obtained in San Francisco. Unfamiliar with the mammal's actual behavior, he pictured it standing on its haunches beside a shoreline boulder, holding a large fish atypical of a Sea Otter's diet (figure 29).

Coincidentally, a young man in San Francisco two years before John W. Audubon's arrival had been so impressed by John James Audubon's Elephant Folio of American birds that he resolved to produce a similar folio on the birds of the Pacific Slope. During his lifetime Andrew Jackson Grayson, who had served as one of Frémont's volunteers, did produce 156 bird paintings, finally published with his notes in 1987. Grayson has become known as the "Audubon of the West," and his paintings enjoy gallery exhibition today.

In 1851 an event proved significant for erstwhile gold digger Dr. Trask. The new California legislature prepared a bill proclaiming the need for a state geological survey and reviewing the potential benefits of such a survey. The next year Governor John Bigler's annual message urged the establishment of a survey; this was followed in 1853 by a joint resolution to the U.S. Congress that emphasized the need for a California geological survey. In early March an "enthusiastic amateur," none other than John Trask, fresh from Sierra mining exposure, was officially requested, as California's "first state geologist," to provide a report on the geology of California, which he accomplished on April 6. A month

Figure 29. John Woodhouse Audubon's painting of a Sea Otter, killed near San Francisco in 1846. (OrHi 86554, courtesy Oregon Historical Society.)

later the legislature charged Trask to "continue his researches." In July his report, "On the geology of the Sierra Nevada, or California Range," not only appeared as a legislative document but was the lead article in the first issue of *The Mining Magazine,* undoubtedly the first major publication on California's geology. In the next issue there was an article, "Mineral Tract of the East Tennessee and Cherokee Mining Company," by another Massachusetts native son, Josiah D. Whitney.

After Trask's geological surveying for the state ended in 1855, he published three comprehensive geology reports in the California Senate journal and two in *The Mining Magazine.* Altogether he received seven thousand dollars from the California legislature for this summary work but was criticized for writing about places he had never studied or even visited. Trask, as someone later observed, "became distinguished as physician, chemist, mineralogist, seismologist, geologist, paleontologist, and botanist." In future years he studied and wrote about California fossil "seashells" and ammonites, freshwater mussels from the Sacramento River, lower animals from San Francisco Bay, and microscopic organisms, and pursued his obsession for earthquakes, past and present.

April 21, 1860, was a landmark day in Sacramento for geology. Governor John Downey approved the legislative act authorizing a detailed geological survey of California and created the office of state geologist, both for a term of four years. The geologist who in 1853 had been writing about Tennessee mineral tracts, Josiah D. Whitney, was appointed to the new position initially at an annual salary of six thousand dollars. Wrote Whitney of the act, it is "a most enlightening measure."

Whitney, born in 1819 at Northampton, Massachusetts, went to Phillips Academy and Round Hill School. He attended Yale University, where Benjamin Silliman's lectures in chemistry especially excited his scientific interest. After graduating in 1839 he did advanced study in chemistry and during 1840 and 1841 assisted chemist Dr. Charles T. Jackson of Boston with a geological survey of New Hampshire and Maine. Young Whitney thought of going to Harvard Law School, but his father and Jackson turned him toward the sciences instead. In May 1842 he was off to Europe, where he traveled from Holland to Moscow, studied at the École des Mines in Paris, and carried out fieldwork with the eminent French geologist and cartographer Professor Élie de Beaumont and with German mineralogist Karl Rammelsberg. Returning to the United States early in 1845, he obtained a position through Jackson as field geologist at the Isle Royale Copper Company in northern Michigan for the summer. In December he returned to Germany, where he spent time in mineralogist Gustav Rose's chemistry lab in Berlin and with Baron Justus von Liebig at his innovative Giessen chemical teaching laboratory. Back home again, Whitney worked as first assistant to Jackson in the United States survey of Lake Superior. Eventually he, and geologist John Foster of Union College, another assistant, replaced Jackson as head of this project, and Whitney started writing articles on Lake Superior's copper and iron regions.

Whitney wrote prophetically to his brother William on December 11, 1848, that Lake Superior geology was passé while "California is all the rage now," and that he was contemplating a geological survey of the Golden State, "where the farmers can't plough their fields by reason of the huge lumps of gold in the soil." Whitney's plan was to be presented to the American Academy of Arts and Sciences, but he was sidetracked by new demands of the Lake Superior survey.

In the early 1850s Whitney was still working as a consulting geologist in the East and by the end of the decade was chief of the Iowa Geological Survey. He gained considerable international recognition for his classic, *Metallic Wealth of the United States,* published in 1854, and by 1859 he was considered by many to be one of the best-trained and most experienced young scientists in the country. His interest in a California survey revived, and he began stirring up kindred interest among influential eastern friends, including Louis Agassiz of Harvard, James Dwight Dana of Yale, Alexander Bache of the U.S. Coastal Survey, and his Yale classmate George Jarvis Brush, with the subtle suggestion that he himself would make a good head.

The California legislators initially knew nothing of Whitney, but Stephen J. Field, a California Superior Court justice, realized the importance of a geological survey and appreciated that its success depended on the quality of its director. Consequently, he sought advice from a number of eastern geologists—Leidy, Meek, Newberry, Dana, Agassiz, Yale paleontologist Othniel Marsh—the American Association for the Advancement of Science, and his brother, Reverend Henry Field of New York City, by coincidence a family friend of the Whitneys. The name foremost in their recommendations was Whitney. Despite the promoting of local favorites by some legislators, Justice Field and a few colleagues including legislator John Conness, chief proponent of the geological survey bill, succeeded in having Whitney appointed California's state geologist. Incidentally, Whitney's married sister Elizabeth now resided in California, and her influential husband, S. Osgood Putnam, secretary of the California Steam Navigation Company, had likewise facilitated Whitney's selection, especially through his friendship with Judge Field. Certainly Elizabeth, since moving to the West Coast, had dreamed that her brother could live nearby and might become head of the California Geological Survey. The rumor even circulated that Whitney himself may have written the job's requirements so that they fitted his own qualifications. The main objections to Whitney's appointment came from fellow geologists William P. Blake of the railroad survey; Jackson, his Lake Superior survey boss; Oscar M. Lieber of the South Carolina survey; and Trask, who did later provide helpful advice for Whitney.

Whitney's all-encompassing charge, with the help of assistants, was not only to make a comprehensive geological survey, with maps, diagrams, and sketches, but also to furnish descriptions and specimens of rocks, fossils, soils, and minerals for a future museum collection, and further to consider California's "botanical and zoological productions."

Whitney wasted no time in picking his staff, who in addition to professional qualifications should be young men, healthy, companionable, adventurous, and unmarried. The foremost candidate was William H. Brewer. Brewer received such a strong recommendation from George Brush, now professor of metallurgy at Yale and like Brewer a favorite of

Professor Silliman, that he simply had to be Whitney's first choice as "Principal Assistant, in charge of the Botanical Department."

William Brewer was born at Poughkeepsie, New York, on September 14, 1828, but he grew up on an upstate farm. At age twenty he went off to study agricultural chemistry with professors Silliman and John Pitkin Norton at Yale, a farm boy traveling for the first time by public conveyance. At the completion of his second year he returned home and became a teacher but in 1852 returned to Yale and was examined for his bachelor's degree in applied chemistry. He graduated in the first class of five students of the nascent Sheffield Scientific School, together with Brush, and William Blake who later served with the Pacific Railroad Surveys. Brewer then returned to academy teaching for three years, applied without success to be a naturalist with the Gunnison railroad survey, and finally decided to go to Germany for advanced scientific training. He studied first with Professor Robert Bunsen in Heidelberg, and then, like Whitney, in Munich under Liebig, who had developed a great interest in agricultural chemistry. For relaxation during the summer of 1856 Brewer hiked six hundred miles up and down the Swiss Alps, particularly examining the flora, and the next year he took a botanical excursion in southern France. Brewer eventually assumed a professorship of chemistry at Washington College in Pennsylvania. During the summer of 1860 his young wife and baby boy died, leaving him at loose ends until he received the unexpected invitation from Whitney, whom he had never met. In Brewer's forthcoming assignment, the young scientist did indeed, as he would title his future classic book, go "up and down California."

Whitney appointed two other assistants, William Ashburner and Chester Averill. Ashburner, from Stockbridge, Massachusetts—"a good fellow," as Brewer characterized him—was a graduate of the École des Mines in France and would serve as "Assistant in the Department of Economic Geology." Ashburner later became a mining engineer in San Francisco and served as honorary professor of mining engineering at the University of California, as regent of that university, and as trustee of its rival across the bay, Stanford University. For sixteen years he was one of the commissioners to "manage Yosemite Valley." Ashburner's wife, whom he married in 1856, was the niece of Justice Field. Averill, also from Stockbridge and a friend of Ashburner's, graduated from Union College in Schenectady and was perforce already a world adventurer, having spent, as a parental "punishment" for a youthful misdeed, eighteen months on a voyage to South America, the East Indies, and China.

Now seeking his fortune with Whitney, he would keep the survey accounts, observe the barometer, and be general factotum. Brewer came to
consider him "a capital fellow."

In October 1860, before leaving Massachusetts with Whitney and his
family for New York embarkation, Brewer was already immersed in the
upcoming adventure. He talked to the publisher of the forthcoming re-
vision of Michaux's *Sylva* by Thomas Nuttall and was promised a set in
exchange for information on California trees. In Cambridge, Brewer visited with Asa Gray and "got much valuable information."

At noon on October 22, Whitney and Brewer were off from New
York for Panama on the *North Star,* together with three thousand dollars' worth of scientific apparatus. They crossed Panama and boarded
the *Golden Age* on the Pacific side, sailing into San Francisco Bay on the
morning of November 14 to a heralded appearance. Newspapers had
earlier announced the forthcoming arrival of the geological survey men,
and now every paper ashore greeted them. They received accolades from
numerous prominent citizens, including a special visit from Governor
Downey and Judge Field. Little did Whitney anticipate that this was
nearly the high point of state support for their assignment; in the future
one of Whitney's totally frustrating tasks involved procuring financial
and logistical support to sustain the work of the California Geological
Survey.

The survey was to commence from Los Angeles, so on November 24,
1860, Brewer left San Francisco by ship for San Pedro, the Los Angeles
seaport. Whitney had gone to John Charles Frémont's Mariposa Estate
in the foothills of the Sierra Nevada with Frémont to meet with French
officials visiting the mining country, so Brewer as first assistant was put
in charge of the party of five, "a heavy responsibility I would like to have
had placed on someone else." On shipboard Brewer's men encountered
an uncouth team of surveyors running an eastern California boundary
line, a "hard set" who were boisterously yelling, carousing, and drinking whiskey. This proved to be the bona fide United States and California Boundary Commission, a party of fourteen men, mostly greenhorns,
using mules and three of Beale's camels to cross the Mohave Desert.
Brewer concluded that the member found dead the next morning had
probably been killed by bad whiskey.

The steamer put in briefly at San Luis Obispo on November 25, a Sunday. Brewer celebrated by going ashore for a bit of collecting. This was
a new flora for him, and in his notebook he cryptically identified the
plants as "composite," "lichen," "small fern," or "growing in tufts,"

and on a few occasions came up with a genus: "*Zauschneria*. Rocky places near landing; *Artemisia*—abundant on dry rocky hills." Brewer's first California plant collection amounted to nineteen specimens. By the end of 1861, his first full year in California, he had amassed 866 specimens, but recognized only a small number to genus and fewer to species.

Once on land at Los Angeles and encamped inland near Mission San Gabriel awaiting Whitney's arrival, Brewer began realizing that California was "still unsettled." There had been about sixty murders in Los Angeles alone the past year, though none in the past two weeks. Brewer kept his small troop well armed, wearing eight-inch bowie knives and heavy navy revolvers, with Sharp's rifles, carbines, and two double-barreled shotguns close at hand. But there were vestiges of civilization. At the nearby ranch of pioneer settler Benjamin ("Don Benito") Wilson, Brewer and Ashburner met "two lovely young ladies," who one day accompanied them during part of a rugged all-day hike. Brewer climbed up twenty-five hundred feet of mountainside and collected some plants.

Arriving from Frémont's Mariposa Estate, Whitney rejoined his crew just as the heavy winter rains commenced. By mid-December the survey team was in the Santa Monica Mountains: the four scientists; a game hunter named Guirado, who was brother-in-law of the California governor and whose sole talent was being able to ride a wild mule; Irishman Mike Eagan, an untrained cook; and Pete Gabriel, a mule driver whom Brewer was pleased at having chosen, "a most valuable man." Among their collecting priorities during this winter season were fossils, which they hoped would provide information on the age of the area's sedimentary rocks.

Between and during the rains of late December and early January, reportedly the worst in eleven years, the surveyors worked in the mountains and valleys surrounding the Los Angeles basin, fortunately not always having to camp out. The Whitney men were delighted to learn that during the heavy rains the rambunctious boundary party they had encountered on shipboard lost everything, except for one camel, when their Los Angeles encampment alongside the river washed away.

Brewer and Whitney were vexed during their surveying by the inaccuracy and even lack of maps. In their own cartographic work Brewer was impressed by Whitney's talent. Whitney used the sextant, Brewer the chronometer, and for altitudes they employed two sets of barometer readings, resulting in impressive accuracy. During his free time Brewer wrote voluminous letters, fixed his plant specimens, sewed buttons on his woolen shirts, and studied Spanish. He visited the already "celebrated asphaltum springs" (La Brea Tar Pits), noting the small birds, mammals, and even cows trapped in the tar.

A celebration on New Year's Eve got the year 1861 off to a good start. It took place at Wilson's ranch with his large family, including those special young ladies. At midnight the "sweet bells" of the mission tolled out the old and in the new. The next day Brewer and Whitney, joined by Wilson, visited some local quarries, and later they enjoyed a sumptuous dinner party at the ranch, the lovely evening concluded by the "even sweeter" nine o'clock mission bells. By now Brewer was aware of Wilson's importance. He had arrived in Los Angeles in 1841 with the overland party of William Workman, during the trip becoming acquainted with the youthful naturalist William Gambel. Wilson began accumulating vast acreages around San Gabriel. During the Mexican conflict, Wilson pushed for Southern California to become a separate state. In 1851 he became the mayor of Los Angeles and then Indian agent for Southern California. Today, Wilson is memorialized by Mount Wilson and its historic observatory; and one of his grandsons was the indomitable General George Patton of World War II.

By mid-January the weather had moderated, though there was snow on the higher peaks and the winds were frequently raw. The men explored the Santa Ana Mountains, San Gabriel Canyon, and far to the east the Temescal Range. On a typical junket Whitney carried the barometer, geology hammer and bag, and water canteen, while First Assistant Brewer had the compass and tripod, level, spyglass, provisions, botanical box, and another canteen of water. Dense chaparral covering lower slopes was sometimes "absolutely impenetrable," according to Brewer, the men often having to crawl through on hands and knees, their shirts, pants, and even drawers often "torn to shreds." There was no chopping a path since both of them had lost their hatchets. On the higher summits there were scattered conifers, including firs and Coulter Pine, and tracks of Black-tailed Deer, Bobcats, and Coyotes were often found, as well as occasional grand views. Now and then they encountered miners scouting for silver and gold, and many in the Temescal Range were absolutely "crazy" about tin ore, the tin mines being the main reason Whitney wanted to visit the area. While Whitney examined the tin "leads" (lodes) and scrutinized the interesting geology, Brewer scouted out the "exquisitely beautiful flowers," sparse and difficult to locate in winter: "small white flower," "lovely little yellow crucifer," "very minute plant of Sedum Family."

On February 7, when Whitney was forced to leave for San Francisco to look after the survey's financial support, Brewer was placed in charge, with orders to prepare for a northward exploration up the San Fernando Valley and along the coast past San Buenaventura (present-day Ventura), with its old mission. On February 19 Ashburner collapsed, having barely recovered from a bout with scurvy on a geological excursion, and had to be taken to Los Angeles for an extended recovery.

During the evening of March 4, 1861, camped near the surf eighteen miles north of San Buenaventura, they heard the boom of a cannon in the distance, signaling the arrival of a steamer at Santa Barbara. They hoped the "Professor," as Brewer called Whitney, would be aboard, or at least some communication from him, since only $3.25 of survey money remained. But neither professor nor money arrived. All Brewer could do was send a letter to Whitney in San Francisco, saying the crew would wait for him—and for funds—in the Santa Barbara area.

During the wait, the weather was abominable most of the time. Their visit to the mission proved depressing because although the building itself was in fine state, the surrounding settlement was in ruins. More appealing were the occasional excursions to the local hot springs, shaded by a fifty-foot California Bay, the men sometimes taking a bath in the sulfurous waters. At a distant hot springs Brewer and Averill visited on muleback, the water temperature was 115 to 118 degrees Fahrenheit. Brewer reflected how valuable such hot springs would be close to a city of health seekers. Along the beach the men encountered a forty-five-foot stranded Gray Whale. On another junket to the shore they found a datura, some fossils, and asphaltum beds. Brewer accompanied a private surveying party to visit a ranch up the coast, where he discovered much asphaltum oozing out of the bedrock and hardening in the sun. He was told that locally the tar was mixed with sand, heated, and used for making roofs. Brewer contemplated that it "will eventually be the source of some considerable wealth."

On March 18, Brewer and Pete Gabriel, employing Cañupio as guide, scaled a 3,800-foot summit, the thermometer registering 120 degrees in the sun at noon. The men suffered excruciatingly from thirst, although Gabriel was able to get relief by chewing on a coin for several hours. The guide did a bit of collecting himself, gathering branches of Mormon tea, informing the surveyors it was "first rate to cure the clap." Impressive during the Santa Barbara stay was the gigantic "king of all vines," a California Wild Grape with a trunk three feet in diameter, extending seventy feet in each direction and reputed to have borne about sixteen thousand bunches of grapes one year.

Whitney finally arrived during the evening of March 19, but only four days later he departed, this time to examine the coal mines around Mount Diablo and the gold mines farther north, leaving Brewer to lead his crew up the coast to a rendezvous with Whitney about mid-June at San Juan Bautista.

On April 2 Brewer's men headed north again, their wagon, with its square-covered body suspended between the wheels on straps, always seeming on the verge of collapse. They traveled along the coast, where Brewer's choice find was a fossilized "mastodon (or mammoth) tooth." Evening's camp was at a lovely spot by the ocean on Dr. Nicholas Den's historic rancho Dos Pueblos, surrounded by thousands of grazing cattle. Brewer remarked on the abundance of what he called "the common 'Pin Clover' of the country," a "very valuable forage" he rightly identified as *Erodium cicutarium,* an introduced weedy geranium. Brewer's party left the Santa Barbara coast by way of Gaviota Pass, the wildest one he had yet seen in California. Having read railroad survey surgeon and geologist Thomas Antisell's interpretation of the pass geology but disagreeing with it, he and Averill devoted a day to examining the pass, collecting some fossils on the way and killing a Southern Pacific Rattlesnake with eight rattles.

At Arroyo Grande, the light wagon escaped on a thirty-degree slope, even with its hind wheels chained, and flipped completely over, throwing one mule on its back as another ran away. The wagon had to be unloaded from underneath: wrapped rocks and fossils from three boxes now in one "promiscuous pile," frying pans, pails, soap, botanical papers, tools, carpetbags, instruments, provisions, and tent ropes, all anointed from a large container of applesauce prepared earlier for the evening supper. Viewing the amalgamated mess, literate Brewer ventured, "Themes for more papers on 'The Distribution of Species' than even the famous antiquarian stone of Mr. Pickwick."

During the second half of April they did fieldwork around San Luis Obispo, filling the time before meeting Whitney at San Juan Bautista in June. A foray was made to the seashore, and a survey of the Santa Lucia Mountains begun. They amassed collections of the abundant fossils, some of them oyster specimens with five-inch-thick shells and up to fifteen inches in length. Altogether sixteen boxes of specimens were accumulated, "enough," as Brewer commented, "for quite a cabinet." All was not necessarily work. One mountain ascent became devoted to rolling boulders, some weighing as much as four hundred pounds, down the grassy slopes. Some would bounce, cavort, and even soar downhill for as much as two thousand feet in elevation, pieces flying off as the rocks struck obstacles, whistling through the air like cannonballs, the sounds audible half a mile away. One that Brewer observed at close hand literally leaped sixty feet off the ground. The arrival of the mail packet about April 27 brought alarming news: Fort Sumter in South Carolina had been attacked, and the United States was on the eve of civil war.

As Brewer's party left San Luis Obispo on April 29 for the upper Salinas Valley, the entire aspect of the countryside changed. Here there were fewer trees, sparse dry grass, the soil parched, and the hills "brown as a stubble field." But Brewer never tired of the beautiful open oak woodlands, with the two species he had observed before but did not recognize, the Coast Live Oak and Valley Oak. The few pines (Foothill Pines) visible on the upper slopes were unknown to him; they somewhat resembled the Italian Stone Pines he had encountered in southern France but had lighter and airier foliage. At an American ranch they visited were some magazines, "rare luxuries for camp," which Brewer, ravenous for literature, spent a whole day reading. Continuing north, they saw "many birds of great beauty," up to fifty cottontails and Black-tailed Jackrabbits a day, snakes of many species, and several kinds of large lizards. Hundreds of "salamanders and chameleons [small lizards, including horned toads] are dodging around every log and basking on every stone," Brewer said, but insects were the most numerous animals, swarming everywhere. There were plenty of ticks, and also the "dreaded tarantula abounds here," with teeth as large as a rattlesnake's. Brewer saved one set of fangs "as a curiosity." On May 11 Brewer hiked to the top of the high ridge west of Soledad mission, where he had a magnificent view of the surrounding landscape. Never had he "been in a land before with so many extensive views": the wide Salinas Valley with mountains beyond to the east (the Gabilan Range), a range of forbidding mountains to the west (the Santa Lucias), while to the northwest lay Monterey Bay.

There was much of geological interest in this area, but with diminishing food and water, Brewer decided to move on to the Monterey Peninsula, where his party arrived on May 15. Although Brewer was still somewhat ignorant of the California flora, he looked the part of a western outdoorsman. He was clothed in smoked buckskins to defy the chaparral, a bowie and a Colt at his belt, his face and hands sunburned to the color of "well-smoked hams," hair trimmed to less than half an inch in length. At night he rolled up in his two heavy blankets on an india-rubber sheet, throwing on an oilcloth against the dew, with his head on his saddlebag. If weather dictated and time permitted, a tepee-shaped Sibley canvas army tent was pitched, using only a middle pole and three guy wires, the latter also serving as clotheslines.

The stay in Monterey was a pleasant respite, in large part because of another coincidence. At El Toro Ranch, the last campsite in the Salinas Valley before the party reached Monterey, the mail stage stopped to water the horses. One of the passengers was a New Yorker whom Brewer had met on shipboard, and even more remarkably, he was a personal friend of Averill's and fellow Union College graduate. Edward Tompkins, a lawyer, had the previous summer received property along the ocean near Mission San Carlos at Carmel. He invited the survey team to camp there.

There was much national news in Monterey via Pony Express, which took only about eleven days to travel from coast to coast. Fort Sumter had surrendered; President Lincoln had declared a state of "insurrection" existed, called for volunteer troops, and declared a blockade of southern ports. The first casualties had occurred—namely, the stoning of four Union soldiers by a mob in Baltimore. Locally, the last steamer leaving Monterey for San Francisco departed fully armed because of a rumor that rebels posing as passengers might try to hijack the ship for the Confederacy.

The news of the geological survey team's presence in Monterey quickly spread. On Sunday afternoon, a few days after their arrival, Brewer and Averill were approached by a man who wanted them to ride out of town a few miles to assess a possible silver deposit. The pair declined the invitation but were soon besieged by other prospectors, one who brought "ore" from a mine and another who reported "indications." To get rid of the ore seekers, the surveyors took the specimens back to camp to blowpipe them. One of the specimens actually did show some silver.

Tiring quickly of old Monterey, with its liquor shops, saloons, cigar and *cigarrito* vendors, gambling, drunks in town, and intoxicated Indians in the outskirts, the surveyors accepted Tompkins's invitation to camp at his El Pescadero Ranch, a large spread—four to five thousand acres—touching the Pacific Ocean (where Pebble Beach is located today). The tent was set up in the ranch yard near a log house the surveyors used

Figure 30. The California Geological Survey party in camp near Monterey, May 20, 1861. The photograph was taken on emulsion-coated leather, and the cracks are due to emulsion shrinkage. (From William H. Brewer, *Up and Down California in 1860–1864*, 4th ed., ed. and trans. Francis Farquhar, 1966. Courtesy University of California Press, Berkeley.)

as headquarters. The former cabin owner, an eccentric, had left an impressive library, including volumes on science, art, astrology, romance, infidelity, religion, and mystery. Brewer persuaded a photographer to come out to the ranch and make pictures of the encampment, four photographs exposed on emulsion-coated sheets of leather. The best photograph, which Brewer won by a draw, shows the white Sibley tent in an open woodland of pine and oak, the men seated and standing in the shade, and beyond the meadow in the background the ridge above Carmel Bay (figure 30).

The survey party visited the nearby Carmel mission, now abandoned and partly in ruins, with broken pillars, the remains of an old shrine, wandering cows, a dead pig in the holy-water font, and a Great Horned Owl flushed from above the pulpit. One morning the men rode to Point Cypress, the picturesque trees reminding Brewer of Cedars of Lebanon. Visits were made to some of the other ranches, Judge Fletcher Haight's having a fine library, a piano, and two attractive daughters who could play it. On May 28 at dawn Brewer, Averill, Gabriel, and a *"buccaro"*

as guide took off with two pack mules up Carmel Valley and into the surrounding hills, Brewer with his botanical box, hammer, and pistol. They
ventured into deep, rocky canyons, had a swim in a cold mountain creek,
and experienced hospitality at isolated ranches, feasts of fresh venison,
a specimen of quicksilver proffered by a rancher named Hitchcock, discovery of a huge fossil fish backbone near Finch's Ranch, and an encounter with a gigantic rattler that had a body as thick as one's arm and
twelve rattles, whose fangs naturally had to be saved by Brewer. The
botanist had finally learned at least the Spanish names of the two common oaks, the evergreen *encina* and the large *el roble*. He came upon "a
most beautiful tree" with foliage richer than a magnolia, undoubtedly
"splendid in cultivation" (Pacific Madrone). On June 1 the adventurers
were back at El Pescadero Ranch, after a junket filled with "much of
botanical and geological interest." Among plants collected were violet-
red penstemons, bushy magenta mints, a greenish-white pea, Big-leaf
Maple leaves, mountain-mahogany foliage, and purple blossoms from
bush lupines. Two specimens identified by scientific name were a shrubby
mint—Woolly Bluecurls, with brilliant blue-violet flowers—and Pacific
Madrone, common in the hills and canyons near Finch's Ranch.

After leaving Monterey, the field party bivouacked at San Juan
Bautista awaiting Whitney's arrival. On June 17, Brewer explored some
striking red sandstone outcroppings southwest of camp where hundreds
of swallows nested in the cliff alcoves, while in the ravine below, "brilliant and fragrant with flowers," there were myriads of Anna's Hummingbirds "flitting like large bees." The next day he visited several "asphaltum springs" at James P. Sargent's large ranch to the northwest, the
hills covered with wild oats, "a grand field of feed, that!" Little could
Brewer anticipate that the Sargent Hills oil field, developed by 1886,
would exist for sixty years, producing more than six hundred thousand
barrels of petroleum. Returning to San Juan Bautista late, Brewer walked
from camp into town to await the mail stage; it arrived at 10:00 P.M. with
a letter from Whitney requesting Brewer to hasten to San Francisco to
confer "on some important business relating to present and future plans
of operation."

Brewer was off at dawn on the stagecoach, going by way of the beautiful Santa Clara Valley, "the garden of California." Brewer stayed at
Whitney's home in San Francisco atop a hillside with a beautiful view of
the city. The two discussed future plans, Whitney describing his
grandiose scheme for a museum to house the survey's collections. In the
city Brewer ordered a new wagon, bought supplies, and met new friends

and old, the latter commenting on "how camp has improved you."
While there he attended an evening meeting of the California Academy
of Natural Sciences.

In late June Brewer was happy to return to the San Juan Bautista
camp: "It seemed like home. There is no canopy like the tent, or the
canopy of Heaven, no bed so sweet as the bosom of Mother Earth."
There were problems in paradise, however. As summer progressed it was
not just hot, but extremely hot. The novelty of exploration was wearing
thin, and the infernal waiting bored the men. Mike Eagan, the cook, became slack in his duties, and the others had taken a dislike to him. But
before Brewer could ask him to quit, Eagan headed to "Frisco" of his
own accord.

Brewer himself was still enthusiastic about continuing up and down
California. On July 1 he and several companions scaled the "very sharp,
steep, bold" three-thousand-foot Gabilan Peak in torrid weather. They
ran out of water and consumed their lunches before reaching the summit, where Brewer plucked a single yellow flower. Going down they
found a fine clear stream in the valley, and a sumptuous dinner was prepared by substitute cook Gabriel. The camp under a large oak was air-
conditioned by a "delicious breeze," and a comet in the west at night
"made us forget the toils of the day."

On the eve of the "Glorious Fourth" at San Juan Bautista, the stage
dropped off a passenger for an overnight stay: Dr. James Cooper, whom
the Whitney men had met in San Francisco when they first arrived in
California. The past winter he had been surgeon at Fort Mohave, doing
some collecting for Whitney, and he brought along a number of desert
plant specimens for Brewer. Now he was on his way to inquire of Whitney about a position as zoologist with the survey.

On July 5 Brewer returned to the tar springs at Sargent's Ranch near
Gilroy and found that Sargent also had an oil works that converted asphaltum to oil for burning purposes. When Whitney arrived back at San
Juan Bautista from San Francisco on July 8, he brought a very important
addition to the survey, Charles Frederick Hoffmann, a German topographer. In his early twenties, Hoffmann had already been topographer
for Frederick Lander in 1857 on the wagon road survey from Fort Kearney, Nebraska, arriving in California in 1858. As principal topographic
assistant, Hoffmann remained with Whitney and the California Geological Survey until it terminated in 1874, except for two years as a professor of topographical engineering at Harvard.

When Brewer, Whitney, and Averill took a short excursion to Monterey and El Pescadero Ranch on July 10, they were told of the discovery of silver mines in the vicinity. Skeptical Brewer only raised an eyebrow at the news. Shortly thereafter, the Whitney survey party left San Juan Bautista and headed southeast to the famous quicksilver (mercury) mines at New Idria. In the desolate landscape the men explored the three principal mines, for hours threading "drifts, galleries, tunnels"; they "climbed over rocks, crawled through holes, down shafts, up inclines, mile after mile, like moles." The miners, mostly Mexicans and Chileans, endured not only unbearable summer heat but also sulfurous acids, arsenic, vapors of mercury, and dust, all of which told on their health. To the surveyors the geology was intriguing, the brilliant red ore beautiful. But as Brewer commented, "Such is New Idria and by such toils and sufferings do capitalists increase their wealth!"

In August the surveyors went north into the Santa Cruz Mountains and on to another famous mercury mining area, at New Almadén, where Frémont and other naturalists had visited. Besides examining the mines, there was a reunion with Whitney and his wife, down from San Francisco; a chance to meet the visiting California governor Downey and his wife; and a round of social affairs. New Almadén's mining engineer was Sherman Day, son of a former Yale president. In fact, while a Yale student Brewer had occupied Sherman's own room in President Day's family home. Sherman had surveyed the so-called Day wagon route over the Sierra Nevada south of Lake Tahoe in 1855, and after the Civil War he became the U.S. surveyor general for California. Brewer enjoyed Day's attractive daughters and their feminine companions. When Brewer departed from New Almadén, his saddlebags were filled with ore specimens.

The remaining schedule for 1861 called for reaching the "geysers" north of Napa before the rainy season. During late summer and fall they surveyed in a crescent around San Francisco: San Jose, Mount Hamilton, Oakland, Mount Diablo, and finally up Napa Valley past St. Helena. Sometimes Whitney joined the field team. They found the Mount Diablo region especially interesting geologically, with its enormous sedimentary strata broken, dipping, sometimes forming high vertical ridges. They discovered a bed of coal four feet thick and many fossils, including shells, wood, and exquisite tree leaves. Whitney and Brewer hiked to the summit of Mount Diablo—"the central mountain mass roasted and baked"—the climb proving easier than anticipated. Here Brewer collected a stonecrop, *Sedum spathulifolium,* near the southern edge of its range, as well as foliage from a California Juniper.

During these months of survey work, two special trips were made to Sacramento, the first because of continuing funding problems, the second to visit the Sacramento State Fair in late September. Brewer spent "two laborious and 'borous' days" there with a committee, making arrangements to send exhibits from the West Coast to the great Crystal Palace World's Fair scheduled for London the next year. He had more fun measuring fruit on display at the fair: an apple over sixteen inches in circumference, a pear almost nineteen inches.

In mid-October the surveyors traveled southeast from Mount Diablo to Corral Hollow, touted in the newspapers for its rich coal mines. They crossed the tedious, dry San Joaquin plains, memorable for Brewer because of the "thousands" of mouse-size tarantulas. At Corral Hollow it turned out that only one mine had ever sold any coal. By the end of October they headed north for Napa Valley and the so-called geysers beyond.

On November 9, 1861, north of Napa and Knights valleys, only four more miles of northern travel remained—in direct line—for the California Geological Survey to conclude its year on schedule. But "*such miles . . . such* a trail," over ridges and through deep gullies. It took two hours to cover a mile of countryside. They went to the head of the Pluton Creek canyon to see the Pioneer Mines high on the hillside, some nine hundred feet above the quicksilver furnaces, smithy, and residences of the mining camp, "in a most picturesque spot." Early the next morning, with foreman Wattles and another miner, the party went to the "Little Geysers," up through a forest and chaparral to the ridge top, where there was "the most sublime views for wild scenery." To the north could be seen a column of steam rising, and shortly the trail led into Geyser Canyon with its "roaring, rushing, hissing steam." Brewer observed that these were not authentic geysers but merely hot springs. He collected a Sweet-shrub, still flowering near one of the geysers, and a grass, Geyser's Panicum (originally named *Panicum thermale*), growing abundantly in the warm, sulfate-impregnated soil. Today this grass is ranked as an endangered species. In one of the hot springs registering 150 degrees Fahrenheit, Brewer obtained specimens of "a low order of plant, like confervae or 'frog spawn,' " but they later proved to be "very unsatisfactorily preserved." After dinner at the mining camp, with the sky darkening and threatening, the men saddled their mules and were off. As Whitney, Brewer, and a few companions reached the ridge summit, "a thick, driving, drizzly fog rolled in from the sea, soon wetting us to the skin, and making our teeth chatter." But the men were lighthearted, since the goal of "the summer's hope" had been achieved, exactly a year after the surveying had started in Southern California.

During this year Brewer had written over a thousand pages of field notes and penned eighteen hundred pages of letters, official and otherwise. At the moment a number of influential men wanted Brewer to act as commissioner for California to the London Exhibition and to accompany California's exhibits. But Whitney felt that the first assistant could not be spared from the survey, and Brewer agreed that he had proved himself "capable of successfully managing a party, carrying it through times of discomfort, and even hardship, accomplishing much labor, and doing it economically."

Surveying ended with winter. In San Francisco, Whitney prepared his first survey report at his office and home, while Brewer had a pleasant room in town and ate his meals at a nearby restaurant. The winter rains outdid themselves. Precipitation for December through February along the coast amounted to more than forty-one inches, while in the Sierran foothills some areas experienced twice that much. Expanses of the Central Valley were under water, including the capital. As the survey's, as well as the state's, financial situation worsened, Whitney personally had to borrow money to maintain the survey. Also, the San Francisco newspaper *Alta California,* encouraged by opponents of the state survey, criticized the amount of survey time and energy devoted to botany, zoology, and paleontology at the expense of economic geology. As the *Alta* editorialized, "And of what benefit will the survey be? Great to science, great to Professor Brewer, and of much credit—among scientific men— to California, but probably of no pecuniary benefit to the industrial interests of this State." The fossil collection was particularly singled out: "It has not the least value or interest for commerce or industry."

THE SACRAMENTO VALLEY AND
MOUNT SHASTA: THE 1862 SEASON

On April 23, 1862, the survey team returned to the field. This season they would finish the map of the greater San Francisco Bay region: north to Petaluma and Napa, south to New Almadén, east beyond Mount Diablo, and west to the ocean. At first Brewer headed up a small group including Averill; topographer Hoffmann; a young French volunteer, Auguste Rémond, who Whitney noted had a particular gift for finding fossils; and a new young paleontologist, William Gabb. Gabb, though

Figure 31. The California Geological Survey party near Mount Diablo, 1862. Left to right: Gabb, Whitney, Averill, Hoffmann, Brewer (in front of the horse), and Schmidt, the cook. (From Brewer, *Up and Down California*. Courtesy University of California Press, Berkeley.)

quite conceited and irritatingly talkative, proved to know his Cretaceous and Tertiary fossils and could make excellent sketches of them.

The five assistants started working alone, but early in May they were joined not only by Whitney and his wife but by a large party of dignitaries, among them the celebrated orator and clergyman Thomas Starr King. The entire group, led by Brewer's "worthies," climbed to the summit of Mount Diablo. One of the choice flowers collected on the north flank was the rare, endemic Mount Diablo Fairy Lantern, *Calochortus pulchellus*. During May Brewer and his crew surveyed southward along the eastern flank of the Diablo Range (figure 31), Brewer collecting deposits from the alkaline pools for later analysis. The surveyors were plagued by superhot temperatures, a superabundance of mosquitoes, centipedes, scorpions, and tarantulas, and myriads of Grizzly Bear tracks. One evening Brewer bet Hoffmann a keg of beer (whenever one could be found) that they would neither hear nor see a bear that night. The men slept so soundly that they missed the grizzly that did track through their camping area, so on a technicality Brewer won the bet. Because the San Joaquin Valley was entirely flooded, the survey party was unable to continue eastward.

During late July the surveyors started from Benicia for the Sacramento River valley, despite flooded areas ahead, planning to proceed all the way north to Mount Shasta. But by the time they made camp on August 6 at Rag Canyon near Putah Creek, Brewer had become critically ill and was incapacitated for days, although he still managed to collect eight specimens, including a Narrow-leaf Milkweed. He finally sent the survey team on, later meeting them by steamer on August 23 just as they arrived at Red Bluff. Then the entire party traveled along the Sacramento River, through country scarred by extensive placer mining, to the village of Shasta, at the edge of the great mountain barrier extending north to Mount Shasta.

On September 2 Whitney joined the party. In a few days they proceeded into the mountains along the river on, as Brewer observed, "the most picturesque road I have ever traveled in this state—in fact, I think that I ever traveled," sometimes at stream level, more often crossing ridges and sinking into narrow ravines. Up the valley were occasional "glorious views" of Mount Shasta, its summit "streaked with glistening snow." Near one camping spot on September 8, Brewer found "a patch of a very curious and rare plant, the *darlingtonia,* a sort of pitcher plant, as yet found in no other locality, the wonder and admiration of botanists." The plants were beside a "running rivulet of pure cold water . . . on the margin in long lines marking the water-course on gravelly soil." Unfortunately, the flowers were already gone, although seeds were ripe and falling. This was the carnivorous California Pitcher Plant first collected hereabouts in 1841 by botanist William Brackenridge of the U.S. Exploring Expedition. Naturally, Brewer filled his botany box with specimens, while his old favorite mule, Blanco, tethered to a nearby tree, made "the woods hideous with her braying." A day later off to the west the "Devil's Castle" (Castle Crags) came into view, with stark white, granitic "pinnacles, minarets, battlements, domes, and peaks." The last two days before reaching Mount Shasta their way lay through a forest of magnificent conifers—pine, fir, cedar, and spruce—"large and grand beyond anything the eastern states know of."

For the Whitney survey the conquest of Mount Shasta was "an item of 'Great Expectations'—it seemed indeed the grand goal of this trip." Special barometers had been constructed more than a year before, and the climb was scheduled for August, but sickness had intervened. Now final preparations for the assault, including testing apparatus and getting provisions ready, were under way on September 10. A guide was found, Mr. Frame, who said he had been on top once and it was "practicable." Two men with their horses from Soda Springs downstream joined the survey party with its three mules. At Camp 99 just below timberline, Whitney woke the climbers at two in the morning, by moonlight. They had scaled the red volcanic cliffs looming above them by eight, and about noon, after traversing wreaths of lava alternating with patches of snow, the party, except for Hoffmann, reached the summit. All the way Brewer and Whitney diligently carried the delicate barometers for checking the altitude. The day was partly cloudy with a brisk temperature of 26 degrees Fahrenheit, but fortunately no wind. The men remained on top for an hour and a half, wearing their goggles against the glare, all experiencing some mountain sickness: fatigue, drowsiness, headache, nausea, bloodshot eyes, blue lips and fingernails. Some of Whitney's fingers were frostbitten, and several of the men threw up. After the ascent Brewer calculated that they had reached the highest point in the state, about 14,500 feet. Mount Shasta is actually 14,162 feet, California's fifth highest peak. Brewer was brief in his opinion of Mount Shasta: "Geologically it is as barren as it is botanically. It is a great cone of lava, nothing else."

The return trip south provided an opportunity to view the east side of the Sacramento Valley, including a look at Lassen Peak, which Brewer described as a gigantic extinct volcano. The last volcanic activity had occurred over a decade earlier, during the winter of 1851–52, producing restricted lava flows and cinder cones. Brewer's "extinct volcano" came back to life again on May 30, 1914.

At Chico the surveyors visited John Bidwell's home. The general was away but had left orders for the men's entertainment. Brewer, having heard of Cretaceous fossil shells up Chico Creek that "might throw much light on the geology of the region," took off with two companions and an Indian guide, only to discover that the streambed was solid basalt. The Indian boy announced that he was leaving, saying this was unfamiliar country to him, but also admitting that he was worried about the local Indians who had recently killed four young people in the vicinity. Naturally Brewer pressed on, until he found a spot where erosion in the creek bottom had cut down through the lava to the underlying sandstone and there were large masses of embedded shells. Ah, "what convulsions of nature that locality must have seen since those animals lived in that ancient sea!"

North of Oroville in the foothills the geology was so intriguing that the party stayed two days. Brewer made cross-section sketches, showing the numerous sedimentary table mountains capped with lava protecting the summits from erosion. He noted that where streambeds were covered by basalt flows, miners had tunneled through the lava to extract placer gold buried beneath.

On October 18, at the "smart city" of Marysville, the surveying party was officially "disorganized," although Brewer noted that there was still unfinished fieldwork. Whitney and Brewer busied themselves around San Francisco for much of the remaining year. Shortly before Christmas the California Steam Navigation Company planned to launch a fine new steamer, the *Yosemite*, at Potrero down the bay. Whitney through his brother-in-law obtained invitations for Averill, Gabb, Hoffmann, and Brewer to travel on the company's *Paul Prey* steamer to attend the launching, which proved a huge success despite the poor weather. On the return to San Francisco, after a sumptuous lunch of cold turkey and champagne, there was a "crash that startled everyone." The little vessel had run into a submerged extension of Frémont's Alcatraz Island. Everyone was rescued, the geological survey men heroically waiting for the last lifeboats. Brewer saw to it that his charge—young Miss Katie, Whitney's niece—went into the first boat, together with all the other women. The next morning at 5:30 San Francisco experienced the most severe earthquake in seven years. Brewer's bed seemed to lift, his house shook, and some feminine screams were heard.

The year ended for Brewer on a positive note, when he and Whitney dined with world-famous German geologist Baron Ferdinand von Richthofen, who had just arrived in California from the Far East. After dinner they repaired to Reverend Thomas Starr King's residence for an evening of conversation. Richthofen stayed in California as well as Nevada for five years, assisting the California Geological Survey and carrying on his own geological studies, which led to the discovery of many new gold finds.

Brewer calculated on December 19, 1862, at the completion of the survey's first two years, that he had traveled 2,067 miles on foot, 3,981 astride a mule, and 3,216 by public conveyance. Along these many miles he had collected more than two thousand plants. Chief Geologist Whitney himself "had seen with his own eyes some portion of at least forty of the forty-six counties of California."

The Geological Survey Continues, 1863–1874

The year 1863 commenced on an optimistic note with the issuance of President Lincoln's Emancipation Proclamation on January 1. While William Brewer waited to resume the California Geological Survey, he delivered a lecture on the third in San Jose, his subject "The Mountain Scenery of California." By now the botanist-geologist was recording secretary of the California Academy of Natural Sciences, and within a month he was offered a consulting position with a Mexican silver mine that he reluctantly declined, since he knew practically nothing about mining. Then one day he visited a "small-fry" college in Oakland, where he chatted with the acting president, Samuel Willey. At the end of March he received word that he had been elected professor of natural science, a post he accepted with the understanding that his survey position held first priority. In 1910 this upstart college, by then the University of California at Berkeley, bestowed an honorary doctorate on Professor Emeritus Brewer of Yale.

In March Brewer spent several weeks in the Santa Cruz Mountains with Charles Hoffmann looking especially at new mining operations, mostly quicksilver. But Brewer, appreciating that the real mineral wealth of the state was in the Sierra Nevada, decided that this field season's efforts should be there. On April 1 he and paleontologist William Gabb set out by stage from San Francisco, with only a few indispensable articles as their outfit for a two-month trip Brewer anticipated would cover over a thousand miles. The pair proceeded to Clayton near Mount Diablo, where Brewer's mule, Kate, and some other mules and horses were picked up. There was a buzz of excitement in the area about newly discovered quicksilver and copper mines, all of which proved worthless.

For a week the two riders slogged across the soggy San Joaquin Valley on a diagonal to Fresno, then south to cross the Kings River, where they visited a tribe of the Kings River Indians, "hard looking customers." Finally on April 10 the pair reached Visalia, beautifully situated in an open oak woodland. They relaxed there for a few days to rest their animals and "to get inspiration for our trip ahead." It was a relief to learn that the Indian troubles in the Sierra had diminished, the ringleader having been killed the previous day.

Brewer and Gabb continued southward along the eastern edge of the San Joaquin Valley. The vegetation that had grown up after the great rains of 1862 was now dying and browning, and the next stretch of plains, from Visalia to the Kern River crossing, was barren. The men stopped at an isolated tavern, probably the last for a hundred miles. The Yale graduate spent the evening there reading his copy of *Les Misérables* and noticed that a miner, on his way home back east, was reading Shakespeare. Brewer suspected that being "far from school, church, or other institution of civilization" in wild California, the next generation would undoubtedly grow up ignorant. As they approached the end of the valley, a black storm came up behind the riders, at first choking alkaline dust, followed by a soaking downpour.

On April 17, after crossing miles of flat, desolate plains, the pair found the greening foothills were a welcome sight. Four miles up a wild canyon they came into a grassy vale surrounded by high mountains, "picturesque, beautiful, quiet." Here were the remains of old Fort Tejon, abandoned as a military post because it was on private land. After a few days of rest, they reconnoitered the surrounding countryside, visiting rich landowner George Alexander at San Emigdio Ranch. Alexander's sister and niece had come out from New York, as Brewer said, "for the romantic life on a California ranch," but obviously the romance had worn off. On April 28 Brewer and Gabb visited the old Tejon Indian Reservation started by Edward Beale. Eventually Beale had bought up an old Spanish land grant in the region, 137 square miles. "That is the way to get land," noted Brewer. The next day the men rode to the top of the new Tejon Pass (old Cañada de las Uvas Pass), seeing many Black-tailed Deer, a large Bobcat, and another of the rattlesnakes, this one still with nine rattles, the newer ones having broken off.

On May 6 they left Fort Tejon, heading eastward into the desert. Brewer was not impressed by the dominant vegetation, Creosote Bush—"every part of which stinks, making the whole air offensive"—and sagebrush. But there were two large, unique plants in full bloom, the Joshua Tree and a century plant, which he described in his journal. He and Gabb went over Frémont's old pass and stopped for the night on Oak Creek. The next day, Sunday, in Tehachapi Valley, they camped under some magnificent oaks and had their usual fare: tea in a tin cup, sugar, crackers, and jerked beef. Half a dozen families now homesteaded there. Brewer heard that earlier that day a visiting Methodist minister had held a service for the mostly Missouri settlers, all southern sympathizers. Brewer and Gabb noticed that all the womenfolk dipped snuff.

The next day the surveyors entered the southern end of the Sierra Nevada over a terrible trail, the rugged granite mountains rising high above green valleys, and continued northward into Walker Basin. They reached the Kern River at Keysville, the largest settlement within ninety miles. It contained only eight buildings, including a saloon, store, and "hotel," but it provided a welcome stop because Whitney's crew had run out of provisions. They got feed for their animals and "square meals" for themselves, and spent the night on bunks (actually boards) in the so-called hotel. There were some gold mines in the area, and Keysville was somewhat of a crossroads between mining districts.

Since Brewer's plan was to "examine the geological nature of the Sierra Nevada," he decided the range should be crossed at as many sites as possible, starting with the old Walker Pass. They left on May 15 and were back on the eighteenth, having gotten all the way east into the desert. Their route was then north, sometimes by rough road, other times by trail, with stops at tiny settlements. One spot they passed was the little green basin at Poso Flat, near where Williamson's Pacific railroad survey had had its depot a decade earlier. On May 21, at the Tule River near the edge of the mountains they were overtaken by a party of men with a band of mules and horses who had come across from Indian Wells at the edge of the desert, where Brewer and Gabb had recently been. A number of their animals had been killed and wounded in an Indian attack, and the survivors were on their way to the safer Tulare Valley.

The following day Whitney's men were back at Visalia, their funds almost gone. They continued for San Francisco, north through Millerton and on to Hornitos near Mariposa. There they received a letter from Whitney instructing paleontologist Gabb to proceed to San Francisco, while Brewer was to join Whitney to the north at Columbia. Thus ended the southern Sierra junket with the mules and horses exhausted, having traveled eleven hundred miles in two months, just as Brewer had predicted. Gabb left immediately, while Brewer remained a day in Hornitos to examine the touted copper mines, which as usual turned out to be overstated. Brewer observed that people were always excited "on the matter of copper, and much money is lost and won in copper speculations."

THE CENTRAL SIERRA, 1863

Brewer never forgot the next two weeks. On June 1, 1863, in the quiet old gold-mining town of Columbia, the placers now completely worked out, the California Geology Survey staff met. Hoffmann went back to Clayton for more pack and riding animals, while Whitney and Brewer perused some of the silver mines during a three-day ride. The group was to head over the Sierra Nevada to the Aurora Mining District in western Nevada by way of Yosemite Valley. But as usual, the excursion never materialized as planned. At the end of the first day at Murphys Camp, reached by a picturesque road crossing the thousand-foot-deep chasm of the Stanislaus River, there was word that Whitney's sister was seriously ill in San Francisco. Whitney left immediately, while Brewer took off for the Calaveras Grove of Big Trees, about fifteen miles above Murphys.

The Calaveras Grove was the original "Big Trees" discovered in 1841 by John Bidwell. Brewer spent several days at the grove, apparently at James Sperry's hotel. The ninety or so Giant Sequoias were scattered among a mixed coniferous forest of Douglas-fir, Sugar Pine, Incense Cedar, Pacific Ponderosa Pine, and White Fir, some of them giants themselves. Brewer measured thirteen of the sequoias, the largest the "Mother and Son," eighty-two feet in circumference three feet above the forest floor. Whitney had counted 1,255 rings on a sequoia stump about twenty-four feet across. Among the few plants Brewer collected in the grove was the scarlet Snow Plant, first found by Frémont and locally called "ice plant."

Upon Whitney's return, plans resumed for the expedition through Yosemite. They expected the trip to take five weeks if they crossed by the Coulterville Trail via Yosemite and returned by the Sonora Trail. The party leaving Big Oak Flat on June 14 included Whitney, Brewer, Hoffmann, a hired man named John, horses, and two pack mules. The saddlebags held one compass, two barometers, and other instruments, plus a few blankets, and out of deference to lightening the load, no tent and very limited culinary wares: knives, forks, and tin cups, a coffeepot, a

pot for cooking beans, a wash pan, and a pan for baking bread and frying meat.

The men went through another stand of Giant Sequoias, the
Tuolumne Grove, before dropping down a steep slope into Yosemite Valley. Passing Bridalveil Falls, they camped at the foot of a massive granite monolith, Tu-tuc-a-nu-la (today's El Capitan). They traveled up
Yosemite Valley, overcome by the glacially carved landscape and the
dashing waterfalls, the "crowning glory of the valley" being Yosemite
Falls. It took Brewer and Hoffmann over fourteen hours of scrambling
to measure the falls, more than 2,600 feet from rimrock to valley floor.
This provoked Brewer to write, "I question if the world furnishes a parallel—certainly there is none known." Today the accepted height of
Yosemite Falls is about 2,425 feet in three cascades, making it the third
highest waterfall in the world. Before the ascent of Yosemite Falls, the
men traveled south past a small lake (today's Mirror Lake) and up the
Merced River main canyon to measure two other beautiful cascades,
the lower, broad one Vernal Falls, the upper, narrow and dashing one
Nevada Falls. During their weeklong stay the surveyors dined with three
New York couples who had packed in. Later when Brewer saw Carleton
Watkins's 1861 pictures of this incomparable valley taken with his cum-
bersome view camera, he thought them "the finest photographs I have
ever seen."

On June 23, 1863, the survey men left Yosemite Valley and headed for
the high country past the two-thousand-foot sheer granitic face of Half
Dome. They climbed the canyon wall beyond North Dome and continued northward to Porcupine Flats, the next day scaling an eleven-
thousand-foot peak, which they named after Hoffmann, and hiking up
through a changing diversity of conifers: Red Fir, Lodgepole Pine, Western Bristlecone Pine, with a single stunted Limber Pine at the summit.
After a night back at camp fighting "mosquitoes with industry," they
skirted the "picturesque alpine" Tenaya Lake on June 25. Despite remaining snowbanks, many flowers were in bloom, including a striking
Crimson Columbine. The party emerged into Tuolumne Meadows the
next evening to camp at Soda Springs. From some road surveyors
camped nearby, the Whitney men were shocked to learn of the invasion
of Pennsylvania by Confederate forces. During the next morning, while
Whitney and company were on their way to Mono Pass, the definitive
battle of the Civil War commenced at Gettysburg.

Brewer and Hoffmann took a side trip east across rocks, ice, and snow
to a towering peak, of whose high-country vista Brewer wrote, "It is not

often that a man has the opportunity of attaining that height, or of beholding such a scene." This peak the two hikers named after James Dwight Dana, "the most eminent of *American* geologists." Whitney, feeling under the weather, had declined to go on the trip, but the hikers' descriptions were so vivid that Brewer returned with Whitney the next day. Whitney agreed that the view from the peak was the most spectacular he had ever seen, transcending those he remembered from the Swiss Alps. On the two trips, Brewer noted "Butterflies, flies, spiders, Coleoptera (Coccinella), Hemiptera and Bees (Yellow Jackets) and winged ants seen to the very summit," with three species of lichens encrusting the very highest rocks.

The four men awoke early on July 1, embarrassed to discover that their previous evening's campfire had started a minor forest fire. Whitney and John (the hired man) returned westward with one pack mule to Big Oak Flat to pick up more supplies, while Brewer and Hoffmann continued their survey of the landscape of peaks. The pair reached one even higher and more spectacular than Mount Dana, whose actual summit thwarted them by its inaccessible pinnacle. This they named Mount Lyell after Sir Charles Lyell, "the most eminent of *English* geologists." Mount Lyell had much more snow, including splotches of "red snow" (pinkish algae in snowbanks), and less vegetation.

On July 4 the surveyors celebrated Independence Day by riding down the Tuolumne River several miles and scaling one of the smooth granitic domes to get an overview of the countryside, in anticipation of making the region's first map. They appreciated that the shaping, polishing, and grooving of these granite mountains was the result of past glaciers. Brewer emphatically noted, "There is no glacier in this state," unaware that the huge snowbank they had hiked upon in the cirque of Mount Lyell was actually a modern glacier. Back at their Soda Springs camp, they enjoyed a holiday dinner of "preserved chicken" by a roaring fire of dead trees, careful to prevent another forest fire. Later in the evening Whitney and John returned, with ample provisions.

The next day was Sunday, a beautiful morning in Tuolumne Meadows. Brewer washed some clothes in the stream, baked two loaves of bread, wrote some letters, and attended to his botanical specimens, amounting to one hundred species collected since the group had left Big Oak Flat. Refreshed, on July 7 the men headed eastward from Yosemite among Lodgepole Pines and over Mono Pass, down the rocky trail known—for mishaps to mules and horses—as Bloody Canyon, with the forest changing to Jeffrey Pines and then Singleleaf

Pinyons. At the "American Dead Sea," Mono Lake, there were neither fish nor reptiles in the borax-laden water, but it teemed with "hundreds of bushels" of Brine Fly maggots, harvested by Paiute Indians and dried for food. Brewer sampled some of the *koo-chah-bee* "worms," finding them "oily, nutritious, and not unpleasant to the taste."

The surveyors were invited to boat out to a volcanic island with an old mountaineer, who during spring gathered the abundant gull and duck eggs for sale. Camping on the island the next morning they explored the volcanic terrain with its steaming vents, hot springs, and cinder cones. The dominant vegetation was Greasewood, with Big Sagebrush and a shoreline of tules. Then they traveled to the mining district of Esmeralda and its city of Aurora—the second most important city in western Nevada—which was originally thought to be in California but proved to be three miles inside the Nevada Territory. Booming Aurora had about one hundred men to every woman or child, and "saloons—saloons—saloons—liquor—everywhere." The group visited one of the mines, looked in at the myriads of gambling establishments, and on July 14 departed "dead broke," not from roulette or faro but because their state funds were exhausted. When the Whitney men discovered that they had two dollars left, they splurged on breakfast before leaving, only to have the remainder extracted from them at a toll gate two miles out of town.

On July 24 the party completed its six-week circuit of the central Sierra, coming down Sonora Pass through hot Sonora, where Brewer obtained some money, and back to Murphys Camp. The next day, Sunday, while Brewer wrote up notes, checked his barometer, and got the horses reshod, Whitney traveled down to Murphys from Calaveras Big Trees, where his wife and young child were spending the summer at a small hotel. Whitney and Brewer conferred about the fieldwork and their next Sierra exploration, due to commence in only two days. The plan was to proceed north over Carson Pass by Lake Tahoe to the Yuba mines, the Feather River, and eventually down to the Sacramento Valley.

Since Hoffmann proved too sick to travel immediately, Brewer rode up to Big Trees, and Whitney and he again went over the late summer's plans, repairing their barometers at leisure intervals. On returning to Murphys, Brewer sent Hoffmann to San Francisco to recruit more personnel. When he checked the evening mail, Brewer found a letter from Professor Brush and one from Professor Samuel J. Johnson of Yale College, each intimating that the next year Brewer would be offered a professorship at Yale. He was ecstatic: "If the place is offered me I shall surely accept it, gladly, and next year will turn my back to the Pacific and my face eastward and homeward once more."

The small party camped in "pretty, grassy" Silver Valley on the evening of July 31 included not only Whitney, Brewer, and John the rehired packer, but also a distinguished visitor. William Hillebrand had been director and head physician of the Queen's Hospital in Honolulu since 1851, as well as a member of the Hawaiian Kingdom's privy council and personal doctor to Hawaiian royalty. But his avocation was botany. Present-day Foster Park Botanical Gardens in downtown Honolulu began in 1855 as Hillebrand's personal garden. In 1888 his classic book on the Hawaiian flora was published posthumously in his native Germany. Hillebrand and his family were vacationing for a few months in California. Planning to accomplish a bit of plant collecting, he had even brought along drying papers.

The men headed into the higher Sierra Nevada to the new mining town of Silver Mountain, sporting forty new houses, the "Restaurant and Lodgings" still without a roof. As a backdrop to the south rose "bold, rugged Silver Mountain," with its three hundred mining claims. In town the surveyors ran into another peripatetic botanical enthusiast, the venerable Britisher Thomas Charles Bridges, fellow of the Linnean and Zoological societies of London and longtime diligent collector of Chilean and Bolivian plants for Sir William Hooker at Kew gardens. He had arrived in San Francisco from South America in November 1856 and traveled to British Columbia about 1857 for two years. After returning to California he made San Francisco his home, and continued his collecting of "Plants, Seeds, Birds, Quadrupeds and in fact any subject of Natural History as time and opportunities will permit me to do so." His collection sites included Mariposa, Yosemite, Mount Shasta, the northern Trinity and Scott's mountains, and Silver Mountain. He shipped his specimens back to his "patrons" in England, except for one shipment of seventy-five Washington Lily bulbs, which went down with the ill-fated steamer *Central America*. Unfortunately, many of his specimens lacked or had misleading dates and localities.

Bridges first resided in San Francisco on Eleventh Street near Market in Chinese House, so named because its wood came from China, and he was a frequent visitor at the new California Academy of Natural Sciences nearby. Later he moved to Oakland. Now at Silver Mountain, Bridges, like too many others, had been lured by Midas and, as Brewer reported, had discarded flora for gold-mine speculation. However, during his gold days Bridges did collect a scarlet penstemon named *Penstemon bridgesii* by Asa Gray. In April 1865, Bridges returned to botanical collecting but in September contracted malaria in Nicaragua and died at sea four days later. His body was brought back to San Francisco for burial at the Lone Mountain Cemetery, no longer in existence. His wife, Mary, donated what remained of his California collections to the Smithsonian Institution. It was said of Bridges that he "left a crowd of friends in California." At its annual meeting on January 8, 1866, the California Academy of Natural Sciences passed a resolution of respect to his memory, while William Dall provided an accolade worthy of all field naturalists: they "who bear the burden and heat of the day, are entitled to honors if not precisely of the same character as those due to the students who in their comfortable libraries work up the results of the collector, still to honors quite as high."

After Whitney visited a few of the local mines, on August 5 he and Brewer scaled Silver Mountain (elevation 10,774 feet; Brewer pegged it at over 11,000, highest in the region). It proved to be a rugged fifteen-mile hike, but the view from the top was superb: Carson Valley to the north, the mountains around Mono Lake and Aurora to the south, the expanse of the Sierra Nevada to the west, and to the east "chain after chain, extending far beyond the state line into the territories." The only trees atop the mountain were a scattering of Limber Pines. Brewer collected some of the alpine flowers and noted again, as he had at Mount Dana, hordes of brilliant red beetles among the rocks. These were Ladybird Beetles, which congregate in great numbers in high places, especially during hibernation.

On August 8 it was back to Calaveras Big Trees to plan for the next Sierra assault. Anticipating that he might not be back the next summer, Brewer ruminated that "we shall skim the cream off from the geology of the state this summer, and, as a consequence, of the sight seeing—so I will have less reluctance about leaving it." Whitney returned to San Francisco on survey business, and Brewer and John set off for Lake Tahoe, passing Silver Lake near Carson Pass. Brewer scaled another high peak south of Carson Pass, taking along his barometer, lunch bag, thermometer, hammer, level, canteen, and botanical box. From atop he had a sublime view to the north of Lake Tahoe's intensely blue water, while around him rose dozens of snowy peaks. Coming back down and crossing Carson Pass, Brewer collected a new reed grass, to be named *Calamagrostis breweri* by George Thurber, formerly with the Mexican Boundary Surveys. Heading north for Tahoe, Brewer got sidetracked by still another high peak visible in the distance, Pyramid (elevation about ten thousand feet). Feeling certain it had never been measured, he and John headed back toward the mountain, cutting across the busy Placerville Road. When John declined to climb, Brewer went by himself, the day proving so fine that he remained on top for three hours, luxuriating in the scenery.

At Lake Tahoe the surveying team camped in a nice grove of trees near Lake House and enjoyed the first quiet Sunday in a month. Brewer was amused at a conversation between two teamsters who were viewing the numerous vacationers. "What are they doing?" asked one. "Nothing," answered the other. Brewer, having seen the lakes of Switzerland, felt that the surroundings of Tahoe lacked its historical and pastoral charm, but with its deep blue water and rugged mountain setting, "its beauty is its own, is truly Californian." Speaking of being "truly Californian," Tahoe was originally named in 1853 after California's third governor, John Bigler, but during the Civil War—Bigler being, as Brewer would say, a "Secesh"—Union Californians demanded that the lake's name be changed back to Tahoe, its Indian name. In 1945 the state legislature finally made the change official.

His spirits raised at Lake Tahoe by the receipt of funds from Whitney, Brewer and party left on Monday, August 24, traveling north along the edge of the lake, on the Nevada side, then on to the Truckee River, and finally into the foothills on the west side of Lake Tahoe at Foresthill near the Middle Fork of the American River. On August 30 Brewer received an emergency message from Whitney summoning him immediately to San Francisco. Whitney wrote that the state owed the survey about $25,000, and that he could borrow no more and did not know what might happen if no funds were received within the next month. Brewer appreciated the situation. His salary was in arrears $2,800 for fourteen months, forcing him to borrow for his personal expenses. "It is infamous—political hacks get their money more regularly. We must wait, as our bills have less 'political significance.'"

On the trip to San Francisco Brewer was to have an encounter significant to the survey's future. On August 31 he rose at 2:30 A.M., rode by stage and then train to Sacramento, and boarded the steamer ready to cruise to San Francisco. Two young men, complete strangers, approached him and asked if his name was Brewer. Once assured, they explained that they had graduated the previous year from Yale's Sheffield Scientific School and this summer had traveled west across the Great Plains. They introduced themselves as James Terry Gardner (who later

in life spelled his name Gardiner) and Clarence King, close friends in boyhood as well as at Yale.

There was a grand reunion of the Whitney California Geological Survey
staff on September 1, 1863, the first since February, with much discussion
of future plans, so dependent on the fiscal vicissitudes of the state legislature. Indeed, much of Whitney's time and energy would continue to involve
confrontations in Sacramento. After the big meeting, Brewer met an eminent world traveler, writer, and mining engineer, J. Ross Browne, who had
first come to California in 1849. In the 1850s, among other official activities, Browne was appointed to investigate western Indian affairs. He wrote
official reports as well as critical articles in *Harper's,* and in 1864 his first
book on his experiences in California and the Nevada Territory would be
published, *Sketches of Adventures in California and Washoe.* After the
Civil War, when he was appointed as federal commissioner of mines and
mining, his *Report on the Mineral Resources of the States and Territories
West of the Rocky Mountains* and *Resources of the Pacific Slope* appeared.

THE NORTHERN MOUNTAINS, 1863

On September 5, 1863, it was time to execute the new survey plans.
Gabb the paleontologist went to Oregon; James Cooper the zoologist,
now reemployed by Whitney, headed to Lake Tahoe; and Whitney and
Hoffmann went to the Sierra, while Clarence King, as "Volunteer Assistant in the Geological Field-work," joined Brewer to explore the northern Sierra, the Mount Lassen and Mount Shasta regions, the Siskiyous,
and the coastal Redwood country. By the end of November the scattered
staff had reassembled in San Francisco and concluded the 1863 season.

During September Brewer and King, with John as packer, traveled
northeast through the foothills gold country from Sacramento, where
there was dramatic evidence of hydraulic mining activities. For hundreds
of yards the bluffs rising above the rivers were washed away, reducing
the picturesque landscape to "miserable, dilapidated" goldless towns—
Camptonville, Galena, and Brandy City along present-day State Highway 49. From there the pair traveled north to Quincy. On September 14,
Brewer's birthday, he detoured eastward, crossing the beautiful American Valley and then Indian Valley, which he considered "one of the
loveliest valleys in the state," and on to Genesee Valley. On this side trip
he looked for fossils said to occur in the auriferous slates of the area.
Brewer was excited when "we found them, a most important matter ge-

ologically." But this anniversary made the naturalist thoughtful: "I am thirty-five—half my 'three-score years and ten' are past." That night in camp, lying in his blanket, he "reflected on it as I watched the stars in the cloudless sky." As it turned out, Brewer lived to be eighty-two.

Returning to the foothills, within a week Brewer and King came in view of "*the* grand feature" of Lassen Peak. On September 26 the two, with three companions, were atop the volcano, where they spent five hours. The anticipated spectacle proved disappointing, the summit not being as high as expected and the weather absolutely miserable, with "a fierce wind, raw and chilly." Undaunted, the pair hiked to the top again a few days later, starting by moonlight. This time, at sunrise, the weather cold but with clear skies, the panorama was spectacular, dominated to the northwest by snow-capped Mount Shasta. Close at hand the many volcanic cones with their hollow centers reminded Brewer of "the great nests of fabulous birds." Mesmerized, the two young men stayed on the summit for over nine hours in warm sun, balmy air, with cirrus clouds high in the sky, the only animals an occasional alpine butterfly and bird "of arctic species." An unidentified sedge Brewer collected at Lassen Crater, and also at Shasta and Dana peaks, eventually was named *Carex breweri* in his honor. On the way down, encountering a lengthy snowbank, King slid a spectacular five hundred feet in fifty-seven seconds, timed by Brewer. The next day the two surveyed the volcano cones on the north side of Lassen; the largest, its top burnt red, was said to have been active as recently as 1857. This was undoubtedly Black Butte, the same cone the Beckwith railroad survey expedition had investigated in 1854, which had actually been active during the winter of 1850–51. Of this volcanic region Brewer observed that "the place is of great scientific interest. These mountains have been *thrust up* from beneath." And on every side of Lassen was evidence of past glaciers.

On October 1, 1863, Brewer and King headed north from Lassen toward the Pit (Sacramento) River east of Hat Creek, on a "route of supposed peril" because of hostile Indians ahead, the terrible "Hat Creek Indians." In three days they were at Fort Crook, having encountered only two braves, who were gathering grasshoppers for food: "Thus vanished the perils and dangers of such a trip." Six days later, traveling west in a beautiful mixed coniferous forest, they came through a pass over the east side of Mount Shasta within two thousand feet of its blanket of snow, catching a sublime view of the sharp peak cone rising over seven thousand feet above them. Continuing northwest through the Shasta Valley,

with its scattered hummocks of lava, they paused to explore a newly discovered lava cave, Pluto's Cave, with a gallery at least fifty feet high
through which basalt had once flowed. Brewer and King proceeded into
the volcano tube about a mile, encountering multitudes of roosting bats,
and near the entrance innumerable skulls of Bighorn Sheep, one of the
largest horns measuring over fourteen inches in circumference.

Reaching Yreka on the same afternoon, October 10, they camped
near town for three days, where they read letters awaiting them—fifteen
for Brewer alone and a number for King as well. From here they headed
northwest, crossed the Klamath River by ferry, and edged across the
Oregon boundary, finding much of geological interest. Autumn was in
the air, the hillside vegetation was taking on color, and storms scattered
a fringe of white on the surrounding Siskiyou summits. At Cottonwood,
their last camp, Brewer sent John back to San Francisco with the pack
animals, while he and King headed west on October 21 through Dead-
wood and Scott's Bar and on to the Klamath River, where the road became a mere trail, edging into Oregon again. Then, after selling their two
trustworthy but jaded horses, the pair arrived by foot on November 2 at
Low Divide, a copper mining community in the new Rockland Mining
District. In this area there was an extensive exposure of serpentine, and
the carnivorous California Pitcher Plant was abundant around creeks
and springs. King and Brewer examined hundreds of pitcher plant
leaves, and the intact ones all contained both water and small insects.
The naturalists recorded that the water tasted disgusting from the dead
insects and also determined that the majority of the leaves had a half spi-
ral twist to the left. King continued to Crescent City and San Francisco
while Brewer spent a week in Low Divide, finding it "a filthier, dirtier,
nastier, noisier place" than any he had yet encountered in California.
During the day he trudged over the surrounding copper country, and at
night he sat in a dirty saloon watching the men "lose their earnings at
poker."

Leaving Low Divide, Brewer walked across to the coast over ridges
of serpentine, south-facing slopes with chaparral of manzanita and silk-
tassels opposite oak woodland and scattered Knobcone Pines. Along the
coast he encountered Lodgepole Pines again, but noticed that they had
persistent closed cones, unlike the ones in the Sierra. He proceeded south
to Crescent City through "the Redwoods . . . and such forests the world
probably does not show elsewhere." The largest tree he measured was
fifty-eight feet in circumference and looked about three hundred feet
high. At Crescent City Brewer experienced a raging storm in mid-

November, the breaking surf sounding like the boom of artillery, tossing great logs into the buildings along the waterfront. Finally the steamer *Oregon* arrived from San Francisco, and Brewer obtained passage south. About midnight on Sunday, November 22, the ship passed through the Golden Gate, "a most beautiful entrance to a more beautiful harbor. The whole scene lay so lovely in the soft moonlight that I stayed on deck until we anchored in front of the city." Thus ended the geological and botanical survey of Northern California by Messrs. Brewer and King.

During the last week of November 1863, all members of the California Geological Survey were in the San Francisco office except Gabb, who showed up a week later. King and Ashburner were ready to leave for the Mariposa Estate, Frémont's vast old mining property, while Brewer took rooms with his old landlady, who now had more fashionable quarters. Brewer returned to office work, lamenting that "to leave the free open air for the confined office and bedroom, and the laborious outdoors work for writing, is a great change and is irksome."

THE SOUTHERN SIERRA AGAIN, AND NEVADA, 1864

On April 9, 1864, William Brewer awaited the arrival of the steamer *Golden City* at the docks in San Francisco. Aboard was a distinguished scientist, Benjamin Silliman Jr., another of Brewer's professors from Yale's Sheffield Scientific School, now a geological consultant planning a year's visit to mining districts in the Golden State and the neighboring Nevada Territory. "The sight of his good, jolly face," Brewer wrote in his diary, "has almost made me homesick." Brewer was delighted to host his mentor, to show him the survey offices in the Montgomery Block, to visit Telegraph Hill and Mission Dolores, and even to attend a theater in Chinatown accompanied by paleontologist Gabb.

Silliman and his secretary, Frank Semple, then left for the Bodie Mining District, just inside the California border near Mono Lake, and on east to Virginia City in the Nevada Territory. On the way back to San Francisco, Silliman stopped by the quicksilver mines at New Almadén. On May 11 at the Occidental Hotel in San Francisco he dined with Brewer and the next day may have talked with Whitney. On May 13, Whitney having sailed for the East with his family, Silliman enjoyed an excursion by ferry with Brewer and Ashburner across the bay to Benicia and back, "a very pleasant ride," as Brewer reminisced.

The 1864 field season for the California Geological Survey com-

Figure 32. An 1864 California Geological Survey party. Left to right: Gardner, Cotter, Brewer, and King. (From Francis Farquhar, *History of the Sierra Nevada*, 1965. Courtesy University of California Press, Berkeley.)

menced on May 24 in San Francisco. There had been an enjoyable get-together at a little French restaurant the previous evening, and then a visit to J. Ross Browne's home, where the writer-cartoonist drew a caricature of the surveyors, complete to their closely cropped hair, woolen shirts, and big boots. The party embarking this time (figure 32) consisted of Brewer, Hoffmann, King, his friend Gardner as volunteer assistant in the topographical fieldwork, and a new young man from Missouri, Richard Cotter, "a very good fellow," who signed on as packer although he was ignorant about packing and had to be taught on the trail.

The survey team traveled southward in scorching heat to San Jose and Gilroy, all familiar territory. Their campsite at 21-Mile House, a hostel twenty-one miles south of San Jose, "awoke pleasant memories" for Brewer, having camped there in 1861 and again in 1862. This night "sweeter sleep came than for many a long night before." Past Gilroy they swung east over Pacheco Pass and camped under the same oak as in 1862, but now the countryside suffered from a terrible drought. They cut across the San Joaquin Valley to Fresno City, on the west the "barren and desolate" Coast Ranges, while to the east the distant Sierra Nevada was barely visible, its glittering snowfields mocking the sweating surveyors.

On June 4 they reached Kings River, with good water, hay, and fine oaks for shade. Spirits were so high that the group sat around their evening campfire and sang songs. Over the next few days they moved into the mountains, eventually encountering magnificent stands of Giant Sequoias in the vicinity of today's Kings Canyon and Sequoia national parks. The trees were impressive, the largest one Brewer encountered being 276 feet tall, with a circumference at the base of 106 feet. The men enjoyed riding on horseback into the trunk of a fallen giant whose heartwood had been burned out in a forest fire. Already there was lumbering in the forest, the wood being used for posts. Brewer reflected "how it takes away the romance of them, using them for fence posts."

From mid-June into early July 1864 the surveyors worked the rugged, beautiful mountain country, hoping to discover the headwaters of the Kings, Kaweah, and Kern rivers. On July 2 they camped near "a beautiful little lake," the headwaters of a creek that would be named after Brewer. The next morning at dawn Brewer and Hoffmann tackled a high granite peak, bristling with pinnacles, visible from camp. It was a terrible eight-hour climb, but when the pair reached the summit, "Such a landscape! A hundred peaks in sight over thirteen thousand feet," deep canyons, cliffs rivaling those in Yosemite, "sharp ridges inaccessible to man, on which human foot has never trod." On the return trip Brewer and Hoffmann slid down an eight-hundred-foot snowbank, in two minutes covering a mountainous slope that had required three hours to ascend.

After Brewer and Hoffmann returned from the peak, later christened Mount Brewer, King and Cotter pleaded with Brewer to let them assault the massifs. On Independence Day the pair filled their packs with six days' provisions, surveying instruments, and blankets, and they were off for the wilderness of granite peaks. Meanwhile, to celebrate the Fourth, Brewer and Gardner climbed Mount Brewer again, planted an American flag on the summit, and left a bottle containing a note with their names and the height of the peak. In 1896, when the bottle was recovered, only one additional name had been added. Later the message was moved for safe keeping to the Sierra Club rooms in San Francisco, only to be burned during the 1906 earthquake.

King and Cotter returned from their escapade after five days, having scaled a fourteen-thousand-foot peak and from its summit spotted and named two much higher mountains, one after railroad surveyor Williamson and the other after their California Geological Survey chief,

Whitney. Cotter had gotten his boots torn off and returned to camp "with an old flour sack tied around his feet." Brewer justified the men's mountaineering with "there is nothing but love of adventure to prompt it after we have the geography of the region described."

Now the survey party headed for the canyons and peaks of the Kings River, accompanied by seven armed soldiers with packhorses and rations for a month in the mountains. This was rugged country that challenged the travelers. On one occasion Old Nell, Brewer's pack mule, fell 150 feet down the canyonside from the terrible trail but was unharmed, although a bag of flour that also rolled down the hill burst. Their descent was over three thousand feet to the South Fork of the Kings River in its picturesque gorge, swarming with trout. Between surveyors and soldiers about eighty fish were "collected." Also plentiful were rattlesnakes, four killed on one day and another encountered on a ridge above eleven thousand feet. The men feasted on trout, and thanks to Cotter's good aim enjoyed one breakfast of Blue Grouse. Brewer, so impressed by the spectacular, wild appearance of this landscape, wrote, "Next to Yosemite this is the grandest canyon I have ever seen." In 1940 this beautiful region became Kings Canyon National Park.

The surveyors found that they could not continue north through the wilderness of mountains, so they headed east into the Owens Valley, through woodlands of Singleleaf Pinyon and onto flats of Big Sagebrush and Greasewood. The Inyo Mountains to the east of the Owens Valley had had some recent mining activity, and on the evening of July 28 the Whitney group camped near Bend City, "a miserable hole" where a large city had optimistically been planned. The next day they passed through the hamlet of Independence, just a year earlier lively but now in ruins.

On August 1, after traveling north up the Owens Valley with temperatures above 100 degrees Fahrenheit, the survey party turned west to cross the Sierra on an old Indian trail. Indian signal fires were visible as the expedition advanced. Over another pass from Mono Lake the trail led down to a fork of the San Joaquin River on the western slope, and the men took side trips to scale some of the peaks. King impressed the others by hiking for twenty-six hours without food.

On September 6 the group arrived at Galen Clark's Wawona Ranch south of Yosemite. They encountered Ashburner's parents and other tourists, as well as the superintendent of the Mariposa Estate, Frederick Law Olmsted Sr., who would become one of America's most notable landscape architects and planners. Olmsted invited Brewer to go north to Yosemite Valley, where Olmsted's family had been staying. Yosemite

was "ever grand, but it was less beautiful" in the fall, the grassy meadows brown and Yosemite Falls but a trickle. The pair spent a week in the
High Sierra on the Mono Trail, enjoying "the scenery of the grand region." Olmsted proved to be a thoroughly genial companion.

Hoffmann continued to complain about a sore leg, even after a rest of
three weeks at Clark's ranch. Finally, on September 10 Brewer, King, Cotter, and Gardner carried Hoffmann on a litter to Mariposa. Eventually
he was transported to San Francisco by steamer from Stockton. Brewer
pessimistically predicted, "He is still very sick and may never recover."

After Brewer and Gardner had spent a few days at the Mariposa Estate with Superintendent Olmsted, Brewer rode down to Stockton with
Olmsted in his private carriage, which was also carrying $28,000 in gold
bullion. Upon arrival in San Francisco, Brewer received unofficial notice
that he had been elected professor at Yale—the chair of agriculture in the
Sheffield Scientific School—and was to report as soon as possible. Professor Silliman Jr. was back in San Francisco, and on October 5 for the
first time he visited the California Academy of Natural Sciences, where
Brewer still served as secretary. Silliman, recently returned from Arizona,
addressed the academy members about his Arizona fieldwork, including
observations of what he thought indicated glacial activity on some of the
mountains there. Geologist Blake, late of the Pacific Railroad Surveys,
noted that this would be the first suggestion of glaciation in that region,
adding that he had seen such evidence as far south as Tejon Pass in
Southern California. Two weeks later quite an assortment of scientists
attended a fifteen-course banquet downtown at Martin's Restaurant, including Brewer and Ashburner from the survey, Professor Silliman and
his secretary, longtime mining engineer Melville Atwood, and Baron Ferdinand von Richthofen, whom Brewer had last seen just after Christmas
in 1862. It was a congenial company, and as Brewer wrote in his journal, "The eve flew rapidly as well as joyously, and we had a most delightful time."

Brewer intended to visit the booming mining camp of Virginia City
and its famous silver mines in Nevada. But this seemed unlikely now, as
the season was late, his health not the best, the expense exorbitant, and
the Indians not at their friendliest. Nevertheless, he could not resist a
quick trip, taking the steamer for Sacramento on November 1, then a
stage to Placerville, and from there the Overland Stage over the Sierra on
a toll road, hardly wilderness anymore with hundreds of freight wagons
and plenty of hotels and stables, even at the divide.

In Virginia City the mining boom was over. During the summer of

1864, Brewer noted, "the vast bubble broke." Gold-mining stock crashed, wildcatting abated, and the population began drifting away. Brewer was taken on a tour of the Gould & Curry Mine by its superintendent, Mr. Bonner, and saw a vein several hundred feet wide. Then Brewer continued south to Gold Hill, visiting several smaller but richer mines. Brewer joined Benjamin Silliman Jr. and Baron von Richthofen, now in the mining town, for an excursion to Steamboat Springs, an extensive hot springs northwest of Virginia City. Since Brewer had received no salary for months from the survey, he could not afford his trip back to New Haven, but Professor Silliman loaned his former student and now academic colleague five hundred dollars for the journey.

During the evening of November 9, his last in Virginia City, William Brewer "took a stroll over the hills a few miles, looking at the rocks and geology of the country." The next noon he was on the stage back to California, by moonlight going over Truckee Pass, where in the near future the Central Pacific Railroad would cross the Sierra Nevada. After a lovely, leisurely steamer trip down the Sacramento River to San Francisco, he spent a few days packing and saying good-bye to many friends. On Monday morning, November 14, rising early, Brewer boarded the steamer, having "bade goodby to dear Hoffmann," the two having been together throughout California for three arduous but memorable years. At 10:30 A.M. the *America* with its 650 passengers moved out into the bay, Brewer in a stateroom by himself through the good offices of an agent he had met in Yosemite. As the ship steamed through the Golden Gate, the countryside of the Golden State was dim through the fog. Brewer calculated he had traveled 15,105 miles in California: 7,564 by horseback, 3,101 by foot, and 4,440 miles by public conveyance: "Surely a long trail!"

William Brewer hiked a few more miles on December 13, 1864, walking from Ithaca, New York, about seven miles west to his home in Enfield, a comparatively easy jaunt for the new professor.

THE SURVEY'S FINAL DECADE, 1865–1874

In its final decade the California Geological Survey shifted direction. The legislative bill for the next four years of the survey, approved on April 4, 1864, was more restrictive than the original bill of 1860, in response to criticism by certain vested interests of the survey's broad-brush activities. There was to be no more investigation of zoology, botany, or agriculture, and not even "general geology." Instead, the focus was to be on "a thor-

ough and scientific examination of the gold, silver, and copper producing districts of this State" and "such scientific and practical experiments
as will be of value in the discovery of mines and the working and reduc-
tion of ores." The appropriation was cut to $4,500 annually for the state
geologist, only two assistants could be employed, and only $5,000 per
year was allotted for expenses.

In late May, while Josiah Whitney's assistants had been enjoying the
California mountains, Whitney and his family were already back home
in Massachusetts. Whitney supervised the printing of the California Geological Survey reports in the East, since it was much cheaper than in the
money-mad West. Whitney was happy to renew at long last his involve-
ment with music and other culture, as well as his friendship with note-
worthies such as Oliver Wendell Holmes, Henry Wadsworth Longfellow,
James Russell Lowell, and especially Louis Agassiz. Whitney still served
as state geologist of California, as well as head of the new School of
Mines at Harvard, but was given a leave of absence without pay for his
California survey obligations. For the state of California Whitney was
chairman of the commission to draft plans for the state university, with
its own schools of mines and of agriculture, and for a state museum to
house, among other things, accessions from the state survey. He was a
dedicated member of the California Academy of Natural Sciences in San
Francisco and became a life member of the premier American Philosophical Society in Philadelphia. In 1863 he was one of the select group
of distinguished American scientists who established the National Academy of Sciences.

In mid-November 1864, William Brewer went to his academic position at Yale. In order to complete the 1864 survey, Clarence King and
James Gardner had returned to Yosemite over a month earlier; there they
met Richard Cotter and William Hyde, a friend and amateur artist with
whom they had come west the previous year. They hired three assistants
and commenced an intensive survey of Yosemite Valley. Since autumnal
storms were already threatening, they had to rush their survey, scaling
and sketching the great granite domes and spires, remeasuring Yosemite
Falls, naming landmarks (including Mount Hoffmann and Mount Galen
Clark), and evaluating prehistoric glacial striations.

The survey and mapping of Yosemite Valley at this time proved to be
of immense significance. During the year there had been an intensive
drive by public-minded citizens—led by Reverend Thomas Starr King,
Frederick Law Olmsted Sr., Jessie Frémont, and pioneer physician Dr.
John Marsh, supported by the California Geological Survey, and even-

tually carried to Washington by Senator John Conness—to have Yosemite Valley and the Mariposa Grove of Big Trees granted to the state of California as a preserve. President Lincoln had signed the final congressional bill on June 30, 1864, granting to California the two unique parcels of public land "for public use, resort and recreation . . . inalienable for all time." The California State Geological Survey's endeavors in Yosemite, especially those of King and Gardner, provided the topographical background for the dedication. In essence, Yosemite represented the first state park in the nation and set the stage for the eventual establishment of national and state parks around the world.

Considering the meager survey budget for the next four years, King and Gardner decided to head home, boarding the USS *Moses Taylor,* bound for Nicaragua, just before Christmas. After some lingering and philandering in Central America, the two arrived in Boston. King reported to Whitney but shortly succumbed to malaria contracted in Nicaragua and was incapacitated for weeks. Nevertheless, Whitney prevailed on the sick King to join him again in California. Eventually both King and Gardner gave in and in late fall traveled back to California with Whitney and his family.

Whitney and his men started to compile the summary reports as the Civil War came to an end. Volume 1, on the general geology of California, was massive, five hundred pages dealing with the history of the survey, the geology of the Coast Ranges, and the geology of the Sierra Nevada. Two volumes by Gabb on paleontology appeared in 1864 and 1869. By 1870 Spencer Baird and ornithologist James Cooper had completed the volume on land birds, and finally by 1876 the first of two volumes on botany by Brewer, Sereno Watson, and Asa Gray was published with private funds. This represented California's first comprehensive flora. Those in California and elsewhere who had periodically raised their eyebrows about the direction of the survey's first four years of fieldwork noticed that there was no volume on California economic geology.

By March 1874 the California Geological Survey was essentially history: as Whitney put it, "The survey has succumbed to the stupidity and malignity of the legislature, backed by the same characteristics on the part of the Governor." Whitney had been accused of giving all the survey collections to Harvard and been charged by some with orchestrating the entire survey from beginning to end on behalf of Harvard. Nonetheless, the California Geological Survey, like the Pacific Railroad Surveys and before them the Mexican Boundary Surveys, had been a proving ground for natural scientists, many of whom became leaders of the

American scientific community during the remainder of the nineteenth century and beyond. Clarence King would head up the 40th Parallel Survey, which did some work in California, and he later was instrumental in the establishment of the U.S. Geological Survey, becoming its first director in 1879.

The California Geological Survey's greatest embarrassment involved California oil. During 1861 the survey parties often encountered asphaltum, primarily in Southern California. Indeed, Whitney sent samples to a renowned Boston chemist for analysis. His report stated that California asphaltum could not be refined into a good illuminating oil. Thus ended the survey's initial petroleum interest, with Whitney stating that if any useful oil were ever discovered, it would be in Northern California.

In late spring 1864, Benjamin Silliman Jr., a consulting geologist for mining companies as well as a Yale professor, was on the West Coast for a year. He had authored in 1855 the first detailed analysis of petroleum from western Pennsylvania, reporting that it could be refined into burning oil. In June 1864, Silliman consulted for his first California oil company, E. Conway and Company, examining oil springs southeast of Santa Barbara and reporting on an "almost fabulous wealth in the best of oil." He visited the Brea and other oil prospects in the Los Angeles area and even took an excursion to Santa Catalina Island, where he did a geological survey for the island's millionaire owner, James Lick. During the summer Silliman's pronouncements on behalf of oil blossomed, as did prospectors, speculators, and investors. Soon there were "oil ranchos" from Pismo Beach to Santa Ana, and Silliman was kept busy consulting on a fee basis.

Meanwhile, having earlier expressed their doubts about the value of asphaltum, Whitney and Brewer became concerned about both their personal reputations and the reputation of the California Geological Survey. Whitney wrote, "If Silliman's reports are correct and his course justifiable, I am an idiot and should be hung when I get back to California." If his reports were wrong, then it seemed Silliman must be involved in some kind of swindle or fraud. When Brewer was interrogated, he emphatically stated, based on his own experiences, that illuminating oil could not be profitably made in California, and he recommended against investing in oil ventures there. Whitney's and Brewer's statements were given wide circulation, and the California oil boom began collapsing, especially after it was rumored that an oil company's promising analysis sample had been contaminated with a lighter oil, possibly kerosene.

Yale colleagues whom Silliman had encouraged to invest early on, including Brewer and Josiah Whitney's brother, William, lost money when the oil bubble burst, and Silliman became involved in some impending lawsuits. During 1867 it even seemed possible that Silliman might be fired from Yale or resign, much to Whitney's and Brewer's satisfaction. Furthermore, Whitney pushed to have Silliman removed from membership in the National Academy of Sciences. This unusual controversy continued into the next decade, with Silliman resigning from Yale. But he would be vindicated for his enthusiasm about California petroleum. Fifteen years after his death in 1885, three and a half million barrels of oil were produced in Southern California.

PART SIX

Institutions and Naturalists

IN 1853 THE CALIFORNIA ACADEMY OF NATURAL SCIENCES was organized as the first scientific academy west of St. Louis. With its meetings and collections, the academy attracted a growing number of residents and visitors interested in natural history and catalyzed the pursuit of scientific endeavor in the Golden State. Among its many active members was Joseph LeConte, who became the first professor of geology, botany, and natural history at the state's new University of California. Affluent academy member James Lick, a devotee of astronomy, underwrote the West Coast's first astronomical observatory. Academy member Edward Lee Greene, a field botanist and pastor filled with provocative ideas about the classification of plants, headed the University of California's first Department of Botany.

The California Academy
of Natural Sciences

It is expedient and desirable that an association be organized
for the promotion of the natural sciences.

> —Resolution at the founders' meeting of the California
> Academy of Natural Sciences

On the evening of June 27, 1821, seven dignitaries met in Sydney, Australia,
at the home of Judge Barron Field and founded the Philosophical Society
of Australasia (later the Royal Society of New South Wales), Australia's first
scientific society, modeled on the Royal Society of London. Although New
South Wales was a new British colony in a raw frontier land and the so-
called tyranny of distance isolated it from the manifold sophistications of
the home country, an early priority was the creation of a scientific society.

In the gold-booming frontier state of California, with its own tyranny
of distance, the same magical number of seven gentlemen met on the
evening of April 4, 1853, in the downtown San Francisco office of real
estate broker Lewis W. Sloat at 129 Montgomery Street. They convened
to discuss "the organization of an association for the development of the
natural sciences" for the new state, then less than three years old. Included among the seven men, five of them physicians, was Dr. John
Trask, who preceded Josiah Whitney in assessing the California geological scene and was particularly interested in medicinal plants. Albert Kellogg, another doctor interested in botany, came to California in 1849
and in 1853 set up a pharmacy business in San Francisco. As time went
on his pursuit of botany intensified, and he contributed plant specimens
to the Pacific Railroad Surveys. Dr. Henry Gibbons had a special interest in meteorology, and his accurate weather records came to be welcomed by the Smithsonian Institution. Lewis Sloat, an amateur conchologist, displayed his shell cabinet in that evening's meeting room. Dr.

Andrew Randall, journalist and naturalist, became the first president of the new academy.

In 1853 the California Academy of Natural Sciences (its name changed in 1868 to the California Academy of Sciences) became the first natural history academy in North America's Far West. As a historical sidelight, at the July 17, 1871, meeting of the academy, the visiting president of the Royal Society of New South Wales "made a few remarks about establishing relations" between Australia's Royal Society and California's Academy of Sciences.

In 2003, celebrating its 150th anniversary, this pioneering California scientific institution could reflect with pride on its lengthy list of distinguished members and visitors, among the early ones Joseph LeConte (member, 1870), James Lick (life member, 1873), and Edward Lee Greene (member, 1874).

"Scientist and Savant" Joseph LeConte and the University of California

Botany, chemistry, geology, physiology, astronomy, philosophy, and the classics were ready at his call.
—Frank Soulé

At the May 4, 1863, meeting of the California Academy of Natural Sciences, Josiah Whitney, the state geologist, spoke about the progress of the California State Geological Survey and was concerned that no provisions had yet been made to provide a permanent home for its collections. He reminded the audience of a recent and very important legislative act authorizing the appointment of a committee including the state geologist, superintendent of public instruction, and surveyor general to report to the next legislature "upon the feasibility of establishing a State University, embracing an Agricultural College, a School of Mining and a Museum, including the geological collections of the State." Whitney's address was timely, because in 1862 the U.S. Congress had passed the Morrill Act, providing land grants to states for establishment of colleges to teach agriculture and the mechanical arts. California's share was 150,000 acres of public land, and a choice site was currently available, at the base of Strawberry Canyon north of Oakland.

There was already a small college in the area, the College of California, on a small, pleasant campus near the San Antonio Estuary (the lower end of present-day Lake Merritt) in downtown Oakland, with a few picturesque buildings but minuscule funds. William Brewer in early 1863 visited it, a "small-fry affair" as he called it, but nevertheless he accepted a professorship of natural science there. When the state legislature on March 23, 1868, finally passed an act creating the University of California, the College of California offered to relinquish its buildings and land to join the new state institution if the university would encompass humanities as well as science.

The first classes convened at the Oakland campus in 1869. The university moved to Berkeley in 1873 when the first two campus buildings were completed, East and West halls (the latter still standing today). The name *Berkeley* was chosen for the university's new site by the university savants in honor of the Irish bishop George Berkeley, who came to America in 1729 hoping to establish a school for "aboriginal Americans" and who provided a model for Columbia College in New York. Reportedly John Charles Frémont, at the beginning of one of his expeditions, stood facing west in the center of the rotunda of the famous Old Courthouse in St. Louis and intoned, "Westward the course of empire takes its way," a line from Berkeley's poem "On the Prospect of Planting Arts and Learning in America" (1752).

By the time the University of California was located at Berkeley, in part a land grant institution with a strong emphasis on the natural sciences, it was an attractive prospect to potential science professors—though not to zoologist James Cooper. Before the university had moved to the Berkeley campus, the university board had suggested that the California Academy of Sciences be incorporated as part of the University of California. Cooper, the librarian of the then-flourishing academy, snorted that the academy members laughed "at the wise professors and regents of the University who kindly informed us that we must be swallowed up in that Asylum for rebel Professors or be extinguished."

Two of those "rebel Professors" were Joseph and John LeConte. John and his younger brother Joseph had applied for academic positions in 1866 when they first heard about the possibility of a new California institution. Both were born in Georgia, at Woodmanston Plantation, into the southern branch of the scientifically distinguished LeConte family. The prominent and affluent Huguenot LeContes came to America during the intolerant reign of France's Louis XIV. The father of John and Joseph, who taught his children about natural history, had a degree in medicine from New York's Columbia College. The family home had bookshelves filled with scientific books and periodicals, a chemistry laboratory, and a botanical garden visited by eminent botanists. Joseph as a youth was especially interested in duck hunting, fishing, camping, and outdoor sports. He graduated from Franklin College (the University of Georgia) in 1841, mathematics and physics having been the only meaningful courses for him there. Joining the Phi Kappa Literary Society meant participating in debates, something he definitely did not enjoy, but by the time Joseph was an upperclassman, the faculty appointed him class orator.

Two years after graduation Joseph started at the College of Physicians and Surgeons in New York City, taking time out for an expedition with his New York cousin, John Lawrence LeConte, into the Old Northwest, where his interest in geology was whetted by features of the terrain around Lake Superior. John L. LeConte, whose father was a distinguished naturalist, went on to become one of America's foremost entomologists and its greatest beetle authority, naming about six thousand species of beetles and refining the entire classification of American insects. He traveled widely in America, did fieldwork in the Rockies, and was with the Union Pacific Railroad Survey from Kansas into New Mexico. He worked under Emory in California in 1849–51, informally attached to the Mexican Boundary Surveys, making an extensive zoological collection, especially insects and reptiles. But because of limited funds his material was not included with the official boundary report. It was John L. LeConte in 1850 who called botanist Dr. Charles Parry's attention to the unusual pines north of San Diego, later to be named after John Torrey. During San Francisco's great fire of May 4, 1851, Dr. LeConte lost a collection of some twenty thousand insects. Following his death in 1883, fellow entomologist George Horn wrote, "He gave to science the results of careful study, knowing that in time whatever was worthy would be adopted."

After the Great Lakes tour, Joseph LeConte and his cousin John L. returned to the College of Physicians and Surgeons, completing their M.D. degrees in 1846. Already Joseph had become acquainted with ornithologists John James Audubon and Jacob Giraud of New York City, Spencer Baird of the Smithsonian Institution, and Professor Torrey at the College of Physicians and Surgeons. The LeConte and the Torrey families had close relationships dating back several generations. In August 1850, now married to Caroline ("Bessie") Nisbet, Joseph went to Harvard to study under Louis Agassiz, which involved a dawn-to-dark field course in Key West, Florida; an expedition to Florida coral reefs sponsored by Alexander Bache's coast survey; and finally a trip to the Catskills searching for fossils with Agassiz and New York paleontologist James Hall. In June 1851 Joseph received a degree in the first class to graduate from Harvard's new Lawrence Scientific School. All thoughts of a lifetime in medicine disappeared as Joseph began his natural science teaching career at several southern institutions, ending up in 1857 at the College of South Carolina as professor of geology and chemistry. His brother John (not to be confused with their cousin John L.) was also there as professor of physics, having earlier taught chemistry at the College of Physicians and Surgeons and chemistry and physics at Franklin College.

Life for the LeConte families during the Civil War was a demoralizing tribulation. John and Joseph at first continued teaching at the College of South Carolina. When it closed in 1863, Joseph produced medicines for the Confederate Army, then as "Chemist and Professional Assistant" for the Nitre and Mining Bureau searched for sources of "nitrous earth" for powder production. John became one of this war corporation's superintendents. After the Civil War, the two brothers, ready to abandon unsettled life in the ravaged South, heard of the new University of California and made application. Joseph's wife was especially supportive of the move, though once in California she never felt completely at home. John was elected professor of physics in November 1868, the first member of the new university's faculty, while Joseph became the university's first professor of geology, botany, and natural history the following year. John served several times in the administration, first as acting president (1869–70), later again as acting president, and then for five years as president (1875–81), while continuing as a professor of physics. Joseph LeConte would become famous as a naturalist, professor, and author.

Joseph, Bessie, and their two young daughters, together with Joseph's sister Jane and her daughter Ada, left New York for the West on August 23, 1869, on an early run of the country's first transcontinental train to Sacramento. Then they boarded a riverboat to San Francisco, and ferried over to Oakland to the luxurious home where his brother's family was house-sitting, the mansion of Charles Webb Howard, president of the Spring Valley Water Company.

In 1870 Joseph LeConte was voted a residential member of the California Academy of Sciences. That summer, after the end of his initial year of teaching, the new University of California professor made his first extended field trip to the Sierra Nevada, at the invitation of a group of male students (figure 33). He later reflected that "I never enjoyed anything so much in my life—perfect health, the merry party of young men, the glorious scenery, and above all, the magnificent opportunity for studying mountain origin and structure." The party of ten rode on horseback from Oakland east into Corral Hollow at the southern end of the Mount Diablo range, across the San Joaquin plains where LeConte recorded a Badger and thousands of horned toads (horned lizards), and into the

Figure 33. Joseph LeConte (back center) and his university excursion party in Yosemite near Yosemite Falls, 1870. (Courtesy the Sierra Club.)

Sierra to the Mariposa Grove of Big Trees south of Yosemite Valley. What particularly impressed Joseph among the Giant Sequoias was the Grizzly Giant, over thirty feet in diameter at its knobby base, its top broken off at a height of 150 feet. Ruminated LeConte, "My own top, with its decaying foliage—made me sympathize with this grizzled giant." On July 31, they enjoyed "sunrise from Glacier Point," with the vista of Yosemite Valley below like a green gem in a white granite setting. Their evening camp was beneath Bridalveil Falls, where in 1851 Lafayette H. Bunnell had been inspired to initiate the movement to preserve Yosemite Valley.

Several days later LeConte's party, during exploration of the valley, paused on the hike to Yosemite Falls to chat at Hutchings's sawmill with "a man in rough miller's garb, whose intelligent face and earnest, clear blue eye" captured Joseph's interest. The workman turned out to be John Muir, who was aware of the university group's visit. It took only a little persuasion to encourage Muir to join the party for the rest of the Yosemite trip.

LeConte and Muir had many a chat during their days together. LeConte observed that "Mr. Muir gazes and gazes, and cannot get his fill. He is a most passionate lover of nature. Plants and flowers and forests, and sky and clouds and mountains seem actually to haunt his imagination." A favorite topic of conversation dealt with the formation of Yosemite Valley, the two men agreeing that the force was glacial, although LeConte felt that before glaciation the river had created the initial valley. This idea of glacial origin contrasted with Josiah Whitney's presumption that the valley was the result of a catastrophic dropped block of the earth's crust, followed by river erosion. Whitney further insisted that no glaciers still existed in the Sierra Nevada, while Muir argued that they did. At the evening campfire in Tuolumne Meadows, Professor LeConte delivered a full-blown lecture about "glaciers and the glacial phenomena of the Sierra." On the morning of October 6, 1871, while hiking in the Upper Merced Basin near Black Mountain, Muir discovered freshly ground "glacier meal" in a streamlet below a vast snowbank, exultantly shouting, "Glacial mud! A living glacier!" He went on to find dozens more. LeConte was the first to put Muir's glacier ideas into print. Today more than fifty "glacierets" still remain high in California's mountains.

At 6:00 A.M. on August 13, 1870, the university group and John Muir went over Mono Pass from "Camp Dana" and proceeded down the east escarpment of the Sierra to an evening spot two miles from the volcanic setting of Mono Lake. The students had received their briefing the previous night from LeConte's lecture, "Salt and Alkaline Lakes." The next morning they visited the most perfect volcanic cone near the lake, where Muir took his departure, a moment of regret for all. Wrote LeConte, "I was really sorry to lose Mr. Muir from our party. I have formed a very high opinion of, and even a strong attachment for, him. He promises to write me if he observes any additional facts of importance." The two mountaineers kept in touch with each other until LeConte's death.

This first of many such trips into the mountains of the West evolved into Joseph LeConte's famous work, *A Journal of Ramblings through the*

High Sierra of California by the University Excursion Party. The young men accompanying "Doc Joe" on this memorable hegira made arrangements several years later for LeConte's journal to be published privately, and it has remained in print ever since. Nineteen years after that original excursion LeConte repeated the exact trip with another group of students. Over his thirty-two years of teaching at the university, with geology his "favorite department," Joseph LeConte fell in love with the natural world of the Golden State, especially Yosemite and the Sierra Nevada, imbuing his many students with this same appreciation.

The year following the original Yosemite excursion LeConte hosted his old teacher Louis Agassiz, now in poor health and in California to recuperate. Joseph's daughter Sallie remembered Agassiz from her childhood and was shocked at how he had aged. With LeConte, Agassiz revived, and Sallie wrote in her diary, "It was very pleasant to see Father and him talking their glacial theories over together—it seemed again the master and the pupil." The two did not agree with respect to Charles Darwin's ideas of organic evolution. Asa Gray in the East and LeConte in the West were major supporters of Darwin, while Agassiz disagreed, though less vehemently by this time. After Agassiz's death in 1873 the diplomatic LeConte suggested that it was Agassiz who had "established the laws of succession of living forms" so necessary for Darwin's theory.

Early in his California career LeConte spent much time doing geological fieldwork, including forays into British Columbia. As time passed, he became more concerned with his teaching, public lectures, writing, and his new interest in physiology. LeConte became an authority on the physiology of vision, writing a book on the subject. At the same time he produced *The Elements of Geology,* published in 1877, which became the major geology textbook used by college students for almost half a century. LeConte viewed *The Elements of Geology*—over six hundred pages in length and illustrated with hundreds of woodcuts—as more appropriate than the current run-of-the mill geology texts. Though less sophisticated than James Dwight Dana's text, it was more readable and useful for the geology student and hopefully for "the intelligent general reader."

In 1892 Joseph LeConte became one of the charter members of the Sierra Club, whose mailing address and archives were maintained by the University of California for many years. In June 1901 LeConte, together with a number of Sierra Club members including Professor Frank Soulé, who had been on the 1870 Yosemite trip, joined for another Yosemite visit. John Muir came along, as did the new chief of the United States Bi-

ological Survey, Dr. C. Hart Merriam, who had recently promulgated
the idea of life zones. The classic 1924 book on Yosemite's wildlife by
Joseph Grinnell and Tracy Storer, *Animal Life in the Yosemite,* focused
on Merriam's life zone concept. LeConte, at seventy-eight, was delighted
to be a member of the group of Sierra outdoor enthusiasts in Yosemite
that summer. He had even optimistically packed his camping outfit for
a possible excursion to Tuolomne Meadows.

On July 5, when the more energetic members of the party climbed up
to Glacier Point from Yosemite Valley, the LeContes and a few others
had a leisurely day in the valley, by coach visiting Mirror Lake, Yosemite
Falls, and finally Bridalveil Falls. That evening Professor Joe joined several of his former students and others around the campfire until rather
late to chat and reminisce, the tall, lean LeConte, in a shabby black suit,
talking in a gentle, husky voice, "his delicate hands clasped behind his
back." After finally getting to bed, LeConte complained of increasingly
severe chest pains. The next afternoon, attended by a physician and resting in bed, Joseph LeConte told his daughter Sallie that he "felt over-
powered with sleep" and closed his eyes. Wrote Bertha Chapman in her
diary, "A light mist actually settled above the granite heights, veiling the
moment of his passing."

Many hoped that this lover of Yosemite could be laid to rest in the national park, "among the monuments of the centuries with the murmur
of the trees to lull, and distant roar of the cateracts *[sic]* to sooth the
sleep," in the words of Chapman. But Joseph LeConte was buried in
Oakland's Mountain View Cemetery, his tombstone a gigantic boulder
from the "Incomparable Valley." Ina Coolbrith, soon to be named California's first poet laureate, penned the memorial poem, which ended,
"No Man was cast in gentler mould, yet stronger none in firm command.
His thoughts our lesser thought controlled, our hearts he held within his
hand." Honoring Joseph LeConte today in Yosemite near Camp Curry
stands the granite LeConte Memorial Lodge, originally built by the
Sierra Club in 1903 and marking in those days the beginning of the John
Muir Trail.

Astronomy Devotee James Lick and the Lick Observatory

If I had your wealth, Mr. Lick, I would construct the largest
telescope possible.

—George Madeira

There is a bounty of sky above the state of California—about 165,000
square miles of it. And in the early days it was easy to see the night sky,
there being few distracting illuminations and little haze in the air except
from dust, humidity, and smoke from Indian fires and occasional ham-
lets. All the early expeditions spent plenty of time observing stars using
the sextant for location, and recording comets, meteors, eclipses, halos,
and other astronomical phenomena.

Members of the U.S. Coast Survey recorded the May 26, 1854, solar
eclipse at Humboldt Bay, Benicia, and Loma Prieta, while three years
later the survey's most significant member, George Davidson, noted the
solar eclipse in San Francisco. Davidson before 1860 published two important western astronomical tracts, "Occultation of Stars by the Moon
on Western Coast of the United States" and "The Occultation of 22 Stars
of the Pleiades, and Solar Eclipse." Davidson's California career was
then just in its infancy.

There were occasional presentations at the new California Academy
of Natural Sciences about observations of the sky. On August 15, 1853,
Dr. Henry Gibbons expounded on "shooting stars" and told the audience that there was always a shower of these early in August (probably
the Perseids, the so-called dog days showers). A few years later there was
a discussion at the academy involving halos. Then on April 17, 1867, a
halo with two "sun dogs" (mock sun refractions in ice crystals twenty-
two degrees on either side of the real sun) was observed. Following an

exchange by academy members about solar and lunar halos, Dr. Gibbons stated emphatically that despite popular notions the presence of a halo around the moon was not an "infallible" forecast of rain.

By 1860 the heavens above the Golden State were being observed through a four-inch refracting telescope set up at the University of Santa Clara, the first for California. The next year at the gold-mining town of Volcano, southeast of Placerville in the foothills of the Sierra Nevada, amateur astronomer George Madeira, with his three-inch refracting telescope, spotted "Comet 1861 II" just hours after it had first been reported in Europe. At age fifteen this Portuguese American youth, interested in both geology and astronomy, traveled from the East Coast to California in 1852. By 1860 he was trekking around the state, delivering lectures on these two sciences. At one talk he gave in San Jose, a man in the audience was so taken by the subject and the speaker that he invited Madeira to his newly completed mansion and gristmill at Agnew near the Guadalupe River, to "remain for a few days" and view the heavens through Madeira's small portable telescope.

The man in the audience was James Lick, who would be elected a life member of the California Academy of Sciences in 1873. He was born in 1796 in Stumpstown (now Fredericksburg), Pennsylvania, a grandson of one of George Washington's Valley Forge veterans. Trained in cabinetry by his father, Lick as a young man sought his fortune in South America, running bullfights and building piano bodies in exchange for silver coins; he amassed wealth by converting his silver into gold doubloons. By January 1848 Lick was in California, initially intrigued by gold prospecting but soon more interested in buying land in San Francisco, around San Jose, at Lake Tahoe, and at one time on Santa Catalina Island. Along the Guadalupe River west of Santa Clara, Lick rebuilt a huge mahogany gristmill at a cost of a quarter million dollars, importing the finest equipment from Boston, to produce the "finest flour in the West." Meanwhile, he himself was living in a shanty, with a mattress on a grand piano for his bed.

As a young man in Pennsylvania Lick had fallen in love with the daughter of a rich local miller. Although miller Snavely's daughter became pregnant, he refused poor carpenter Lick permission to marry until he owned as prosperous a mill as Snavely's. Lick departed for South America with the retort "Some day I will own a mill that will make yours look like a pigsty." When the grandiose mill on Guadalupe River was completed, Lick sent a photograph of it to Stumpstown, but it apparently arrived too late to impress Snavely. Lick never married, but his illegitimate son, John Henry Lick, came to live with him shortly before James Lick's death.

When stargazer Madeira visited Lick in 1861 at his mansion near the magnificent gristmill, Madeira was undoubtedly impressed by Lick's accumulation of scientific and metaphysical books and the beginnings of his extensive garden. A few years later Lick and Madeira were together again, this time supposedly in Volcano, where Madeira now had a six-inch reflector set up. Madeira talked to Lick about some of Europe's large telescopes and what astronomers such as William Herschel and William Parsons had accomplished with their instruments. Madeira supposedly exclaimed to Lick, "If I had your wealth, Mr. Lick, I would construct the largest telescope possible to construct."

In 1871 Secretary Joseph Henry of the Smithsonian visited San Francisco. Henry had been named chairman of the Federal Lighthouse Board and was touring both coasts. On September 4 he delivered an address at the California Academy of Sciences. He concluded his lecture with comments on his activities for the board while in California and urged that "the wealth of this country should be generous in aiding and promoting science." During Henry's stay, from August 23 through September 18, he was at the opulent three-story Lick House hotel on Montgomery Street that James Lick had built during the Civil War, doing the fancy interior woodwork himself. Henry chatted with the affluent Lick, emphasizing the national needs of science, and not too subtly suggested that Lick could perpetuate his name by providing for a scientific institution just as British chemist James Smithson had (unwittingly) done for America's future Smithsonian Institution.

In 1872 Professor Louis Agassiz of Harvard visited San Francisco with the *Hassler* oceanographic expedition. At the California Academy of Sciences on September 2 he delivered an impromptu lecture praising the academy, noting that California had already made many contributions to science, and like the eastern states should continue to support "the growth of scientific study." On September 25, the eve of his departure, he addressed a special meeting of the academy, emphasizing the present aims of science. As commendable as Agassiz viewed the existence of the academy, in its isolation on the west side of the continent, he was distressed by its cramped, inadequate facilities and its empty treasury. He challenged that "I hold it one of the duties of those who have the means, to help those who have only their head, and who go to work with an empty pocket."

By 1873 cabinetmaker and entrepreneurial landowner Lick was

worth about $3.5 million, making him the richest man in San Francisco. On February 15 he signed the deed of donation to the California Academy of Sciences for a valuable lot on Market Street for building purposes, in consideration of his desire "to promote the diffusion of Science and perpetuity of the California Academy of Sciences." Two days later at the academy's weekly meeting the gift was announced, and James Lick was elected a life member.

Lick had been working with close friend David Jackson Staples on a Lick will, including some interesting items such as funding for towering statues of Lick's father, mother, and himself; a five-hundred-foot marble pyramid to be built on the San Francisco shoreline; and an allotment of three thousand dollars for his illegitimate and sometimes estranged son. Also in Lick's proposed will was half a million dollars to build a major telescope. Academy president George Davidson, who had his own personal observatory in San Francisco, reputedly had given Lick a small telescope to view the stars from his home, and Davidson himself dreamed of mounting a large telescope atop one of the High Sierra peaks he had climbed. When Davidson came to thank Lick for the gift of the Market Street land to the academy, he also commended Lick's interest in underwriting a telescope, preferably in the Sierra Nevada. Lick had known and admired Davidson for some time, and after the thank-you visit Davidson was invited for further discussions about a big telescope.

On April 1, 1873, Lick suffered a paralytic stroke and was moved from his homestead in San Jose to a room at Lick House in San Francisco, where Davidson continued to visit and discuss the telescope project. Lick was finally convinced to abandon his more outrageous proposals, and to Davidson's delight he made the commitment to invest a large part of his fortune in building a telescope "superior to and more powerful than any telescope yet made." At the academy meeting of October 20, 1873, President Davidson, after delivering a lecture on the discovery and progress of spectrum analysis of stars, announced to the large audience the Lick pledge for an observatory in the Sierra Nevada, concluding his remarks with "A thousand years hence, the James Lick Observatory, endowed with perpetual youth, will continue to unfold the mysteries of the cosmos, and to search for new worlds to conquer." Proclaimed the headlines of the next day's *Daily Alta California*: "THE LICK OBSERVATORY, THE GREATEST SCIENTIFIC WORK OF AMERICA ABOUT TO BE COMMENCED. THE NOBLE AMBITION OF JAMES LICK."

Of course, the project experienced some difficulties. Various sites were proposed: Lick suggested his downtown property (pointed out as not dark enough), his property on Lake Tahoe (too much snow, but once optimistically called Observatory Point), and Mount St. Helena up the Napa Valley. He insisted on visiting the latter site, although he was still bedridden from his stroke and the trip would involve camping. Unfortunately, the horse-drawn wagon, in which he lay on a mattress, tipped over on the steep, rutted road near the summit, throwing Lick and his mattress out. Thus ended the convalescent's enthusiasm for on-site visits.

The recommendation of Mount Hamilton by Lick's homestead foreman, Thomas Fraser, was finally chosen. It was ideal: a high mountain, yet close to home and civilization. This was the then-unnamed "high peak" east of San Jose that William Brewer and Charles Hoffmann had decided to tackle on August 26, 1861. In this first ascent by white men, the small survey party with their mules had been joined by Laurentine Hamilton, minister of the San Jose Unitarian Church. After abandoning the mules, the men covered the last three miles on foot, reaching the summit shortly after four in the afternoon. Wrote Brewer, "The view was very extensive and the day very clear." Josiah Whitney's stalwarts wanted to name the mountain after him, but Whitney declined. So this "high peak" was christened Mount Hamilton after the minister, who was also San Jose's superintendent of schools.

In the summer of 1876 Lick agreed on Mount Hamilton as the observatory site, with the provision that Santa Clara County build an access road, "first class in every particular." The eventual thoroughfare became known as Lick Avenue. Unfortunately, even as Lick lay on his deathbed, controversy continued over the disposition of his Trust, even including claims that Lick had become demented. However, when James Lick died on October 1, 1876, in his small room at Lick House, a new group of trustees—middle-class professionals rather than the earlier politicians and bankers—was at his side. Flags flew at half-mast in San Francisco as Lick's body lay in state at Pioneer Hall, while citizens filed by the open coffin. On October 4 the funeral cortege, the hearse drawn by four black horses, proceeded to the Mechanics Pavilion for funeral services and temporary interment in the Masonic vault.

A decade later, on January 10, 1887, Lick's casket was borne by the Mt. Hamilton Stage to the top of Mount Hamilton. The next day the coffin was placed in a brick vault with a stone cover beneath the iron pier prepared to support the great Clark thirty-six-inch reflector telescope that had arrived on December 27, 1886, a beautiful day. The installation of the largest reflector telescope in North America was completed in

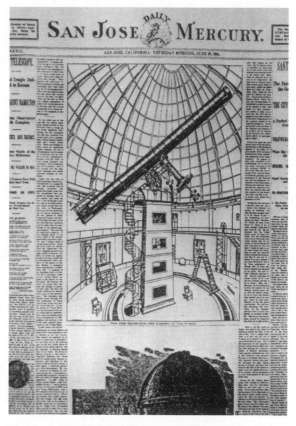

Figure 34. The Lick Observatory dedication reported
on the front page of the *San Jose Daily Mercury,* June
28, 1888. (Courtesy California Academy of Sciences.)

1888 (figure 34), replacing the twelve-inch instrument first placed on the
mountain summit, and the Lick Observatory was donated to the University of California. By its fourth year the Lick Observatory had reported the fifth moon of Jupiter, the first new Jovian satellite since
Galileo and Simon Marius saw four moons in January 1610. The Lick
Observatory was to add more Jovian moons in the future, as well as en-
abling manifold other astronomical discoveries.

James Lick also made contributions in the discipline of horticulture.
In those early days at his mansion near Agnew, he worked all day with
his garden and exotic fruit trees. He obtained rare flowers and shrubs
from around the world, as well as eucalyptus trees from Australia, im-

ported with Australian dirt in which to plant them. On one occasion, to test the obedience of a new group of gardeners, he ordered them to plant trees upside down! Bones turned up on his ranch were ground up to use for fertilizer. After years of flooding by the river, Lick moved to his land in San Jose, where he built Lick Homestead and put in three large hot-houses for plants brought from Agnew. Although Lick was frequently somewhat short with people, his biographer, Willard Farwell, said of the millionaire that "with his plants he was gentle and tender. His days were spent among them, hovering over the plants like they were capable of love. The flowers bloomed luxuriantly, as if to thank him" (quoted in Wright 1987, 5). Later in life he ordered replicas of the white conserva-tories at Kew gardens near London for the Lick Homestead, but they were never erected there. After Lick's death they became the impressive centerpiece for the new Conservatory of Flowers in Golden Gate Park (rebuilt after a fire four years later). This 1878 Victorian conservatory edifice remains the oldest building in San Francisco's Golden Gate Park.

The total amount of seven hundred thousand dollars was awarded from the Lick estate for the Lick Observatory on Mount Hamilton, a lasting mountaintop memorial for the amateur stargazer. George David-son, scanner of the skies in his position with the U.S. Coast and Geodetic Survey, missionary for a California observatory, and the scientist who gave James Lick his first telescope, served as honorary professor of geo-detics and astronomy at the University of California from 1870 to 1905.

Edward Lee Greene

The University of California's
First Bona Fide Botanist

A botanical problem attracted him, and he rode at full tilt like
a medieval knight.

—Willis Jepson

In late July 1883 an unorthodox man of the cloth executed an ecclesiastical assault in Berkeley, California. Locked out of his church by a disgruntled vestry because he was injecting too much Catholic doctrine into his ministrations, Reverend Edward Lee Greene, with a small band of zealous supporters, charged down Bancroft Way to St. Mark's Episcopal Church. With the axe he had been carrying over his shoulder, Greene smashed the padlocked doors and stormed into the church. Donning cassock and surplice, he intoned morning prayer, two long passages from Job, selections from the Psalter, and an unorthodox sermon to a growing audience. After the Litany and Ante-Communion, he climbed back to the pulpit and delivered another sermon. It was Reverend Greene's last presentation for the Episcopalians. He was subsequently to march north across Bancroft to become the first instructor in the University of California's new Department of Botany.

Edward Lee Greene, born in Hopkinton, Rhode Island, on August 20, 1843, was early imbued by his gardener mother with an interest in nature, and he never forgot the potted scarlet geranium on his windowsill that soothed him as a peevish and fretful baby. At six among his earliest primers was Mrs. Almira Hart Lincoln Phelps's *Familiar Lectures on Botany,* which had a strong religious slant. After a brief stay in Illinois, the family moved to Wisconsin. Greene entered Albion Academy in 1859 and became acquainted with Dr. Thure Kumlien, professor of natural history, a graduate of Linnaeus's University of Uppsala in Sweden and

the Old Northwest's most famous naturalist. At the end of Greene's junior year, in 1862, he began serving in the Union army, together with two
brothers and his father. He came to detest the war for two good reasons:
the horrible bloodshed and "no chance to collect anything." Actually, he
did collect ferns and other plants, as well as study birds, along the Cumberland River in Tennessee, carrying extra army clothing in his hands
while his knapsack held plant specimens and Alphonso Wood's *Class-
book of Botany*. His flowers from the Fort Donelson battlefield were
later displayed at a Chicago fair. After the war Greene returned for his
final year at Albion Academy, graduating in 1866.

Greene taught in rural Illinois schools for three years, and though a
Baptist he occasionally preached in Methodist churches. In 1869 he published his first botanical article, four paragraphs on white and double
flowers, in the *American Naturalist*. His last publication, number 565,
appeared posthumously sixty years later in the first volume of California's new botanical journal, *Madroño*. Greene had written it years earlier but never published this article extolling the state flower of California, the California Poppy.

Greene became excited about western botany by reading the Pacific
Railroad Surveys with Professor Kumlien and subsequently corresponding with George Engelmann and Asa Gray. On April 8, 1870, Greene
arrived at Denver in the Colorado Territory by stage, to discover that
springtime had not yet arrived in the Rockies. With the swirling snow or
blowing dust, he initially did well to find a dozen plants worth collecting, and the only plains trees were cottonwoods and willows. He de-
jectedly wrote to Kumlien that he "should never like this country for a
home." When the blooming season finally arrived, Greene's opinion of
the region improved, but now there were the Indians to contend with.
Up one foothills valley he ran into a band of well over a hundred, led by
an imposing chief who demanded to see the bag Greene was carrying.
When Chief Colorow, a feared Mountain Ute, looked in the bag and saw
nothing but plants, he snorted in disgust, "Ugh! Medicine Man." The
six-foot botanist was left to go on his way, but not before he challenged
and threw to the ground one Colorado brave, and himself was thrown
by another.

Greene, always religious, visited in Denver with Episcopal bishop
Randall and decided to become an Episcopalian. Encouraged by Randall, he entered the Collegiate and Divinity School of Jarvis Hall in
Golden, Colorado, as a divinity student and instructor of botany. By the
spring of 1871 he was teaching botany at Wolfe Hall, an Episcopalian

girls' school in Denver, and reputedly gave the school one thousand plant specimens, half of which he had collected in Colorado during the past year. Between summer teaching stints, he spent ten days on a botanical foray in Colorado's high country. Early in September, then, Greene became deacon at Greeley, Colorado's Trinity Church, extending his peripatetic preaching and botanizing to include Wyoming.

Eastern botanists began requesting Rocky Mountain specimens from Greene. When in August 1872 Asa Gray of Harvard attended the dedication of Grays Peak, named for him by Dr. Charles Parry, Greene was invited to join the party. On August 12, Parry, Gray, and Greene, accompanied by some "unscientific gentlemen" and Mrs. Gray, hiked to the summit. The adjacent peak honored John Torrey, but Torrey was too infirm for such an adventure. However, by accident Greene ran into Torrey six weeks later at Denver's railway station. Greene thought he recognized the old botanist by the plant press he was carrying. When Greene asked the elderly gentleman's name, the venerable botanist replied, "I am called Dr. Torrey." Torrey and his daughter were bound for the mountains, where Maggie would climb Grays Peak; Greene accompanied them on the train.

Ordained in January 1873, Reverend Greene moved to St. Peter's Church in Pueblo, on occasion holding services elsewhere, including nearby Canon City, where another botanist, Townshend Brandegee, was the county surveyor; like Greene, he later became a California resident. In July, as Greene headed for a mountain vacation, the Pueblo newspaper enthusiastically acclaimed, "The reverend gentleman is among the leading botanists of the country." During this period, when Professor Kumlien inquired if Greene were married, he responded, "It seems with my preaching and botanizing I have enough to keep me busy and happy. Happy as any mortal ought to expect to be." He remained a bachelor throughout his lifetime, plagued by veiled allusions of "sexual perversion."

In February 1874, Greene was assigned to the parish at Vallejo, California, and by mid-April was overwhelmed by the new plants he had added to his collection. He visited the California Academy of Sciences, comparing his specimens with those in Dr. Albert Kellogg's herbarium, and in May he was elected a resident member. By September, however, despite a good salary and a large congregation, Greene concluded that his new parish was "too much for me." In particular, demands of pastoral work interfered with his "scientific studies." Faced with the death of his mother in Colorado and his father's declining health, Greene returned to Colorado in April 1875. During that spring and summer he lived at the mining camp of Georgetown, near Gray and Torrey peaks, preaching there and at nearby Empire. Then in October the Northern California district bishop called for Greene's return, this time to Marysville, up the Sacramento River. The move failing to materialize because of disagreements, Greene spent another winter at Georgetown, worked over his plant collections, added to his botanical library, and rationalized that "I ought to be very happy."

During March 1876, Greene was called to an isolated church at Yreka in Northern California, where with free time on his hands he relished his floral excursions into the nearby Klamath Mountains. His local discovery of invasive European weeds resulted in his first published article on California plants. But in late January 1877 he departed on an extended junket to San Diego, then east by foot to Yuma through the Colorado Desert. Missionary Greene preached several vigorous sermons in Yuma, Arizona Territory, where the newspaper informed its readers that "you will like him and will find that he is no chicken-eating bummer." He continued into the heart of the Southwest where he remained for some years, especially in New Mexico. Though a minister, he was not so staid as to turn down a mug of mescal. On one occasion when he was thirsty, he accosted other itinerant missionaries and asked for a drink. They had apparently been warned about this crazy six-footer, who carried nothing but an umbrella and a "huge book" (his plant press).

In 1881 Reverend Greene was delighted to receive a call to serve as rector at St. Mark's Episcopal Church in Berkeley. Asa Gray was not excited about this move, writing that "I had rather have you in some vast wilderness." The University of California was nearing the end of its first decade and still had no botany department, such subjects being taught in the College of Agriculture by its head, Dr. Eugene Hilgard, and by a quartet from the California Academy of Sciences: forestry by Henry Bolander, field botany by Walter Gibbons and Albert Kellogg, and lower plants by H. W. Harkness. Charles Bessey, a distinguished visiting lecturer from Iowa State College, taught economic botany. In 1882 Greene reputedly began teaching botany part-time, and on May 7, 1883, he was again elected a resident member of the California Academy of Sciences. At the academy herbarium, initially working amiably with Mary Curran, he was appointed joint curator of botany. His stint as rector of St. Mark's survived three years of turmoil, at which time he resigned to become a Roman Catholic. Meanwhile, for the 1885–86 school year Greene was hired as "the first strictly botanical appointee in the history of the University of California" at seventy-five dollars a month, an instructor with no scientific training in the field but with plenty of ideas of his own.

Greene approached his teaching with enthusiasm and also continued his study of the California flora with extensive collecting trips throughout the state, including to offshore islands. By 1887 he had published landmark papers in the *Bulletin of the California Academy of Sciences* ("Studies in the Botany of California and Parts Adjacent"), considered to be "the first noteworthy contributions to western botany published in the west by a resident botanist" (Bartlett 1916, 164). He became a full professor in 1893, recognized as the foremost West Coast botanist, though not always appreciated. Asa Gray tended to be a conservative classifier, while Greene was a vigorous splitter of species. His on-and-off friend Kate Brandegee chided that "not more than one in ten" of his new species was tenable. In Gray's caustic opinion, Greene should be taking "more time for elaboration" and have "less confidence as to specific distinctions, and a more restrained judgment about genera." The eastern botanists may also have frowned on what Willis Jepson called Greene's "successful struggle to assert the rights of independent investigation and publication on the part of western botanists," and his habit of identifying plants by smelling them. But after Gray's death in 1888, some viewed Greene as the leading American plant taxonomist.

In 1890, when the College of Natural Sciences was established at the University of California, the Department of Botany came into being, with Greene as head. Under him the staff increased, with some valuable additions. Greene initiated a botanical journal, *Pittonia,* and later founded the serial *Erythea,* both of which were filled with Greene's prolific articles, ranging from "A Botanical Excursion to the Island of San Miguel" to "Dates of Publication of Nuttall's Compositae." Though a botanist of the old school, Greene developed new ideas about the classification of plants and in 1892 was one of three American botanists named to the International Committee on Botanical Nomenclature. In the University of California president's biennial report for 1893 it was noted that "in the department of botany much good work has been done with the students and still more in classification and correspondence." In this year Greene was elected president of the International Botanical Congress by acclamation, and the next year he received an honorary doctorate from Notre Dame. In part, Greene's ideas about classification were the result of his religious beliefs, since he firmly accepted the biblical concepts of special creation and fixity of species, rather than evolution.

To students Greene was a striking figure: tall, with ruddy features and

Figure 35. Professor Greene's botany laboratory and herbarium in the University of California, Berkeley's South Hall, 1893. (Courtesy University and Jepson Herbaria, University of California, Berkeley.)

a shock of white hair, articulate, somewhat autocratic, ordinarily in good humor but with a wit that could bite. He would commence his university day about six, lecture for general students before nine, retreat briefly to his book-lined study, then return to work in his herbarium lab (figure 35) with advanced students and staff, many of whom later became famous. After lunch Greene pursued botanical research of his own, and by midafternoon was off to the newly developed Garden of Native Plants below North Hall. Weekends he headed for the field, often accompanied by a favorite student. But his tour at Berkeley was not always tranquil, especially his relations with the administration.

During the first semester of the 1894–95 school year Greene took a leave of absence to visit the great botanical sites and herbaria in Europe. Upon his return, he handed in his resignation, to be effective at the end of that academic year. The *San Francisco Call* mourned his departure from the university as "Berkeley's great loss." Collecting together his vast personal library and herbarium, not to mention appropriating some staff members' specimens, Greene left for Catholic University of America in Washington, D.C., as professor of botany, later becoming an "honorary figure" at the United States National Herbarium of the

Smithsonian Institution. In 1914 Greene, disaffected, accepted an invitation to move with his herbarium to Notre Dame, where he was promised a lifetime annuity, lodging, and publication of everything he might write.

While many of Professor Greene's important botanical publications were inspired by his California experience, at the Smithsonian he turned his interest and talent toward writing *Landmarks of Botanical History,* about the history of botanical endeavor from its earliest days. The first volume appeared in 1909 as Smithsonian Miscellaneous Collection no. 54. The manuscript for the second volume was almost completed when he died but was not published until 1983, in an edition including both volumes. After writing more than eight hundred pages, Greene was just barely through the seventeenth century. Left unwritten were two additional volumes on the history of British and American botany. The Smithsonian initial volume was long hailed as having "no rival as a source of information on the development of botanical thought and botanical practice in Europe" (McVaugh 1983).

Edward Lee Greene died in his sleep in Washington, D.C., on November 10, 1915. During his lifetime he insisted, against much criticism, that he had described more than three thousand new species. Today, modern taxonomic techniques are increasingly supporting some of his claims. As far as Greene the teacher is concerned, more than forty years after sitting in Greene's classroom at Berkeley, Dr. Morris Carpenter, visiting Professor Jepson's office on the Berkeley campus, spied Greene's portrait and exclaimed, "Ah! There is old Greene!" Reflected Jepson: "I am sure few worthwhile students ever forgot him."

PART SEVEN

The Postwar Naturalists

AFTER THE CIVIL WAR, California increasingly was a mecca for individual naturalists from around the world, some who just visited and others who became permanent residents. Drawn by the uniqueness and diversity of California's natural bounty, visitors repeatedly pursued the natural sciences throughout California. Notable among them was Charles Christopher Parry, who—though he made his living as a doctor in Iowa—frequently traveled the state alone and with other visiting botanists. At the same time California acquired an increasing number of naturalists calling the Golden State home: individuals like John Muir, zoologist James Cooper, and the two naturalist couples John and Sarah Lemmon and Kate and Townshend Brandegee. As residents they did their field collecting, research, and writing, as well as making contributions to natural history institutions and activities within California. But many discovered that the naturalist's life, despite its many attractions, could not provide an adequate livelihood.

The Latter-Day
Peripatetic Naturalists

THE ULTIMATE PERIPATETIC:
PARRY AND HIS NATURALIST COLLEAGUES

A familiar figure to hunters, prospectors, mountaineers,
and all sorts of outdoor people . . .

—C. H. Preston

During the February 17, 1868, meeting of the California Academy of Sciences there were two presentations about ongoing western surveys. One was by geologist Clarence King, who now headed the government's 40th Parallel Survey. In the winter of 1867–68 King's survey men had stayed in Nevada, the geologists in boomtown Virginia City and the rest of the scientific staff in Carson City. Although the previous field season's discoveries needed writing up and the collections needed organizing, King took time for a quick trip over the snowy Sierra to San Francisco. At the academy meeting—where his former superior, Josiah Whitney, presided as president—King presented a résumé of the survey's accomplishments. Also on the program were two of his former colleagues, geologist James Blake and zoologist James Cooper. But the main lecture was by one of the academy's earliest corresponding members, who had just returned to California with another survey expedition after a sixteen-year absence: Dr. Charles Christopher Parry (figure 36).

From 1849 to 1851 Parry had served in California as assistant sur-

Figure 36. Dr. Charles Christopher Parry. (Courtesy
University and Jepson Herbaria, University of California, Berkeley.)

geon, as well as chief botanist and geologist, with the Mexican Boundary Surveys. He then returned east to work with John Torrey on those
surveys' scientific report. In 1861, leaving his medical practice in Davenport, Iowa, Parry visited the Colorado Rockies, where he botanized on
and off for a number of years. During the summer of 1861 Parry named
some of the Front Range's high peaks after naturalists—before they
could be named for politicians. There was James Peak, after Vermonter
Edwin James, first geologist-botanist to climb a fourteen-thousand-foot
peak in North America (Pikes Peak) and collect the alpine flowers; Engelmann Peak, after the great St. Louis botanist George Engelmann;

Mount Guyot, after geologist Arnold Guyot of Princeton; and Mount Flora, after Colorado's wildflowers. Parry especially selected a pair of fourteen-thousand-footers on the Continental Divide southwest of Denver, naming the higher one after Harvard's Asa Gray and the slightly lower one after John Torrey of Columbia. In the summer and autumn of 1862, Parry energetically pursued the Colorado flora, particularly the high country species. He climbed Pikes Peak, on whose flank he collected what has become one of the world's most famous ornamental conifers, the Blue Spruce; it eventually became Colorado's state tree. Two years later Parry returned, accompanied by ornithologist Dr. Jacob W. Velie, and climbed a peak named for Parry by the Colorado Territory's surveyor general, Francis M. Case. Hiking up a high, dominant, unnamed Front Range mountain, Velie suggested that Parry name it after John James Audubon.

Parry was to collect more than eighty new species from Colorado, but he often neglected to record important collection data for his specimens, such as specific localities and dates. However, he did relish writing narratives about his western botanical activities, many published in his hometown Davenport newspaper and other daily and weekly newspapers across the country.

In February 1868, Parry was back in California, becoming the only frontier naturalist to have experienced the Golden State during the Mexican Boundary Surveys who would now be a regular visitor for more than two decades. His academy lecture dealt with his experiences in 1867 as botanist in a surveying party for the Kansas Pacific Railroad. He presented "an account of some of the natural features along parallel of latitude 35 degrees, and the probabilities of coal being found on it," reflecting his expertise in both botany and geology. This survey was supported by railroad entrepreneur General William Palmer of Pennsylvania. The proposed route Parry traveled started in eastern Kansas, came into Colorado along the Arkansas River, then turned south into New Mexico, and eventually entered California. In addition to Parry as botanist, Palmer's party included geologist Dr. John LeConte, then a professor of natural and mechanical philosophy at the University of South Carolina. The next year LeConte was to become the first faculty member at the new University of California. Dr. Lewis served as entomologist, while Dr. William Bell was invited by LeConte as surgeon.

Palmer's survey team had commenced in the spring of 1867 and reached Fort Mohave on the east bank of the Colorado River in midwinter. They spent several days there before crossing the river downstream at Needles, continuing west through the Mohave Desert to the San Joaquin Valley and proceeding north to San Francisco. Some distance beyond Needles they crossed an extensive salt flat, named after Parry but misspelled as "Perry Sink." When the Palmer party arrived in San Francisco, Dr. Bell, who had preceded them, reported that the men were "five of the shabbiest-looking fellows I ever saw. Their coats were torn, their caps washed into shapeless mushrooms of felt, their faces tanned and branded, and their figures covered with mud." Parry had collected fewer than two dozen plants in California, many of them near the Colorado River, whereas in Kansas Parry's rambles, as Bell wrote, were "among the prairie flowers, and very beautiful were these heralds of spring, all quite new to us, although well known to that experienced western traveler."

After Parry's academy presentation on the survey, his paper, "Botany of the Region along the Route of the Kansas Pacific Railway, through Kansas, Colorado, New Mexico, Arizona, and California," eventually appeared in General Palmer's official report and in William Bell's *New Tracks in North America* as an appendix. Parry's treatise included ten pages on plants he had collected, many of which eventually went to the British Museum.

Starting in 1869, for three years Parry served in Washington as a botanist for the U.S. Agricultural Department, whose new building next to the Smithsonian Institution housed the U.S. National Herbarium. Professor Torrey had recommended Parry for the position, which primarily involved arranging specimens collected over the years by the various government expeditions. In 1871 Parry made an official visit to England's Kew gardens, where he developed a close friendship with director Sir Joseph Hooker, son of William Hooker, and took time to visit his own Gloucester birthplace. Then without explanation Parry was dismissed from his Washington position, much to the shock of the botanical community. Apparently the cause was some conspiracy in the commissioner of agriculture's office.

In 1872, back in the Colorado mountains, Parry escorted a large party of dignitaries, including Asa Gray and his wife and botanist Edward Lee Greene, to the summit of Grays Peak for the formal dedication of Grays and Torreys peaks (Torrey was too infirm to participate). The next year Parry joined a reconnaissance under Captain William Jones of the Army Corps of Topographical Engineers to the Yellowstone country, followed by some botanizing in Colorado with Greene. During 1874 and 1875 Parry explored southwestern Utah, pursuing Frémont's earlier footsteps.

PARRY, EDWARD PALMER, AND
LEMMON IN SOUTHERN CALIFORNIA

By the end of 1875 the peripatetic Parry was again back in California. After a visit to the California Academy of Sciences, he informed Asa Gray that the herbarium there included a number of interesting new plants but the provincial staff was reluctant to share the specimens with eastern savants. About this time Parry became acquainted with plant collector John Lemmon of Sierraville, north of Lake Tahoe, with whom he had corresponded. Lemmon was one botanist who actually overwhelmed Gray with western specimens. Parry wrote to Spencer Baird that he was on the track of some Indian mounds he planned to excavate and looked forward to a visit the next spring by Edward Palmer, another peripatetic English naturalist, who was spending the winter at St. George, Utah. "Doctor" Palmer's medical prowess was derived from dispensing medicines and caring for the sick aboard the *Water Witch* with the La Plata Expedition in South America, and from casually attending the Cleveland Homeopathic College. Palmer's avocation was collecting plants and animals. For several months in 1861 he worked as zoologist James Cooper's assistant with the Pacific railroad survey along the California coast. Then during the Civil War he served as an army surgeon. When Parry started as U.S. National Herbarium curator in 1869, Palmer was employed as a general agricultural explorer for the Agriculture Department, the Smithsonian, and the Army Medical Museum.

In mid-March 1869, Palmer left Kansas for the Southwest on a collecting assignment. He was at Fort Mohave on the Colorado River by August 10, went down the Colorado by steamer with a stop at La Paz to ship two boxes of plants and bird skins, and then continued on to the mouth of the Colorado River, where three more boxes were shipped and apparently lost. Parry worked with some of Palmer's Colorado River plants in Washington, noting that there were "several nice parcels of plants from the mouth of the Colorado River."

The next year the Agricultural Department charged Palmer to "make collections of Natural History in the Western Territories of the United States." Palmer returned to the Southwest, collecting along the Colorado River and across the desert to Los Angeles. Again Parry worked with Palmer's desert specimens at the National Herbarium, but since Parry preferred boreal species, he commented curtly that "I do not find much of interest in Palmers [sic] last set the Alpine plants are very meager."

For several years thereafter, Palmer worked in Washington for

Spencer Baird, who was now assistant secretary of the Smithsonian as well as U.S. commissioner of fish and fisheries. Much of Palmer's assignment involved collecting and preparing marine invertebrates and vertebrates. In the autumn of 1874, after reading a magazine article about the isolated island of Guadalupe off Baja California, Palmer became obsessed with the idea of being the first naturalist to make collections there. He was in San Diego by mid-January 1875 and on Guadalupe by about the first of February, with provisions for a six-week stay. When the boat scheduled to pick him up did not appear, he and a few starving companions lived on goat meat and the sparse vegetation until finally rescued in late May by worried friends. Yet Palmer managed to collect several sets of about 140 different plants for distribution in America and Europe. Palmer stayed in the San Diego area and Baja, finally going east to St. George, Utah, for a season of archaeological work, where his field investigations were the first for Pueblo ruins, and his artifact collections, in the view of William Holmes, the first "of importance known to have been made by exhumation" (McVaugh 1956, 68).

During 1876 Palmer returned to Arizona and California to complete his collection of ethnological artifacts from the Indians along the Gila and Colorado rivers. He visited the Mohave Reservation, shipped those collections back, and made a plant-collecting excursion into central Arizona. In spring of that year, he headed west into California, taking the stage along the Mohave River route and arriving at Crafton east of San Bernardino in late May. Here he met Parry and quickly accepted Parry's invitation to join a grandiose expedition conjured up by Parry and John Lemmon the previous year. This enterprise would commence with a botanical exploration of the San Bernardino Mountains and end up on the Columbia River. Trip headquarters were established at Crafton Ranch near today's Crafton Hills, which was sheep country in 1876. Here Mr. and Mrs. Myron Crafts had started a resort, one of the few local places to accommodate tourists. At nearby Zanja Creek Parry had already discovered a curious little plant that proved to be a new species of spineflower, *Chorizanthe parryi,* a buckwheat then common on the rocky plains but today listed as potentially endangered.

The threesome never made it to the Columbia River, but they did collect in the San Bernardino Mountains and into the desert from late spring into summer of 1876. Their first venture was east over San Gorgonio Pass to the Mohave around Agua Caliente (Palm Springs). On this trek Parry's Beargrass, *Nolina parryi,* was discovered on a dry, rocky ridge northwest of Agua Caliente. On May 29 the three botanists assaulted

San Bernardino Peak with eleven other hikers. Parry named a new lavender plant from here after one of his fellow hikers, *Collinsia childii.* Later the naturalists went over Cajon Pass and collected into the desert as far as the Mohave River. Gray named a Parry specimen from this trip, a showy little purple relative of gilia, *Linanthus parryae,* after Parry's second wife.

On one of the trips in June, Palmer was thrown from his horse and injured, temporarily putting an end to his arduous fieldwork. He was finally able to make his way by steamer to San Luis Obispo, whence he leisurely collected plants on his inland trek to San Francisco. Meanwhile, Parry and Lemmon continued their survey of the San Bernardino Mountains, finding several new species that were subsequently named after Parry. On a hike up Edgar Canyon, Parry discovered the exquisite Lemon Lily, *Lilium parryi,* now rated as rare because of loss of habitat. Incidentally, Lemmon would shortly advertise Lemon Lily bulbs for sale.

In July 1876 Parry made a collecting trip north into the Sierra, and then finally returned to Davenport. Here he discovered Palmer, awaiting assistance in working up over 650 specimens from the Southwest. About half were from California, especially from the Mohave Desert, the San Luis Obispo region, and the Redwood country around San Francisco.

In 1877 Parry remained back east but wrote Lemmon about how to plan a field trip for the California visit of Asa Gray and Sir Joseph Hooker in the autumn. Meanwhile, Palmer spent much of the year in southwestern Utah collecting plants, forty-one species of butterflies, and Indian artifacts for Harvard's Peabody Museum of Archaeology and Ethnology. Then in 1878 Palmer and Parry carried out a very productive excursion into Mexico to Mexico City, and particularly to the silver-mining country around San Luis Potosí.

PARRY AND ENGELMANN VISIT THE GOLDEN STATE, 1880

In 1879 Parry's father, at home in upstate New York near Lake George, became ill, and Parry traveled east from Iowa to be with him. The senior Parry died, but Charles Parry, his annual calendar disrupted, made no extended field trips during the remainder of the year. However, Parry planned to visit California again, the next time, he hoped, with his long-time friend George Engelmann of St. Louis. Parry wrote to Engelmann in January 1880 that the Southern Pacific Railroad was being extended eastward in Arizona from Yuma to Maricopa, reportedly to connect eventually with the Santa Fe Railroad. Parry urged that "before they became too old, the two of them should go over that interesting travel route together." But Engelmann's son was ill and his wife, Dorothea, was in poor health, so he did not reply.

Later that year, after Engelmann's wife had died and his son's health had improved, Engelmann answered Parry's second letter, which related progress of all the western railroads and included the comment "Pity we were not young again." Parry was near sixty, while the venerable Engelmann of St. Louis, the botanical gatekeeper for the West, was just past seventy. Parry proposed that Engelmann join him on the train to Salt Lake City, travel up and down Utah so Engelmann could settle some questions about what would become a new species of fir, the White Fir, and then explore the West Coast.

After Parry and his wife had visited Engelmann in St. Louis in March 1880 to excite the old botanist about the junket, Engelmann wrote to Gray that he was seriously considering the California and Oregon trip "with Parry." The decision was made. Parry, on a visit to the East Coast, chatted in New York with paleobotanist John Newberry of the Pacific Railroad Surveys about good botanizing places in Oregon and made arrangements for complimentary railway passes on the Union Pacific line. He later was promised passes for traveling in California from railroad magnate and former governor Leland Stanford. Parry visited Harvard's Arnold Arboretum, where he talked to Director Charles Sargent, learning that his projected junket tied in nicely with Sargent's preparation of the government's first official forestry report for the tenth United States Census. Sereno Watson, curator of the Gray Herbarium at Harvard, who was going by himself to the Northwest, would later join Parry, Engelmann, and Sargent, while Edward Lee Greene and John Lemmon might meet them in California. The trip was becoming a botanists' extravaganza.

In mid-June 1880 Engelmann joined Parry in Council Bluffs, Iowa, to head west. Engelmann wrote to Asa Gray that "I could perhaps not have gone without such a trusty friend as Parry." Arriving in Salt Lake City, the pair met the somewhat eccentric and egotistical field botanist Marcus Jones, whom Engelmann judged to be "a good fellow who will learn and improve." Jones had corresponded with Parry at Davenport in the early 1870s about identification of some plants Jones had collected. Although Jones had never met Parry, he had recorded his opinion that Parry "was a self-advertiser who kept his name before the public as much as possible."

In Salt Lake City, Parry and Engelmann spent a day looking over

Jones's plant collection at the Salt Lake Academy, where he had been teaching. Engelmann was especially interested in the cacti and Mormon tea specimens. The following day the three took a field trip up City Creek Canyon so that Engelmann could see Utah Junipers in their natural habitat, and the following morning Parry and Engelmann visited Jones at his academy room to peruse his plant collection once more. Parry and Engelmann then decided on an excursion out to Saltair Beach to swim in the Great Salt Lake and invited Jones along. After procuring bathing apparel large enough to encompass his ample girth, Engelmann, who could not swim, was off for the briny lake. He waded out until the water level passed his middle, whereupon he was floated off his feet, facedown, by the buoyant salty water. Parry, his "trusty friend," being otherwise distracted, Engelmann almost drowned before someone noticed his dilemma.

Leaving Salt Lake City, Parry and Engelmann went up to Ogden. There they met Charles Sargent and his colleague, Skinner, and the four were off on the train to San Francisco, where, as Engelmann wrote to Gray, "Botany as you know is very unsatisfactory." The party proceeded by steamer to Portland, where they ran into Yale professor William Brewer, and then pursued their fieldwork in the Northwest. In British Columbia they encountered the British surgeon-naturalist Dr. W. F. Tolmie of the Hudson's Bay Company. Tolmie, who was about Engelmann's age, showed the eastern scientists a copy of the David Douglas journal printed in Hawaii, but he did not recollect that a genus of northwestern saxifrage, *Tolmiea,* had been named after him.

The party considered meeting Lemmon at Mount Shasta, but Parry at any rate was flagging. He wrote to Gray, "I may well be excused and take my own slow 'puttering' course which if it does not accomplish much has its attractions." Sargent himself was eager to return to San Francisco by September 1 and to Boston by October 1. Forty-year-old Sargent had worn Parry and Engelmann out. As Engelmann informed Gray, Sargent rarely allowed the others more than five minutes' rest before moving on. But, joked Engelmann, if the leader were Asa Gray or Sir Joseph Hooker, the time "would have been reduced to 2 or 3, I fear."

Engelmann and Parry arrived at San Francisco in September 1880. Engelmann continued to San Bernardino to join amateur botanist Samuel Parish for a brief junket eastward through San Gorgonio Pass into the desert at White Water (northwest of present-day Palm Springs). Here Engelmann saw his first live Teddy-bear Cholla, *Opuntia bigelovii,* a cactus he had described and named for its collector, railroad survey botanist John Bigelow. In Dr. Engelmann's excitement, circling the cactus with joyous cries of "Oh, Bigelow! Oh, Bigelow!" the ponderous botanist backed into another cactus. As Parish embellished, "I spent the afternoon prying Engelmann loose." Engelmann continued by train to the Sonoran Desert at Tucson, viewing along the way many cacti he had described from specimens but had never seen alive. The greatest thrill was the Saguaro. His original description of this largest cactus in the world was based on a letter and sketch from Lieutenant William Emory, and snippets of the plant.

From Tucson in the Arizona Territory Engelmann had decided to return to St. Louis until Parry's enticing letter from San Francisco reached him: "If you are to come back here I shall see you and arrange for a *winter campaign!* . . . I have arranged a trip to Mount Tamalpais," and to go with California Academy botanist Albert Kellogg to Monterey, and perhaps up the coast to Mendocino, "& meet Mrs. Parry at Truckee." However, this tour never materialized.

At the California Academy of Sciences meeting on September 20, 1880, Parry was introduced and made a presentation on salt-marsh Arrow-grass and the big-leaved Elk Clover of the Ginseng Family. After Engelmann's return to San Francisco from his desert trip on October 3, he was lauded on October 18 at the academy, ten years after having been elected an honorary member. He presented an account of his travels in Arizona, and also lectured about California oaks and their economic value.

By now Parry was thoroughly captivated by the Golden State. In mid-October 1880 his wife, Emily ("Goodness"), joined him in San Francisco. After Thanksgiving they dropped in on newly married botanists John and Sarah Lemmon at their rooms in Oakland, and by the end of December were settled in Colton south of San Bernardino, not far from where Samuel Parish and his brother William resided, as well as William G. Wright, an avid butterfly collector. Wright, who operated a planing mill, would eventually publish the first popular work on California butterflies. Wright in 1876 had introduced Samuel Parish to his first professional plant collector, John Lemmon, for whom Parish developed an immediate dislike. Parry observed, "Lemmon made so unfavorable an impression that Mr. Parish's ardor for botany was a bit damped." It was getting to know Dr. Parry that improved Parish's opinion of plant collectors. Lemmon, who had an equal dislike for Parish, was urged by Parry that "the paramount interests of science call for harmony & good feelings."

At Colton Parry busied himself with his collections and "potboiling" newspaper articles on his West Coast adventures: in the *San Francisco Bulletin* "The Gardens of Riverside" and "The Desert Palm," and for folks reading the *Davenport Gazette* in snowy Iowa, "Midwinter in Southern California." Parry took several winter field excursions: across to Yuma where he had been with the Mexican Boundary Survey in 1849, west to the coast near Los Angeles, into the Mohave Desert with Samuel Parish and Wright, and numerous trips with Parish to the ocean beaches and the local mountain slopes.

Parry and his wife visited John Muir during the winter at Muir's home east of San Francisco. In June 1881 a party was put together to visit Yosemite, where Parry "got drenched 3 times in the spray of the falls hunting *drip* plants." Emily Parry and some other ladies were so taken by Yosemite that they stayed on for a month. Parry wrote up his experiences for the *Davenport Gazette* readership, and in a letter to Engelmann, now back in St. Louis, he explained that "I ought to tell you more about Yosemite, but am not equal to any description." He then reminisced, "I shall think *sadly* of the anniversary of our start on that glorious Western trip *June 23d 1880,* when you seemed so well and accomplished so much."

Though headquartered in San Francisco, after mid-1881 Parry was off to San Bernardino and Tehachapi with his Colorado botanical acquaintance Edward Lee Greene, who was visiting. When Parry continued north to camp and collect in the northern Sierra, Greene headed south to collect tarweeds. In October Parry met Engelmann's favorite cousin, the agriculturalist Dr. Eugene Hilgard of the University of California in Berkeley, and traveled up the Sacramento Valley by train to visit General and Mrs. Bidwell at Rancho Chico, where Parry surveyed the general's great agricultural development. By December 4 Parry and his wife had "settled down at last in winter sunshine" at "botanical head quarters" in Colton "to overhaul my whole collection" and take an occasional trip into the desert with Parish.

THE INFAMOUS PARRY-JONES-PRINGLE-ORCUTT
TREK INTO BAJA, 1882

In March 1882 Parry and his wife headed south to San Diego, where they booked into the Bay View Hotel. Also at the hotel was botanist Cyrus Pringle, who had been collecting wood specimens for the American Museum of Natural History's famous Jesup collection, making gen-

eral collections for Asa Gray, and as agent for the U.S. Census Department exploring forests of that region to collect data for the report that
had been started by Charles Sargent of Harvard's Arnold Arboretum. It
was Sargent who procured railway passes for Pringle's various junkets.
Pringle, by the way, was becoming the most successful of the growing
congregation of professional plant collectors who made a living selling
their specimens.

Also arriving at the Bay View Hotel, coincidentally, was Marcus
Jones, whom Parry had seen most recently at a California Academy of
Sciences meeting the previous summer. Jones, after finishing his initial
teaching year at the Salt Lake Academy in June 1881, made his first trip
to California, accompanied by Miss Lena A. Wakefield, who was in
charge of the academy's primary school. After sightseeing in San Francisco, Jones headed south for Santa Cruz and spent time with Dr.
Charles Anderson, an amateur naturalist who had made major botanical collections in Nevada for Gray. Then Jones traveled to Oakland for
some botanizing. On July 18 he attended a meeting of the California
Academy of Sciences, where he renewed acquaintance with Parry and
met Albert Kellogg, the academy's botanist, whom he characterized as "a
genial red-headed old bachelor" and was shown some of Kellogg's oak
drawings. In August Jones returned to Salt Lake City and his wife and
several-month-old baby.

By February 17, 1882, Jones was back in San Francisco for what became a prolonged California stay. He visited Lemmon in Oakland and
Kellogg in San Francisco, headed south for Santa Barbara and Los Angeles, and ended up in San Diego at the Bay View Hotel on March 9,
where the Parrys were staying and where he met Cyrus Pringle for the
first time, though the two had corresponded. In fact Pringle had recently
written Jones asking what he thought of Pringle's coming west and
botanizing in Mexico. Jones had written back, warning "that he better
keep out." But here Pringle was, on the Mexican border, and viewed by
Jones as "a prince of plant-collectors, whose name was well-known to
botanists everywhere."

Jones busied himself collecting both plants and shells in the vicinity
of San Diego until early April, getting into the northeastern foothills
(Temecula Canyon) and down to the Mexican border. Once Jones, who
had never sailed before, raced by himself under full sail across San Diego
Bay to uninhabited Coronado Beach, where amid the weeds he found a
new plant species. Continuing in his sailboat past Point Loma he found

himself four miles offshore and becalmed, with the sun going down and the tide going out. It was a lucky breeze that brought him back to shore. Sometimes Jones and Parry would go botanizing together. And in San Diego Parry became acquainted with Daniel Cleveland, who dabbled seriously with seaweeds, ferns, and other plants. Cleveland, a wealthy New York attorney, had arrived at San Francisco in 1867, moving to his permanent home in San Diego two years later, where, active in public affairs, he remained until his death in 1929. It was in this vicinity that most of Cleveland's plant collecting was carried out. His particular interest was ferns, and many of his specimens went to Asa Gray.

On April 5, at Parry's suggestion, Jones, Pringle, and he departed on what became an infamous collecting trip to Ensenada along the coast south of the border. Parry had met gawky teenager Charles Orcutt of San Diego, a protégé of Cleveland's who was interested in natural history, especially plants and shells. Parry rented a buggy from Charles's widowed mother and invited Charley along as driver and cook for Jones and himself, while Pringle had his own driver, team of horses, and wagon. Jones and Pringle enjoyed, as Jones wrote, "many pleasant visits around the camp fire on the trip." Jones was less friendly toward Parry, since he considered Parry, as he once snidely commented, "a typical toady," like others who had obtained a government job, "a great feeder of hot air, who slobbered over the great to keep in their good graces." Pringle, a quiet-spoken Quaker in his early forties, was married but, unable to get along with his wife, had turned to western botanizing instead. The opinionated Jones quickly developed a dislike for Charley Orcutt, whom he judged to be lazy, professing knowledge of "a little more about everything than the Almighty," and believing that "there was not room enough in California for himself [Orcutt] and other botanists." Two years after Jones declared that Orcutt "never amounted to anything," Orcutt started *The West American Scientist*, the first magazine in the West to publish solely natural history articles.

These disparate naturalists headed south from San Diego, realizing they had gotten into Mexico when they encountered some Mexican customs officials. They camped on the Tijuana River, the next day botanizing through the Valley of the Palms. Jones was already irritated with Orcutt, who paid more attention to Parry. Finally, Jones pulled Orcutt aside and told the lad that he, Jones, was paying for half the expenses, and if Orcutt did not do more of his bidding, Jones would beat him up! Shortly, Jones was delighted when Orcutt fell out of the wagon, breaking his shotgun. They stayed at Ensenada, camping first on the beach for botanizing and shell collecting. On April 12 they traveled inland to the hilly country around San Rafael, where Pringle left the party.

Before Pringle departed, however, Jones had espied a beautiful new endemic rose with rose-pink flowers near the shore. When he showed the flower to Parry, Parry immediately had to collect some for himself and called Pringle's attention to the plant. Parry would send his specimens to Engelmann, who named the rose *Rosa minutifolia,* the Small-leaved Rose.

Exactly what materialized on the return to San Diego has become a bizarre botanical legend, revolving in part around who should get credit for finding the Small-leaved Rose. On Sunday, Parry supposedly raced back to San Diego with his rose specimens in the wagon with Orcutt, leaving Jones stranded in Baja. Orcutt and his brother brought the wagon back down on Wednesday, but, ignoring Jones, they left the wagon and went off to eat dinner. Apprehensive, Jones strapped on his freshly oiled and loaded six-shooter, hitched up the Orcutt team, and was ready to leave alone for San Diego when the boys returned. They told Jones he could not go until he had paid his trip bill, including the cost of Orcutt's broken shotgun. Jones leveled his pistol, declaring he would not pay, and ordered the boys out of the way. He took off for San Diego in their wagon, leaving them to walk twenty miles home.

In the aftermath, Parry facetiously complained to Engelmann about Jones, "who may swoop down on us from the 'Sierra' any Monday with loaded revolver and force us into a thicket of *Rosa horrida* or impale us on a cushion of *Cereus maritimus!*" Jones in turn referred to Parry and Orcutt as "nothing but riff-raff" and singled out Parry as one of the "worst grafters" he had ever met. He declared that Orcutt was so patronized by San Diego's Cleveland that "he swelled up like a toad and— burst." Engelmann's final warning to Parry was "when you are with active young fellows like Jones, you must look to your laurels."

In June 1882, Parry, back in San Francisco, was requested to give a talk at the California Academy of Sciences on his trip in Southern and Baja California, describing "the most interesting plants he had met with." One of these was the California Adder's-tongue Fern near San Diego, reminding him of when, as a botanist with the Mexican Boundary Surveys, he had first discovered it at San Diego in 1850.

PARRY'S FINAL VISITS TO CALIFORNIA

The Parrys were back in San Francisco from Iowa by early fall of 1882, "away from work and hay fever," then in Colton in November. During January, Charles Orcutt and William Wright joined Parry for a return trip to Baja. The Parrys spent the 1883 summer in San Francisco, with Parry "under the weather," and in mid-September returned to Davenport. Next was an extended visit to England (1884–85) for research at Kew, where Parry renewed his friendship with Joseph Hooker and relaxed at Ilfracombe, a developing port town and resort on the Bristol Channel. California again became the Parrys' home away from home in November 1886, when they celebrated Thanksgiving with the Bidwells in Chico. Parry was well along on research he had begun in 1882, important monographic studies on California-lilacs, manzanitas, and alders. Professional articles appeared in the *Proceedings of the Davenport Academy of Natural Sciences* (manzanitas and California-lilacs) and the *Bulletin of the California Academy of Sciences* (alders and manzanitas). At the California Academy of Sciences meeting of March 7, 1887, Parry read a paper on California alders, and on June 20 he presented his paper on the manzanitas. In July he lectured at the Chautauqua Assembly in Pacific Grove on the topic "What a Botanist Saw in Europe."

In June 1888, Parry—with encouragement from John Muir's wife, Louie—invited Muir to accompany him on a trip to Lake Tahoe to help identify certain species of manzanitas and California-lilacs growing thereabouts. Despite late spring work demands at the Muir Ranch, Muir accepted the invitation, eager to get away from "the eternal grind, grind, grind" and back into the wilderness. The excursion was a refreshing interlude for both naturalists. Muir would fondly recollect "more than a week with Parry around Lake Tahoe in a boat; had him all to myself— precious memories."

Parry's final weeks in the Golden State were spent house-sitting for the Lemmons at California Hall in Oakland. He later reminisced that "I often fancy myself again laid back in an easy chair & enjoying the prospect or hiding my flowing beard in the Tea tank." Parry left California in mid-June 1889, and his train was met in Davenport after midnight on July 15 by "Goodness." That fall the couple paid an extended visit to northeastern Canada, New England, New York, and Philadelphia. The day after returning to Iowa, January 4, Parry was stricken with pneumonia. On February 20, 1890, as friend and biographer Dr. C. H.

Preston eulogized, Dr. Charles Christopher Parry "died at his pleasant home near this city, one to whom the Davenport Academy of Sciences was deeply indebted, and whose memory, fragrant and pure as the flowers he loved, it will ever cherish." Wrote John Muir, "It seems as if all the good flower people, at once great and good, have died now that Parry is gone."

On February 27, 1877, Sir Joseph Hooker had written a congratulatory letter to Parry, calling him "already king of Colorado botany" and then expressing great interest in the results of Parry's explorations in California. By 1890 the Golden State could appropriately be added to peripatetic Parry's kingdom.

Frontier Naturalists
Calling California Home

JOHN MUIR: THE PREMIER RESIDENT NATURALIST

Man of Science and of Letters, Friend and Protector of
Nature . . .

—University of California doctor of laws citation for
John Muir (1913)

John Muir first arrived in San Francisco on the morning of March 28,
1868, coming from the Panama Isthmus bound for "California's weeds
and flowers." He died on Christmas Eve 1914 at the California Hospital in Los Angeles, after becoming ill at the Van Dyke Ranch in the Mohave Desert.

University of California professor Willis Jepson said of Muir: "He
was many things: a geologist, a geographer, and a zoologist—but he
liked best to be thought a botanist." Muir indeed developed an intimate
appreciation for western flora, especially Giant Sequoias. But ornithologists laud Muir's observations of the American Dipper, the "Hummingbird of the California Water-Falls," and mammalogists his account of
Bighorn Sheep and his record horn from Mount Shasta, sixteen inches
in circumference. Geologists praise Muir's "extolling this stupendous
landform," the Sierra Nevada, his "Range of Light," and credit him with
first discovering in Yosemite's high country in 1871 a "living" glacier,
"swooping down from the gloomy precipices of Black Mountain." Muir

discovered at least sixty-five more, although Josiah Whitney had insisted that none remained. Also, Muir correctly attributed Yosemite Valley's sculpting to glaciers, rather than Whitney's geological cataclysm.

But Muir's role as a naturalist transcended scientific specialties. He became the advocate for the state's natural bounty. It was Muir who guided botanists Gray and Hooker up Mount Shasta, hosted Presidents Taft and Roosevelt, as well as Ralph Waldo Emerson, naturalist John Burroughs, and other eminents in Yosemite. He was one of those responsible for the establishment of the Sierra Club, Yosemite and Sequoia national parks, the U.S. Forest Reserves, and the federal Antiquities Act, and he fought to save Yosemite's Hetch Hetchy Valley from being dammed. In 1908 Muir Woods National Monument was set aside in his name.

Muir never completed his studies at the University of Wisconsin, leaving it, as he wrote, for "The University of the Wilderness." He later received honorary degrees from Harvard, Yale, and Wisconsin. When President Wheeler of the University of California conferred a Doctor of Laws degree on Muir in 1913, it was for the "Widely travelled Observer of the world we dwell in, Man of Science and of Letters, Friend and Protector of Nature, Uniquely gifted to Interpret unto others Her mind and ways."

ZOOLOGIST JAMES COOPER CLAIMS CALIFORNIA

So well known as an indefatigable and accurate
naturalist . . .
                                            —Elliott Coues

The Cooper Ornithological Society held its centennial meeting at Sacramento in April 1993, with an international symposium titled "A Century of Avifaunal Change in Western North America." More than thirteen decades earlier, the Cooper Society's namesake, Dr. James Graham Cooper, presented his first paper for the California Academy of Natural Sciences in San Francisco on July 7, 1861 (figure 37). His topic was animals new to California, including Whitney's Elf Owl and Lucy's Warbler, both of which Cooper had collected and named. Probably few in the audience were aware that Fort Mohave, where Cooper had shot the two birds, was not in California but across the Colorado River in the New Mexico Territory (Arizona). When Cooper died in 1902 he was among a select group of naturalists who were not merely visitors to California but made contributions in the Golden State as longtime residents.

James Cooper was born on June 19, 1830, in New York City. His fa-

Figure 37. Dr. James Graham Cooper, acting assistant
surgeon, U.S. Army, 1865. (Courtesy Cooper Ornithological Society.)

ther, William Cooper, studied zoology for three years in Europe and became a founder and leading light of New York's Lyceum of Natural History in the 1820s. It was William, not his son, after whom the familiar
Cooper's Hawk was named. When James was seven, the Coopers moved
across the Hudson River to a farm at Guttenberg, New Jersey. Growing
up in a naturalist's family, Cooper later wrote of his "budding of ornithological tastes" arising from "huggling" gosling geese to death.
While a youngster, he searched for bird nests, shells, and reptiles. He
stuffed specimens of Eastern Gray Squirrels and kept Virginia Opossums
and flying squirrels as pets. But James's obsessions were hunting and fish-

ing; he thought that if he "ever went into a wild country, the hunter's life would be my choice." Instead, as a young man he attended the College of Physicians and Surgeons in New York City, graduating as a doctor in 1851 and going to work in metropolitan hospitals for two years.

Then the chance for wild country arrived. One of his father's friends was Spencer Baird of the Smithsonian Institution. Baird, knowing of young James's interest in field zoology, suggested a position "as either surgeon or savant" with one of the pending government expeditions: Captain Page's exploration of the La Plata River with the steamer *Water Witch;* the North Pacific Exploring Expedition under Captain Ringgold; or Governor Isaac Stevens's Northern Pacific Railroad Survey. Thanks to Baird, Cooper was appointed surgeon and naturalist for the last, at a salary of seventy dollars a month and all travel expenses.

On April 28, 1853, with instructions from Baird, Cooper departed by steamer from New York, bound for the Pacific Northwest with a brief pause in California. He was with Stevens's survey from that spring until the autumn of 1855, serving with a cadre of young men destined for fame, including fellow naturalist George Suckley; John Evans, the geologist of the Oregon and Washington territories; ethnologist and geologist George Gibbs; artist-photographer John Mix Stanley; assistant engineer Frederick Lander; and Captain George B. McClellan, under whom Cooper and Gibbs would serve in the survey's western division. Lieutenant Ulysses Grant was quartermaster at Fort Vancouver in the Washington Territory when Stevens's survey party arrived for a month's stay to acquire supplies and equipment, and Cooper struck up a friendship with Grant.

Initially at Fort Vancouver Cooper was frustrated because the trunks containing his preservation materials had been left behind in San Francisco. But later there were months of far-ranging and productive exploration in Washington and Oregon. On one occasion Cooper was honored to meet Hudson's Bay Company's Dr. William Tolmie, namesake of Tolmie's Warbler (now MacGillivray's Warbler), and several days later he actually collected a specimen of the small bird.

On October 7, 1855, with the Stevens survey completed, Cooper was invited to sail to San Francisco Bay aboard the USS *Active.* During his two-month stay in California, he attended several meetings of the new California Academy of Natural Sciences and was elected a corresponding member. He met several members of the North Pacific Exploring Expedition, including New York naturalist William Stimpson. In late October Cooper traveled south into the Santa Clara Valley, found a board-inghouse in Mountain View, and had a marvelous month of observing and collecting California birds and mammals before returning to San Francisco on December 1 and finally sailing for Panama on December 6. His stateroom companion on the *Golden Age* was Dr. John Newberry, geologist with Williamson's 1855 Pacific railroad survey. At Christmastime in Panama City Cooper encountered George Suckley, his fellow Northern Pacific Railroad Survey naturalist, who was returning to the Washington Territory; and the two discussed their preparation of the survey's natural history report. In Panama Cooper also met conchologist Reverend Joseph Rowell and collected some seashells for his father, William Cooper. Incidentally, William would prepare the section on Mollusca for Governor Stevens's Northern Pacific Railroad Survey.

Back east, with a touch of "Panama fever" (malaria), Cooper worked on his zoology report for the survey and briefly practiced medicine in New Jersey. In February 1857 he learned from Baird of a possible naturalist position with the Northwest Boundary Survey. In mid-March he headed to Washington, D.C., spending a day in Philadelphia with entomologist John Lawrence LeConte, cousin of Joseph and John LeConte. Cooper admired the exquisite illustrations of the railroad survey insects LeConte was preparing. A letter from Spencer Baird awaiting him at LeConte's assured Cooper he would have housing in Washington, but informed him that Caleb Kennerly had been appointed naturalist for the Northwest Boundary Survey.

The Washington rooms into which Cooper moved belonged to William Stimpson, the naturalist with the North Pacific Exploring Expedition whom Cooper had met in San Francisco. Their boardinghouse was facetiously dubbed the Stimpsonian, since the occupants all worked at the Smithsonian Institution. Living here also were Newberry; William P. Blake, the geologist from the 1854 Williamson-Parke railroad survey; and Ferdinand Vandiveer Hayden, who had been naturalist for G. K. Warren's Sioux Expedition. Spring was arriving in Washington now. In early morning Cooper and Stimpson would stroll in a nearby woodland and collect land snails, then work until noon at the Smithsonian. On occasion Spencer Baird stopped by the Stimpsonian for a visit, and Cooper spent a delightful evening at the Bairds' home, meeting daughter Lucy.

On February 17, 1857, President Pierce signed a bill for the construction of federal roads under the direction of the Department of the Interior, rather than the War Department. Cooper was appointed by the new secretary of the interior, Jacob Thompson, to serve as surgeon, with additional duties as naturalist, for the Wagon Road Expedition from Fort Kearny, in the Nebraska Territory, to Honey Lake Valley in California. But under the disorganized Colonel Magraw this survey fell into complete disarray. By early October Cooper, totally disgusted, abandoned the expedition and returned to Washington. He had gotten no closer to California than a distant view of Laramie Peak in Wyoming.

Arriving in Washington on November 21, 1857, Cooper went to work again at the Smithsonian on scientific reports and specimens, with five other naturalists. The young men were a congenial group, occupied all day but spending "an hour in eating, drinking, and laughing." This congregation, together with other Washington naturalists, evolved into the Potomac-side Naturalists Club, which existed into 1866. Meanwhile, the group at the Smithsonian working for Baird—"Baird's Boys"— eventually constituted the Megatherium Club (named after a giant fossil ground sloth). The motto of this august scientific body was "Stand up to your liquor and face danger when it comes, and above all never let your evening's amusement be the subject of your morning's reflections." Among Cooper's companions was "a wild Illinois naturalist," Robert Kennicott, at the beginning of a short but brilliant career that ended with his premature death in Alaska.

Cooper was in good spirits, anticipating payment for his work on the Northern Pacific Railroad Survey report. He was also developing a forest biogeography classification for North America, including associated maps, that impressed Interior Secretary Thompson. During this period at the Smithsonian, Cooper and Suckley were preparing in addition a lengthy manuscript on northwestern natural history. When published in 1860, it was the first major treatise on the subject. Cooper also wrote a discourse for the Smithsonian, "The Influence of Climate on the Growth and Distribution of Vegetation."

On August 7, 1860, Cooper, as contract surgeon for three hundred army recruits in twenty-two wagons, departed from Fort Benton (Montana) for Fort Vancouver on the Columbia River. Upon arrival, since his obligations as surgeon were terminated, he joined another group of soldiers bound by sea for San Francisco. Arriving in California on November 5, he was delighted to learn that he could continue his army contract. He was assigned for the coming winter to Fort Mohave east across the Colorado River, where surgeons were badly needed. He traveled by ship to Los Angeles for the inland trek to Fort Mohave. When informed that there would be a two-week delay, Cooper caught a cut-rate steamer back to San Francisco, little anticipating that this brief return would be a turning point in his professional career.

Fortuitously, Whitney's California Geological Survey men encountered Cooper shortly after they arrived at San Francisco in late November 1860. The man they met was tall, "spare, a man of soldiery bearing, with a dark beard, well silvered, clear blue eyes, delicate hands, a voice slow and not given to a ready flow of language." Cooper had likely learned of their arrival through local newspaper publicity about the new geological survey. At first somewhat reticent, Cooper warmed up and told the Whitney staff that he had been over most of the United States, traveling back and forth across the plains, had served for two years in the Northwest with the railroad survey as surgeon and naturalist, and had already seen much of California. Cooper mentioned that he had written a large treatise with naturalist Suckley on northwestern natural history. William Brewer, Whitney's assistant, subsequently judged Cooper to be knowledgeable and "of more than ordinary intellect and zeal in science." However, Brewer frankly suspected that he would not be a "very companionable fellow in camp." When Whitney learned that Cooper would be surgeon at Fort Mohave during the coming winter, Whitney induced him to procure animals and plants and make observations in the desert for the California Geological Survey, intimating that there might be a position for him when his duty at Fort Mohave ended.

Cooper returned to Los Angeles, and on December 4, 1860, he was on his way to the Colorado River over Cajon Pass on muleback, traveling with an army wagon train. Near the pass Cooper heard from chaparral slopes "the loud ringing trill" of the Wrentit, "the chief bird-music at this season," and shot his first specimen of the crested Phainopepla. Arriving at Fort Mohave on December 19, he found a deciduous woodland along the Colorado River, but the surrounding upland had "a most barren and desolate aspect." At Mohave Cooper collected several new species of birds. He wrote Baird about a new warbler, asking if Baird's daughter Lucy would consent to his naming the warbler after her. He went on to write, "I hope she may never have to live in such a country as the bird inhabits, but if she does, her presence like the bird's would go far to make a garden of the desert." With respect to a new owl, Cooper noted with tongue in cheek that he called it *whitneyi* "merely from its habit of saying 'whit whit,' nothing else I assure you." Cooper dissected the tiny owl and found in its stomach small feathers and insects. Life at Mohave was monotonous, but Cooper kept busy collecting birds, mammals, and eventually reptiles, as well as occasionally doctoring; and he enjoyed dinners with the commandant and his wife. It was a pleasant surprise when Lieutenant Joseph Ives, explorer of the Colorado River, showed up in February 1861.

With the start of the Civil War, there was imperative need for troops elsewhere, so Fort Mohave was abandoned on May 28, 1861. One of Cooper's last observations was of another Lucy's Warbler, this time at its nest. Cooper left the fort with the heavily laden wagon train carrying the entire army command back over Cajon Pass and on to San Diego. Cooper then headed north to Los Angeles along the shoreline, picking up California seashells and stimulating his excitement for conchology. In Los Angeles Cooper received a letter from Spencer Baird reporting that Whitney intended to hire Cooper as zoologist with the survey. This was welcome news, since he had received no word from Whitney.

William Brewer next encountered Cooper in early July 1861 at the Brewer survey camp near San Juan Bautista, where Cooper's stage stopped for an overnight stay on its way to San Francisco. Reputedly, he brought Brewer four to five hundred plants from the desert. Brewer anticipated that Whitney, up in San Francisco, would probably add Cooper to the survey's naturalist staff at least temporarily, which Whitney did.

In August Cooper—assisted by the Whitney survey's general factotum, Chester Averill, who had a knowledge of sailing—carried out shoreline collecting and dredging for four days in Carmel Bay. They procured 140 marine specimens, and 111 more from dredging and collecting in Monterey Bay into September. Cooper also added to the survey collection some living corals fished up by Chinese boatmen near Point Pinos.

On one occasion when Cooper was fishing along the shore, he encountered a comely lass with two sisters. The four of them took a moonlit walk, then sat and talked beneath a tall sycamore; "She was sweet as sugar or more so." Later he observed her, on horseback, watching him dredge in the bay. He was "dead in love" and suspected that the girl knew it. When Cooper returned to San Francisco, the sisters, on their horses, were on hand to bid good-bye. As the three young ladies began to ride away, "the dear little girl" reined her horse back and shook his hand. He thought, "Oh my heart!" In San Francisco, Cooper brooded about his love for three days. Whitney, a great matchmaker and feeling sorry for his zoologist, promised to introduce Cooper to "three of the finest and richest young ladies in the city." It proved a hollow promise.

From fall to spring of 1861–62, Cooper was in Southern California continuing his diversified collecting for the survey, though with limited funding. In San Francisco he had picked up an assistant, Edward Palmer, about thirty years old, who would later be one of Dr. Charles Parry's associates. Palmer, born in England but raised in Cleveland, Ohio, was a protégé of zoologist and physician Dr. Jared Kirkland, professor at the Cleveland Medical College, who possessed a large natural history library. From 1853 to 1855 Palmer was with the La Plata Expedition. Cooper and Palmer arrived by steamer at San Pedro, where they found some Tertiary fossils in the beach cliffs and did some dredging, over a two-week period collecting ninety-seven species. In late October 1861, with two other men, they attempted to collect on Santa Catalina Island, but high winds prevented a landing, and they spent four days drifting helplessly about.

Continuing on to San Diego for an extended stay to do more dredging and other collecting, they first lived in an army cottage in town and then for a month and a half near the mouth of San Diego Bay. That winter Palmer primarily procured marine invertebrates for Cooper, who considered him a diligent collector. Among the shells Palmer found was one erroneously thought by Cooper to be a new species of triton, which he named *Tritonia palmeri*. Cooper reported on the find at the California Academy of Natural Sciences meeting of November 3, 1862. While with the survey Cooper also bestowed, inadvertently, an appropriate species name on a new Cretaceous brachiopod shell honoring the survey's extremely loquacious paleontologist, William Gabb. The survey men guffawed when "the serious, the unbending" Dr. Cooper announced the name, *Ligula gabbii*.

In mid-November 1861 the two naturalists moved to the little settlement of La Playa, north of Ballast Point on the eastern shoreline of San Diego's Point Loma. Here Palmer was stabbed in the foot by a stingray and had to return to San Francisco. Shortly before Palmer left, he helped Cooper pack survey specimens for shipment. Noticing that Palmer was secretively tucking some of the best ones among his personal possessions, Cooper waited until Palmer was away for a moment. Then he removed the pilfered specimens from their wrapping paper in Palmer's baggage, replacing them with stones and other rubbish.

After Palmer's departure, Cooper hoped to venture along the coast and into the backcountry to collect vertebrates, and also to visit the Coronado Islands. However, heavy winter rains in the lowlands and snow in the highlands delayed his trips until the end of February 1862. In San Diego on March 28, he received depressing news from Whitney: his salary with the survey was to be suspended as of April 1. Discouraged, he returned north in mid-May, stopping again in San Pedro and getting out to Santa Cruz, one of the Channel Islands.

When the survey's 1862 summer field season commenced, Cooper, though still without salary, continued to work on his collections for them. The fossilized mollusks from around San Diego elated Whitney because some of the shells indicated Cretaceous-age sedimentary formations, something not known before for that region. In late July Cooper packed up a large number of specimens—mammals, birds, shells, crustaceans, and insects—to be shipped to the Smithsonian and the British Museum, with the admonition that at least some new species be saved for naming by California naturalists. The boxes went into the hold of the *Golden Gate,* a wooden side-wheeler steamer, which caught fire off Manzanillo, Mexico, and was totally destroyed, with the loss of 223 lives and over a million dollars in gold, not to mention Cooper's specimens. Philip Carpenter, an English conchologist who was going to describe the shells, wondered what future shell collectors at Manzanillo would think when they came upon these exotic northern specimens in their local beachcombing.

During September Cooper moved to Oakland. He would have preferred the Monterey Peninsula because of its natural diversity, but felt the climate there less beneficial for his health. Hearing from Whitney that the state might provide another survey appropriation for the next year, Cooper splurged and bought two parcels of lovely land in San Mateo, across the bay, with Redwoods, beautiful flowers, and abundant birds "rich in song and plumage." He daydreamed about working for the survey in summers and practicing medicine here during the winter.

The Civil War was now well under way, but Cooper optimistically calculated that it might be over by winter. During 1863 on and off he was in the field on salary for the California Geological Survey again. From late April to August he explored the coastal islands from San Clemente north. The collections from the southern coastline and island trips resulted in fourteen large boxes and one keg of "wet specimens." He apologized to Whitney for having obtained only 130 additional species, but added that "their novelty and interest surpasses any collection made during any other three months." In early autumn Cooper collected in the vicinity of Lake Bigler (today Lake Tahoe), returning a month earlier than expected because Whitney reported that his pay must stop. Even though Cooper was piqued with Whitney, "who has treated me badly in this matter of salary," he ended the year in San Francisco, continuing to work with the survey collections. In January 1864 he opened a medical office for a short time, until survey funds became available in April. Though suffering from what was probably recurrent malaria, Cooper spent the rest of that year collecting, mostly throughout the San Francisco region.

In May 1865, with the Civil War finally ending, Cooper volunteered for a three-year army tour. He was commissioned assistant surgeon with the Second Cavalry, California Volunteers, and reported to Camp Union near Sacramento, whose climate definitely did not agree with him, and which he hated. Despondent, he wrote to Baird that "I begin to think it is time to provide for the downhill of life, which with me may not be a very long journey either." At the end of August his regiment went to Southern California, where at least he had the good fortune to meet fellow ornithologist Elliott Coues, who was delighted to find in Cooper "an encyclopedia of interesting biographies of the birds of the Pacific coast." Cooper now feared he might be transferred for duty to hot southwestern Indian country, mourning that he could not "withstand the hardships of the field which were once a pleasure to me." Out of consideration for his health, he turned in his resignation and was discharged in San Francisco just before Christmas 1865.

Throughout his army period Cooper pursued his zoological interests in his spare time but increasingly realized that it was impossible to make science "a self-supporting business." From time to time he asked Baird to approach Cooper's old colleagues from the Northern Pacific Railroad Survey, especially Lander and Generals Grant and McClellan, with respect to employment opportunities. Cooper even asked if Baird could get him a place with the Mexican consulate.

Out of the army, Cooper was on occasion back with the California Geological Survey as Whitney's zoologist. He prepared the material on California birds for the survey's volume 1, *Ornithology: Land Birds*. Baird, who edited Cooper's manuscript, called the treatise "the most valuable contribution to the biography of American birds that has appeared since the time of Audubon."

On January 9, 1866, James Cooper married Rosa M. Wells, an Oakland minister's daughter. He began a medical practice at Santa Cruz early that year. By February 1867, the practice was failing and nothing had come of Cooper's idea of writing zoology textbooks, so he and his young family moved to San Francisco, where he complained about the climate but did produce the detailed "Geographical Catalogue of the Mollusca Found West of the Rocky Mountains" for the California Geological Survey. Baird tried to interest Cooper in a collecting trip to Alaska, but Cooper, fearful of the weather, turned down the offer.

Cooper, with an insecure financial future, had good reason for his frequent pessimism. In 1868 when he applied for a teaching position at the new University of California, his application was late, and the opening was filled by Joseph LeConte, his friend John L. LeConte's first cousin. This year Cooper was ill with lung trouble (tuberculosis) and had a light case of smallpox, but he still managed a ten-day collecting trip up the Sacramento River. And he completed a long chapter on zoology for Titus Cronise's tome *The Natural Wealth of California*, published by Hubert Bancroft.

In 1870 Cooper's interests turned increasingly toward conchology, but he still did scattered fieldwork in the Sierra Nevada for Whitney. During 1871 Baird recommended Cooper as surgeon and naturalist for Lieutenant George Wheeler's southwestern expedition, but Cooper turned it down because it paid too little. In the spring of 1872 he was again on Whitney's payroll for a number of months, first doing fieldwork in the Santa Cruz Mountains. The last week of April Cooper embarked on what would be, for him, one of his most delightful excursions, this time to examine the geology of "a region scarcely known to naturalist," the Cuyamaca Mountains in Southern California. The small party—comprising Cooper, California Academy of Sciences geology curator Watson Goodyear, and Mr. Fox of the Southern Pacific Railroad—headed northeast from San Diego past the old mission. By the end of the month they were well up in the mountains, having traveled through lovely oak woodlands "alive with the songs of migrating spring birds," and then above four thousand feet into a mixed coniferous forest where travel was hazardous because of ice and fresh snow crashing down from branches of the tallest conifers. On the way back the men were literally blown into San Diego by a hot desert wind—the "Santa Ana."

Soon Cooper was off by wagon for Los Angeles, where he examined the geology of some of the shoreline hills and collected fossils. Here he encountered a huge California Condor, standing quietly on a nearby hillside. As Cooper approached, the bird, a fine-looking specimen in apparent good health, never moved except to open and close its bill. Cooper thought about picking the condor up and carrying it back to camp, but it was a several-mile trek, and he was loaded down with fossils. So he wrote that he "left it to fulfill its destiny."

Cooper moved his family down to a ranch at Saticoy, east of Ventura, in June and went back to collecting animals. In August 1872 he asked Baird about a surgeon's berth with a forthcoming naval expedition in the Gulf of California but learned he would have to become a naval officer first and undoubtedly could not pass the physical exam. Cooper on one occasion complained to Baird about the hard life of a field naturalist, "the pursuit of science as a private business is a losing game. . . . Almost all the 'enlightened' people of this city know me as a 'naturalist,' which is the title of all the taxidermists also, and . . . they avoid employing me professionally as they would a bird-stuffer."

By the summer of 1873 the Coopers had moved back to San Francisco, James still involved with fieldwork, collections, and reports for Whitney, including trips with Charles Hoffmann and others to Mount Diablo, Antioch, and the coal town of Somersville, and work on the map of San Francisco Bay. In October Cooper visited the mercury mines at New Almadén. In mid-1874, while doing geological fieldwork at Livermore, he learned that the California Geological Survey had been suspended. The next February Cooper and his growing family finally moved to a permanent residence in Hayward, where the climate was more moderate. He returned to the practice of medicine, was in relatively good health, and began earning a respectable income from medicine and selling specimens to museums and personal collectors. One of his bird specimens with an unusual history came from botanist Sarah Plummer (shortly to become the wife of botanist John Lemmon). The Yellow-rumped Warbler had fallen dead at her feet on her cousin's porch in Alameda, and she skinned, preserved, and sent it on to the ornithologist.

Since presenting his first paper for the California Academy of Natural Sciences in 1861, Cooper had demonstrated his broad interests as a naturalist, his reports ranging from new and rare terrestrial mollusks and specimens of "little-chief hare" (Pika) from the High Sierra, to new genera and species of California fish and Oregon Indians' use of the root of an Aspidium fern as medicine. One new freshwater mollusk he found from the Feather River was, at the time, only the second species in its genus, the other occurring in Cuba. Cooper served the California Academy several times as curator of zoology and invertebrate zoology, and even as paid curator of paleontology. He was on occasion the academy librarian, corresponding secretary, vice president, and during the last four years of his active participation (1888–91) the director of the museum. In his mid-sixties he was on the academy payroll as a researcher on mollusks. Though thought of as an ornithologist, Cooper published more then forty papers on shells. The last paper he read at an academy meeting, in December 1875, was "New Facts Relating to California Ornithology," but he continued to present papers, by title only, through 1896.

With the passing years Cooper, chronically suffering from rheumatism, malaria, and tuberculosis, found getting into the field more taxing. However, as railroads proliferated, Cooper took to this new form of transportation, often using complimentary passes. For the Atlantic and Pacific Railroad he was even employed to report on the geology between Mohave and Needles. In 1890 Cooper included geological descriptions of some California train routes for the second edition of *The American Geological Railroad Guide*.

During the 1890s there was little money from the academy or elsewhere for the zoological collector. In 1893 Cooper received a federal pension for general debility, six dollars a month, and the next year applied unsuccessfully as an examining surgeon for pensioners. His health continued to fail, and during the last four years of his life Cooper was "debilitated," partially paralyzed, barely able to speak. On July 19, 1902, the indefatigable field naturalist died in Hayward. He and his wife, Rosa, who died on June 9, 1909, are buried in Oakland's Mountain View Cemetery, along with many other early naturalists. As California's first and longest resident natural scientist, he left a permanent mark. The minutes of the California Academy of Sciences announcing his death observed, "He wrote many valuable papers and in all branches, to which he directed his attention, he was widely and favorably known for his scientific attainments."

On June 22, 1893, the Cooper Ornithological Society was organized. The enfeebled Cooper was able to attend only one meeting of the society, held at his own home. As a tribute to Cooper after his death, Otto Emerson wrote in the society's journal, *Condor,* "Three years ago last March he sat among us, for the last time in public, for the study of his favorite work, ornithology." He was "a man who could lead you in paths of Nature to the haunt of birds, to the lurking places of shells, or to hidden fossils, and could name for you the rocks, trees, and plants of the mountains, hills and plains." Dr. James Graham Cooper was the last of "Baird's Boys."

A BOTANIST COUPLE IN RESIDENCE:
"J. G. LEMMON & WIFE"

Both are scholarly, gifted, and write beautifully.
    —*Santa Barbara Weekly Press* (December 4, 1880)

In October 1866, Union army veteran John Lemmon awoke early on his
first California morning, south of the little town of Sierraville northwest
of Lake Tahoe. Looking out of his brother's farmhouse window, he was
excited at what he saw: "a practically unknown world," one filled with
"strange new flowers, bushes, and even trees." Lemmon was in his mid-
thirties, tall, emaciated, with sunken but pleasant blue eyes and a prominent Roman nose. Toward war's end he had endured three weeks in the
notorious Confederate prison at Andersonville, Alabama, where daily
rations were a piece of soggy corn bread. He was then transported to the
prison stockade at Florence, South Carolina. Despite several attempts to
escape, he became a "parole of honor," was "promoted" to steward of
Hospital Ward One, and was ordered to play flute in a prisoner ensem-
ble performing for Rebel socials. Finally liberation came on February 28,
1865. Lemmon, weighing only eighty-five pounds, was the first to discover the prison gates open, the Rebel jailers vanished.

John Lemmon was born on a farm near Lima, Michigan, on January
2, 1832. At nineteen he began teaching at village schools for eight years,
then attended lectures at University of Michigan until the Civil War. A
fervent abolitionist, in 1862 he joined the Fourth Michigan Cavalry and
fought in thirty-six battles before his capture on August 29, 1864, during General Kilpatrick's ill-conceived raid south of Atlanta as part of
Sherman's famous "march to the sea."

After the war John's mother, Amilia, had the invalid transported from
Michigan out to his brother Frank's farm either to recuperate or to die.
Here the "genial climate," the diversified landscape, and the distinctive
western flora reawakened Lemmon's childhood interest in natural history, especially for plants and insects. "Imagine," he thought, "a terra
incognita, a paradise, an open field of opportunity; all in sight!" He was
shortly "groping" about the yard, one hand on the fence rail and picking flowers with the other. Helped by his mother, he ventured farther
afield, finally able to be put on "the good Dr. David Gould Webber's"
horse and to be escorted by his doctor on short collecting excursions.
Later, Lemmon spent much time in the elderly Webber's cabin at the
Webber Lake resort, with the widowed doctor or alone. As Lemmon's

health improved, he became acquainted with Eliphalet Case, who had started a small coeducational "academy." From 1870 to 1874 Lemmon occasionally taught at the village school, although his perceptive mother continued to feel that her son "was born a botanist, perhaps inheriting the reincarnated spirit of an ancient weedpuller."

As Lemmon's plant collections increased, Case, now his hiking companion and also interested in natural history, suggested that Lemmon should ship his thirty or so specimens to botanist Henry Bolander at the California Academy of Sciences. Unable to identify many of the plants, Bolander sent some to Professor Asa Gray at Harvard, with an accompanying note about Lemmon: "His specimens are poor, but still they may interest you. In future he may do better; he is quite an enthusiast, and a good mountaineer; he may be able to find many new plants yet in those mountain recesses." After perusing Lemmon's specimens, Gray's friendly reply, including the new species names, launched Lemmon on his plant-collecting career: "Lots of new plants; but don't work too hard. . . . The plants will wait for you. I congratulate you upon the change of a rebel prison pen for a California paradise." After reading Gray's letter, Lemmon was so excited that he ran out into the yard, shouting out the names of the new species, many honoring him. The curious five-leafed clover John had collected on his brother's doorstep had become *Trifolium lemmonii* (not to be collected again for well over a century); the Indian paintbrush "flaming up beside the gate" *Castilleja lemmonii;* the plant across the road with its purple flowers and large leaves *Asarum lemmonii.* . . .

During Lemmon's first decade in California he began exploring his new state and adjacent Nevada for plants, on occasion accompanied by "Professor" Case. There were collecting trips by foot, horse, and wagon in summer, while during winter Lemmon read in Webber's cabin, "buried in 8 feet of snow," or sorted his plants on the tables in the now-empty Webber Lake Hotel. He especially liked to have his new species named by Asa Gray or Gray's assistant, Sereno Watson, in honor of Lemmon or one of his personal friends. At his request a Sierra Valley milkvetch was named for Dr. Webber, another under sagebrush near Beckwourth Pass after Case. When a plant honored one of Lemmon's acquaintances, he considered it "a grand festival" and often celebrated by printing a special herbarium label.

Lemmon visited the Lassen Volcano area in 1873, and in late May 1875, he and Case trekked to Pyramid Lake in the Great Basin desert, "to toil for weeks through its sands, alkali and prickly bushes, revel in its curious and rare flora." A "day of adventure and accident" occurred when their borrowed rowboat nearly swamped during a sudden gale. Finally reaching shore, they had to hike five miles back to camp with their gear. During this expedition Lemmon collected more than fifty plants new to him. In the early 1870s Lemmon combined his floral excursions with selling subscription books, a great opportunity for getting acquainted with numerous settlers, particularly women, whom he interested in collecting plants. He commenced writing up his "strange scenes and striking adventures," as well as his Civil War experiences, for a variety of publications, especially the *Pacific Rural Press.*

One of Lemmon's early paying botanical jobs was gathering wildflowers and arranging them in vases for a Sierraville church. As his collections increased, he put together sets of dried specimens, as well as bulbs, seeds (especially from conifers), and herbal medicines, for sale. He even contemplated moving to Yosemite and selling pressed wildflowers to tourists. Eventually, Lemmon's letterhead and business card advertised him as "Botanist, Lecturer, Microscopist and Collector in Natural History."

By now Lemmon was corresponding with some of the nation's important eastern botanists, especially Asa Gray and the peripatetic Dr. Charles Parry of Iowa. In 1875 he provided information about and specimens of the Sierra Nevada White Fir for George Engelmann, who was comparing the White Firs of the Rockies and the Sierra Nevada (subsequently both *Abies concolor*). And Parry and Lemmon enthusiastically began planning a joint exploration that would take the pair from the Mohave Desert north to Canada. In late spring 1876, indeed, Lemmon, Parry, and Edward Palmer botanized in the San Bernardino Mountains and the desert beyond. At the Mohave River headwaters Lemmon collected a new species of Purple Mat waterleaf that Gray named *Lemmonia californica* (now *Nama californicum*), "to send Lemmon's name down to posterity along with his adopted state." During the trip Parry's assessment of Lemmon was that he was active but "excessively *nervous & fidgety,* does not like to stick to steady work, and likes to make a display of what he does besides being short of funds. He is a thoroughly *good fellow* and I would like to see him do well."

About this time Lemmon expressed his own opinion of himself: "I neither drink, smoke, chew, swear, play cards, or lose my temper, yet I am not a member of any religious order, I am a zealous Odd Fellow and a Patron of Husbandry. In science I adhere to Darwin and Gray. I am unmarried but not a woman-hater. Mean to marry the best woman I can

find—after I can travel no more." But after he met artist and botanist
Sarah Allen Plummer in 1876, he could wait no longer!

Sarah (sometimes spelled Sara) was born in New Gloucester, Maine,
on September 3, 1836. She attended the "Female College" and Normal
School in Worcester, Massachusetts, receiving a certificate. Moving to
New York City, Sarah started teaching at Grammar School No. 14 under
the tutelage of the famous Caroline Whiting. In addition, she apparently
cared for wounded Civil War soldiers with Dr. Henry Bellows's newly
formed U.S. Sanitary Commission. When she contracted a severe case of
pneumonia, her physician recommended a move to a healthier climate.
On advice from a friend, in 1869 she took ship bound for California.
Once in Santa Barbara, barely having recovered her health, Sarah was seriously injured in a buggy accident, her first newspaper publicity being
her erroneous death notice in the local paper.

Miss Plummer, starting a little book, stationery, and novelty shop,
quickly became "prominently identified with the intellectual, scientific
and social interests of the place" and developed "hosts of warm loving
friends." Missing the cultural stimulation of the East, she turned her
shop into a meeting place for "local literata and artists" as "a haven for
intellectual development." Obtaining two hundred books from her New
York Unitarian friend Dr. Henry Bellows, Sarah established a circulating library, reading circle, and art gallery. With $1,500 of contributions
she visited San Francisco and bought more volumes, her collection eventually becoming Santa Barbara's public library. Sarah was active in the
new Santa Barbara Natural History Society and contemplated doing a
watercolor book on California flora.

John Lemmon likely first became acquainted with Miss Plummer in
February 1876, when he stopped in Santa Barbara for a week on his way
to the Southern California excursion with Parry and Palmer. And on his
way home John apparently stopped again. He had been contracted to
procure California plants for the U.S. Centennial Exposition in Philadelphia, and at Santa Barbara he encountered many helpful enthusiasts,
among them teachers from "Santa Barbara College"; members of the
Natural History Society, including Sarah Plummer; and Charles Parry
and his wife, Emily. All were excited to have a "genuine live botanist"
pay a visit, and as Mrs. Elwood Cooper, fern aficionado and wife of the
college's superintendent, innocently but portentously suggested to Sarah,
"I will meal him if you will bed him." Unfortunately for Lemmon, after
his Southern California trip the Department of Agriculture reported that
the federal treasury was so depleted there was no money to pay for his

Centennial Exposition collections and expenses. John and Sarah now began corresponding and seeing each other, and Lemmon commenced tutoring her in botany, sometimes at the California Academy of Sciences while she stayed with her cousin, Captain William Plummer, and his wife in Alameda. When Sarah collected a new shrub related to Coyote-brush in Glen Loch Ravine near Santa Barbara, Lemmon sent the specimen to Professor Gray, who named it *Baccharis plummerae* in her honor, "an ardent botanist whose name it is a pleasure to commemorate."

In mid-1877 Lemmon, learning that Asa Gray would be bringing Sir Joseph Hooker, director of Britain's Kew gardens, to the Rocky Mountains and on to California, enthusiastically began planning a weeklong field trip (hopefully including Miss Plummer) in the northern Sierra for the visiting botanists. Gray finally wrote that an already tight schedule would permit only a brief meeting with Lemmon, if any, and it would be impossible for Lemmon to accompany the planned Gray–Hooker–General Bidwell excursion with John Muir to Mount Shasta. Despaired Lemmon: "All plans fail."

By August 25, 1877, the distinguished visitors were in San Francisco at the Palace Hotel, having come by wagon across the High Sierra, through two groves of Big Trees and Yosemite Valley. The following week they enjoyed a tour to Monterey, met the LeContes and other faculty at the new state university across the bay, and visited the California Academy of Sciences. On September 2 Sarah Plummer, busy at the academy with her plant collections and hoping she might be lucky enough to meet Hooker and the Grays, was excited to learn from Dr. Kellogg, with whom Sarah had been conferring, that Hooker was downstairs, packing cones. When Sarah was introduced, Sir Joseph said, "Oh, I have heard Dr. Gray speak of you." He invited Sarah to come to the Palace Hotel and meet Gray, saying, "If he is not there I will be & you can call for me & wait till his arrival." Sarah walked over to the hotel, where Sir Joseph "entertained me for over half an hour" in his room before the Grays showed up. Professor Gray drew his chair next to Sarah's and explained why the party would probably be unable to see "Professor" Lemmon. Whereupon Sarah exclaimed, "You must in some way arrange to see my good soldier boy & the most enthusiastic botanic worker on the Pacific coast. Your great admirer." The next day at the academy Sarah and Gray chatted again about visiting Lemmon, Gray finally saying, "Do you not think it would be well to invite Mr. Lemmon to meet us at Lake Tahoe when we could have one day with him?"

Thus did Miss Plummer engineer the memorable meeting of Lemmon,

Hooker, and the Grays, although she did not join them. Lemmon and his diminutive mother came south from Sierraville on September 11 to the Truckee train stop near Lake Tahoe to meet Gray's party, returning from their Mount Shasta excursion. It was midnight when the train finally reached Truckee, with botanist Dr. Harvey Harkness of Sacramento in charge of the little party, which now included another British scientist, Sir David Wedderburn. The imposing Hooker greeted John with "Glad to meet you, Lemmon. I feared you would be too sick to join us. When "Professor Lemmon" was introduced to Asa Gray and Mrs. Gray, he felt obliged to tell "the greatest of American botanists" that the title of professor he was wont to use "had not been conferred in any regular manner." Gray quickly put his hand on Lemmon's shoulder and told him, "You are quite as worthy of the title as any who regularly have it, and I shall call you Professor." Lemmon treasured this tribute, recalled Willis Jepson, "in hallowed memory all his days."

The next morning Harkness arranged for a special train to Donner Summit, where there was half a foot of fresh snow. Then after breakfast at Truckee, the party proceeded cross-country to Mount Stanford, Hooker busy collecting bark, wood, and lichens with his huge knife. Gray and the two Britishers actually reached the top of the peak. The following morning a stage took the party to Lake Tahoe, where they boarded a steamer, with free passage. Once away from shoreline vistas, there was time for talk. Hooker and Lemmon discussed the different conifers of North America. Anon, Lemmon told Gray of his hope to lead botanical expeditions or teach botany. Dr. Gray replied that John could undoubtedly handle expeditions but needed more knowledge of "structural botany" before he could teach.

Returning through the forest to the hotel in Truckee, the visitors again busied themselves collecting conifer wood, cones, and foliage. Before Hooker and the Grays departed on the eastbound train, Lemmon, who was suffering from catarrh, had not slept for two nights, and felt "under a cloud" the entire time, apologized for his fogginess by telling Professor Gray that "the honor of your visit quite upset my head." But he had still managed to correct Dr. Gray's misidentification of *Helianthella californica*. In 1885 when Lemmon's aged mother died, Hooker wrote him a letter of sympathy about his "most worthy and affectionate parent," reminiscing that "especially do I recall incidents of our trip over the basin of the wonderful Lake Tahoe accompanied by Mrs. Gray and your sprightly mother."

At the California Academy of Sciences meeting of January 21, 1878,

John and Sarah were elected resident members, Sarah being one of the first two women ever proposed for membership. At the same meeting Lemmon presented a paper on the insectivorous California Pitcher Plant, one that Hooker had hoped but failed to see. In June Lemmon was back in Santa Barbara to join a lengthy excursion to Yosemite that would include Sarah Plummer, whom he now called "Amabilis" while she called him "Lemmonia." Six campers went, but Amabilis, "though getting well again is still too weak to join us." The batch of plants Lemmon collected on the trip was shipped to Gray on September 1.

In the spring of 1879 Lemmon made an extended trip from Sierraville north to Nevada's Pyramid Lake, and by mid-May on to the Modoc Lava Beds. This time he was accompanied by Sarah Plummer, with his "lively & most entertaining" seventy-six-year-old mother as chaperon, the trio camping in a two-horse covered wagon. In late April at Pyramid Lake Lemmon, a "sentimentalist" (as one friend noted), dedicated a flower he collected "Above Poet's Camp" to the poetess "Miss Sara A. Plummer." Later, a clover that she picked at the lava beds would bear her name as collector on Asa Gray's herbarium label.

During July Lemmon scaled Mount Shasta with a party of mountaineers from Sisson's Tavern at Strawberry Valley, hiking at first up through coniferous forest and collecting plants along the way. On the evening of July 9 they camped at timberline, to begin the assault of the summit early the next morning. Hiking across snowfields taxed Lemmon's endurance. "Every few rods," he wrote, he was "obliged to drop down on my face and rest a few seconds." Above the snowfields were icy slopes, but John was determined. And by early afternoon the party reached the top. On the trip down he revived sufficiently to do more collecting.

During this period California had been ravaged by outbreaks of the "Devastating Grasshopper," "leaving the ground bare and sere." Interested in insects, Professor Lemmon was employed to research the scourge and write a lengthy report, gaining fame as "the highest authority on the coast," and was interviewed by newspapers. According to one enthusiastic but misinformed reporter, Lemmon also "goes to Alaska next summer with John Muir."

Late in 1879 Lemmon moved from Sierraville to Oakland. Harry Hollister of Truckee had run into Lemmon as he was moving and wrote friend Sarah Plummer that "Prof. L. is both lonely and disgusted. What he may do in his desperation I cannot imagine." John's old colleague Case, just married as well as becoming a Mason, now had new obligations, and John's application to join Clarence King's 40th Parallel Survey had been turned down, King having decided he did not need a botanist. Under the circumstances, the next year, with Miss Plummer returning from a trip east to visit her family and Mr. Lemmon returning from a Sonoran Desert excursion, the pair concluded that the time had come to get married.

The wedding took place on Thanksgiving Day 1880. The couple's marriage received a warm tribute from the *Santa Barbara Weekly Press* beneath the headline "MARRIAGE OF DISTINGUISHED SCIENTISTS." "Both are scholarly, gifted, and write beautifully. May their life be full of happiness and usefulness." John Coulter, editor of the *Botanical Gazette*, received notice of the marriage, which ordinarily he would not have publicized. But this time he made an exception, noting in the January 1881 issue that "this one is so botanical that its mention seems very appropriate," one well-known botanist marrying another well-known botanist, and "they will reside this winter in Oakland, and having united fortunes and herbaria, are ready to welcome their friends in their new herbarium rooms."

For the time being the couple occupied Lemmon's "very cozy" rooms at Blake House in downtown Oakland, and John's mother came from Sierraville to live with them. John and Sarah diligently made up sets of their collected plants for sale to eastern and overseas botanists as well as "lovers of California flowers," thus hoping to "keep the wolf from the door." Thursday afternoon was set aside "to receive any social calls," and among their first visitors were Dr. and Mrs. Parry, on their way south in December for their traditional winter's stay in Southern California.

The Lemmons' protracted honeymoon of 1881–82 was in part Sarah's adventurous choice, a "grand botanical raid into Arizona." John had already botanized in the Arizona Territory in 1879 and 1880, making his exciting discovery of native potatoes. Also, he had collected large bundles of Mormon tea to sell as medicine. The Lemmon honeymoon trip involved not only scaling an unnamed peak near Tucson (which would be named after Sarah Lemmon), but also exhausting plant collecting while avoiding outlaws, yellow fever, and, in particular, marauding Apaches. Ina Coolbrith, Oakland's librarian and poetess, wrote the couple on September 7, 1881, at Fort Bowie in the Chiricahuas, "I think, perhaps, you don't hear as much there as we do here, of the extent of this [Apache] outbreak." Actually the Lemmons had heard more than enough. A number of Apaches who escaped from the San Carlos Indian Reservation were now hiding nearby in the rugged mountains. Furthermore, on the Lemmons' return to Arizona in 1882, they were stopped in the Huachuca Mountains by a band of war-painted Apaches, whose chief carefully scrutinized the botanists' plant presses, then simply tapped his head and departed with his raiding party.

After the "honeymoon," on October 16, 1882, there was a reception for the couple at the Harmon Seminary in Oakland, with seven hundred tickets issued. The pair gave addresses on the perils of botanizing in Arizona, there were musical presentations, and Sarah showed her watercolor sketches of desert flowers. A newspaper reported of the event that "certainly all who heard went away with new ideas of the great wonderland to the southeast of us." For their twenty-fifth wedding anniversary in June 1905, the Lemmons returned to Tucson and, accompanied again by rancher Emerson Stratton, and by Mrs. W. R. Kitt, revisited the summit of Mount Lemmon. There, by the conifer where they had carved their initials and the rock cairn they had erected "during our Wedding trip" in 1881, the couple recelebrated the first scaling by a white woman of the peak, which Stratton had named after Sarah.

With Oakland established as their permanent home, the Lemmons energetically pursued their profession (figure 38). The pair made numerous presentations at the California Academy of Sciences, Sarah reading one of the first major papers on Pacific Coast ferns at the March 7, 1881, meeting just before they left for Arizona. The Lemmons, as John wrote, now "quite thoroughly explored the Pacific Slope from Mexico to British Columbia, discovering many hundreds of new plants and other objects of natural history" and abetted in their travels by free railroad and steamer passes.

In late 1884 John was appointed to superintend the Northern Pacific floral exhibit for the forthcoming World's Industrial and Cotton Centennial Exposition at New Orleans (1884–85), with Sarah in charge of the California exhibits for the Woman's Department. This became the most extensive fair ever held in the United States, and the floral displays were housed in the largest greenhouse in the world (six hundred feet long). At exposition's end, the Lemmons received $190 with their "first-grade gold medal diplomas," including for the "largest and best exhibit" of North American and of Pacific Slope flora, grasses and sedges, ferns and flowers, and Sarah's eighty paintings "of rare and beautiful plants," plus ten dollars for the exhibit of Arizona indigenous potatoes. The following year Sarah's "Marine Algae of the Western Shore" was published.

Continuing his search for a salaried position, in April 1886 John Lemmon was appointed special agent for the Forestry Division, U.S. De-

Figure 38. John and Sarah Lemmon in the field. (Courtesy University and Jepson Herbaria, University of California, Berkeley.)

partment of Agriculture, especially to collect and report "all the facts known of the Pitch Pines of the Pacific Coast." This project was in preparation for exhibits at the 1889 Paris World's Exposition (for which the Eiffel Tower was built). Lemmon received directions from Bernhard Fernow, chief of the division, with emphasis on "original observations of importance on the biology" of the conifers, including their turpentine capacity, with a specific and formal final report. When John turned in the completed treatise, after the deadline, it was rejected. Fernow wrote Lemmon that it "is not germane to the subject under consideration," there was an "absence of original observations," and furthermore the de-

partment could hardly print what was more appropriate for the newspapers to which Lemmon regularly sent popular articles. Fernow's reprimand stunned and depressed Lemmon.

For a period in 1887, while John was researching and writing and
Sarah painted, the couple lived on a ranch at Cholame. They optimisti-
cally attempted to organize a flora class for people in San Luis Obispo
County, but to no avail. Toward the end of this year it appeared that
John might get a position with a proposed museum at San Diego near
Hotel del Coronado. The Coronado Beach Company had contracted
with Henry A. Ward of Rochester, New York, to develop a large museum
with special exhibits such as he had presented earlier in San Francisco.
San Diego's Daniel Cleveland wrote John that the proposed museum
"would be a capital place for you and Mrs. Lemmon, where you would
find congenial work and appreciation and congenial people." Unfortunately, Ward liked San Diego so much that he decided to spend part of
the year there himself, putting the museum under his personal attention.

On January 2, 1888, Lemmon received another job offer, from
Charles Orcutt, now publisher and editor of *West American Scientist*.
Lemmon was invited to travel Northern California for a year to, as Orcutt wrote, "canvass for my magazine at every country town and farm
house, botanizing and forming collections of whatever comes in your
way, along the route." Lemmon would get seventy-five dollars a month
and could keep half of the plants collected, turning the other half over
to Orcutt; any month Lemmon sold more than seventy-five dollars in
subscriptions, Orcutt would retain the excess. When Lemmon turned
down the offer, Cleveland wrote him, "You are safer 'paddling your own
canoe' without having him on board as the Captain."

The Lemmons returned to Oakland that year and moved into California Hall downtown, where they would reside for a decade. This housing, the top floor of a business college, though not spacious, became a
favorite haunt for many visitors. During the 1888 summer the Lemmons
were scheduled to lecture for the Chautauqua Literary and Scientific Circle in Pacific Grove, John on grasses and Sarah on "Consider the Lilies,"
and the next summer Sarah talked again on lilies while John lectured on
California pines. From 1888 to 1892 the couple worked for the California State Board of Forestry, John as forester and Sarah as artist. If not
afield with collecting equipment, notebook, sketch pad, and camera, including forays into Oregon and Washington, John was busy at the
forestry board's office in downtown San Francisco or for extended periods in Sacramento. In 1888 his *Pines of the Pacific Slope* appeared and

in 1890 *Cone-bearers of California,* both receiving wide recognition. In-cluded were photographs by John and some of Sarah's exquisite cone paintings. In 1889 when the Lemmons were away from Oakland, Dr. Parry volunteered to manage their herbarium.

One of Sarah's interests involved her for nearly two decades. In 1884, when the Lemmons attended the New Orleans exposition, there was dis-cussion among participants about each state having an official state flower. At the 1893 World's Columbian Exposition in Chicago, with the Lemmons in attendance, John occupied himself in the Forestry Building before visiting relatives in Michigan. At the magnificent Woman's Build-ing Sarah delivered a lecture on forest conservation, and there was much discussion about voting for a national flower and consideration of state flowers. The National Floral Emblem Society was officially created, and to her surprise Mrs. Lemmon was appointed to chair the California State Committee, charged with "confirming the selection of the State flower by act of the next State legislature."

Although the California Poppy was already informally accepted as the state floral emblem, for the next decade Sarah proved an inspired activist on behalf of legislative recognition of the poppy, corresponding, sending out hundreds of publicity circulars, and traveling around the state giv-ing lectures. She noted that the wife of California's first "governor," Jessie Frémont, had earlier said, "Why, the Eschscholtzia *[sic],* of course; the Golden Poppy was born for the State of California." Upon Sarah's urging, in 1895 and again in 1899 a bill for adoption of the poppy al-most unanimously passed the legislature but was left unsigned by the governor. Governor Henry Gage's comment in 1899 was "I do not think the adoption of a State flower is a proper subject for legislation." Dur-ing May 1901 Professor Lemmon entered the fray, offering a resolution at the meeting of the Pacific States Floral Congress to support the Cali-fornia chapter's recommendation of *Eschscholzia californica* as Califor-nia's state flower.

On March 3, 1903, the *Oakland Evening Enquirer* ran a front-page headline, "THE GOLDEN POPPY IS OUR EMBLEM," describing the cere-mony the previous evening in Sacramento after the signing of the Cali-fornia Poppy bill by Governor George Pardee. Senator Smith, who had introduced the successful bill, told the audience that Mrs. J. G. Lemmon of Oakland "was the one largely responsible" for the statewide cam-paign, and presented her with the Bald Eagle quill-feather pen used by the governor in signing the bill.

At the end of the century the Lemmons finally moved to 5985 Telegraph Street in upper Oakland. Their first real and fitting home was a several-story white wooden house, with a sign above the porch announcing in large capital letters, "LEMMON HERBARIUM." As Professor Willis Jepson of Berkeley later contemplated, perhaps a bit enviously, "In the 80s and 90s if you asked any chance person in California the name of a botanist, if he knew the name of a botanist the chances are very greatly that he would have known the name of Lemmon and only Lemmon." The Lemmons' home was like a museum. One end of the parlor was a research alcove, floral paintings hung on the walls, and conifer cones crowded on high shelves. Sarah's work table was along one wall with John's across from it, both cluttered with plant specimens, papers, hand lenses. . . . The main part of the parlor, with its scattered wooden and wicker chairs and packed bookshelves, was separated from the study by heavy sliding drapes.

In 1902 Lemmon, having completed two demanding years on Berkeley's city council, was ready for relaxation. During the summer he joined two hundred Sierra Club aficionados to explore the southern Sierra Nevada. Lemmon's exciting discovery was a tall, yellow-orange lily with recurved petals, with a second specimen collected in a difficult defile of Bubb's Creek. The old botanist had to be helped by young San Francisco naturalist Lynwood Kelley, after whom John named the lily *Lilium kelleyanum*. Two years later Lemmon was honored with an invitation to the California Promotion Committee's fourth excursion, bound for a week in the Redwood country. By stage, railroad, and steamer the party of forty-nine important bankers and businessmen, and "Professor J. G. Lemmon of Oakland," traveled north to Eureka, dined and wined along the way, with a tour of Luther Burbank's garden. Amid the conifers Lemmon was in his element. As he wrote to Sarah, "The crowds seem to remember what I tell them about the noble trees and the lovely flowers all the better for my extravagant praise of them. . . . I've nearly lost voice in talking." Then for seven months in 1905 and 1906 the Lemmons were traveling and collecting in the woodlands of northern Mexico. And the next year, when John was seventy-five, he and Sarah were appointed delegates to the fifteenth National Irrigation and Forestry Congress, meeting in Sacramento. The two, in declining health but appreciating the value of irrigation and forests to the West, felt "we had to consent, though it will be difficult to leave our desks."

Through the years both John and Sarah had experienced poor health, even wondering before they married if two such chronically ill people should wed. In 1894 John confided to a friend that Sarah "is suffering with painful neck and left shoulder, but is as sprightly as ever, her friends never suspecting the effort she makes to be agreeable." Was it any wonder that many Lemmon friends wrote notes of sympathy to John when in March 1907 local newspapers reported the death of an Oakland-area Mrs. Lemon ("PIONEER DIES AT ADVANCED AGE")? No one bothered to read the obituary details about Mrs. Elizabeth Jane Lemon (spelled with one *m*).

Despite Sarah Lemmon's chronic weak constitution, she determinedly pursued her many activities, writing about topics from orchids to nursing schools and a tribute to her friend Clara Barton (who founded the American Red Cross), providing flower illustrations for assorted authors' works including the State Department of Education's *Third Reader,* and giving many public lectures in and out of the state on a variety of topics, particularly forest conservation. She was California's representative at the Louisiana Purchase Centennial in Chicago; chaired the California State Red Cross Board, being especially involved during the Spanish-American War; and was active with the California Federation of Women's Clubs, the California Press Club, the Women's Christian Temperance Union, the Ladies' Relief Corps of the Grand Army, and the Woman's Suffrage League. Furthermore, one admirer called her the "Nature Poet of the Sierras"; another hailed her "as one of the most accurate painters of nature in the State; her soul is in the art."

John Lemmon died of pneumonia in Oakland on November 24, 1908. His prime regret was that he and Sarah had never enjoyed a sustained source of income. In 1916 Harvard's Gray Herbarium had the Lemmon Herbarium appraised and was shocked to learn that about "40 percent of the specimens are without data." After Sarah moved away, the Lemmon home on Telegraph, still housing the herbarium, was rented for some time. In 1927, when the house was advertised for sale, University of California professor Willis Jepson rushed to remove the huge herbarium of at least ten thousand specimens of flowering plants, ferns, algae, and cones, among them Sir Joseph Hooker's wedding-present floral gift; there was even an insect collection. Jepson was dismayed to discover the historic collection crammed into a small, cold, damp, unlighted storage room. It took about thirty packing cases to transport it to the University of California. John Lemmon's monumental masterpiece on trees of the Pacific Slope still rests in its storage box today, an uncompleted manuscript.

Sarah, the "& wife" of the Lemmon Herbarium specimen labels, never really recovered from Lemmonia's death. Suffering a nervous breakdown as early as 1913, she moved to Stockton, entering the Stockton State Hospital on April 22, 1916. Visitors reported that Sarah remained in fairly good health, spending a good deal of her time reading. She died there of "general arteriosclerosis" on January 15, 1923, "aged 83" according to the hospital record (which was incorrect). In Kate Brandegee's words, Sarah's "firm and aggressive attitudes had a powerful lifelong influence upon her gentle, ungainly husband."

In the centennial year of Sarah Lemmon's state flower (2003), a gravestone for the couple was erected at Oakland's Mountain View Cemetery, bearing the simple inscription "COMPANIONS IN BOTANY."

MARY KATHARINE LAYNE CURRAN BRANDEGEE

My botanical trips are too numerous to catalogue.

—Kate Brandegee

Most of California's famous resident naturalists after the gold rush arrived from the East as adults. Enchanted by the Golden State's diverse natural history, they then remained as citizens. But Mary Katharine "Kate" Layne arrived as a child and remained a permanent resident well into the twentieth century. She was born in the wilds of western Tennessee on October 28, 1844, and her wanderlust father moved his large family to the gold-mining foothills of El Dorado County, California, when Kate was nine. In the Sierra Nevada she grew up "in much isolation," as she wrote. At twenty-two she became Mrs. Hugh Curran in Folsom, up the American River from Sacramento. But after Curran, a confirmed drunkard, died in 1874, Kate moved the next year to San Francisco and matriculated at the University of California's two-year-old medical department, the second oldest "medical school" in the West.

Katharine Curran received her M.D. in 1878 but found that her new practice left plenty of leisure time. From childhood she had been interested in natural history, especially insects and birds, and luckily she discovered a haven at the California Academy of Sciences. In February 1879 she was elected a resident member of the academy and in January 1882 read her first paper, on "caecums" of the Black-tailed Jackrabbit. Although Dr. Curran admitted that "with me botany was accidental," she was already helping at the academy herbarium, which was "in a shocking condition." During 1882 she started her own collection of the local flora and a year later joined Justin P. Moore to curate the academy's botany collection. Curran totally abandoned medicine for plants, "giving up my small practice and devoting the succeeding ten very active years to the service of the Academy."

In addition to modernizing the academy herbarium, Curran began extensive "botanical rambles—Sierras, valleys, Coast Ranges, and deserts," and over the border into Nevada. Taking advantage of free train passes, she traveled sometimes by Pullman and even in a freight locomotive. During the decade of the 1880s she published more than two dozen botanical articles, ranging from descriptions of plants she collected during 1884 to a flora of Yosemite, Sierra Nevada plants occurring in the Coast Ranges, and a catalogue of the plants and ferns of San Francisco. Her unfulfilled life ambition was to write a flora for the entire state.

Curran shared her interests with numerous old-timers of the academy, among them Dr. Hans Behr, who tutored her in botany; Dr. H. W. Harkness, whom she assisted with his fungi research; and Dr. Albert Kellogg, as well as her young assistant, Alice Eastwood. But the two men who had the greatest influence on her scientific—and personal—life were Edward Lee Greene and Townshend Stith Brandegee. Both men had earlier pursued botany in the West, Brandegee with the famous Hayden and others, and Greene as an itinerant preacher and science teacher.

Edward Lee Greene, now minister at St. Mark's Episcopal Church in Berkeley, spent much time at the academy herbarium. It was probably Greene who encouraged Curran to send her plants to Asa Gray and others. In 1885, after Curran and Greene were jointly appointed curators of botany, Greene egotistically bragged to Gray that he was the academy botany curator, with Kate as his assistant. Both Greene and Curran were strong, independent personalities who generated sparks wherever they went. After Greene departed for a position at the University of California, he spitefully referred to Kate as a "she devil," while she "of the bitter pen" attacked his prolific naming of new species. Greene countered that "since he was an ordained minister, any attack on his acts was a sin against God, and the act was therefore devilish." Nonetheless, Greene named several new species of California plants *layneae* in Kate's honor, using her maiden name.

Relations between Kate Curran and Townshend Brandegee—according to Marcus Jones "the oldest and most accomplished botanist in the country"—proved considerably different (figure 39). Brandegee, born in 1843 in Connecticut, graduated as a student of William Brewer's from Yale's Sheffield Scientific School in 1870, with broad interests including birds, mammals, and plants. He first came to California for Professor Charles S. Sargent during the winter of 1886–87 to obtain more tree

Figure 39. Katharine and Townshend Brandegee. (Courtesy University and Jepson Herbaria, University of California, Berkeley.)

trunks for the historic Jesup log collection at New York's American Museum of Natural History. When Kate Curran, whom Jones called the "husky and red-blooded woman of tireless energy," met Townshend, the "little, quiet, refined man," at the academy, she wrote to her sister that she fell "insanely in love" with her "Dolly," as she came to call him. The two were married in San Diego on May 29, 1889. For their honeymoon they set off on foot for San Francisco, botanizing along the way. They headquartered at the California Academy of Sciences for five years. Kate had earlier encouraged publication of the academy proceedings, and now her husband underwrote a new journal, *Zoe,* primarily for her benefit. In 1892 Mrs. Brandegee started the West Coast's first botanical club.

The adventures and anecdotes generated by the Brandegees during their California residence are legion. Shortly after marriage, they made a collecting excursion around Ukiah, accompanied by a female companion. Although they had a wagon for the specimens, the three walked. Dolly and Kate "were very demonstrative, continually stopping to kiss each other." Their companion, embarrassed, kept shouting over her shoulder at them, "My face is set sternly to the front." When the Brandegees were botanizing in the Santa Lucia Mountains, 150-pound Kate broke her leg and had to be carried on a stretcher for two days by four ranchers through rough country. During an extended trip into Baja California in 1893 with her husband and Dr. Gustave Eisen, Mrs. Brandegee returned a month early by ship. It was a foggy night when the vessel reached San Pedro, and it wrecked on the rocks. When she later told Townshend about the accident, his first question was "Did you save the precious plants?" Mrs. Brandegee was fond of a glass of beer, and on a hot day in the eastern Mohave Desert, she dropped into a saloon and ordered a mug. When she was struggling to find money for the drink in her ragged work skirt, the barman said, "No, no, no, poor Lady! I'd never think of taking a dime from you for a glass of beer." While collecting in the Tehachapi Mountains, Kate stayed at a boardinghouse, arranged her plants in tissue paper in the tiny front parlor, and slept on the floor with her revolver beside her. In reply to her landlady, who once said, "God bless you," Kate promptly retorted, "There is no God!" She did possess a sharp tongue, and on one occasion she used it on a male companion with "I've dissected better men than you." But it was the railroad agent at Barnwell in the Mohave Desert who captured the essence of the Brandegees when he said, "They went out after plants on the hills yonder. And the old lady 'ud post the old man on a rock and she'd scout around and bring him plants for to look at. The old lady was boss. That was plain to see."

In 1893 Kate Brandegee turned the academy herbarium over to Alice Eastwood. The next year she and Dolly left for San Diego. There they developed a delightful site on a mesa overlooking the town, with a brick building to house their herbarium, a spacious botanical garden filled with rare natives, a pet guinea hen, "mockingbirds, quail calling, and other native songbirds making the air musical with song," as well as a California Thrasher that would come at Kate's beck. Initially, the couple themselves occupied a wooden shack with a tent in front. They continued their plant studies and field explorations, together or alone, including a number of excursions into Mexico.

In 1906 the Brandegees donated their botany library and herbarium, called "one of the richest private herbaria ever assembled in this country," to the University of California. More than doubling the size of the university herbarium, it included 76,166 specimens. As was true for many other collectors, scores of the Brandegees' specimens were poorly labeled. Kate and Townshend, homesick for the San Francisco Bay region, moved to Berkeley, where at the university herbarium Townshend was appointed honorary curator—without salary—and Kate worked as a volunteer. She loved to stroll about the wooded campus in the most rustic attire. Students, mistaking this shabbily dressed elderly woman for a vagrant, often pressed money into her hand, saying, "Take it to help you, dear."

Kate Brandegee, despite her vigorous mien, suffered from diabetes for much of her life. Because of her affliction, she abandoned a planned trip east for herbarium visits at St. Louis, New York, and Harvard to examine western type specimens. In 1920, when she did attempt the excursion, she collapsed in St. Louis and nearly died before returning home. Her condition worsened, and she passed away on April 3, 1920. At first, Townshend continued to work every day at the university herbarium. As someone reported, he "has 'picked up' a lot. He seems very cheerful and contented and really appears to be enjoying life since he was freed from the intellectual dominance of Mrs. B." But as time passed, he deafened, was blind in one eye, suffered from a debilitating paralysis, and became withdrawn. Yet he continued his work with the Mexican collections of German botanist Dr. C. A. Purpus until contracting pneumonia. A few months later, on April 7, 1925, Townshend Brandegee died.

This distinctive couple is buried in Kensington's Sunset View Cemetery, near Berkeley. Mrs. Brandegee had been a California resident for nearly seventy years.

Postlude

The End of California's "Frontier"

The frontier has gone, and with its going has closed the first period of American history.

—Frederick Jackson Turner

The World's Columbian Exposition, commemorating the four hundredth anniversary of Columbus's arrival in the Americas in 1492, opened on May 1, 1893, in Chicago. At a special meeting there of the American Historical Association, a University of Wisconsin history professor read a landmark treatise titled "The Significance of the Frontier in American History." Frederick Jackson Turner began his paper with a quotation from the superintendent of the 1890 census: "a frontier line" no longer exists for the nation and consequently the frontier "cannot, therefore, any longer have a place in the census reports." The frontier opened by Columbus had—supposedly—come to an end.

THE DEATH VALLEY EXPEDITION, 1890–1891

Among the vicissitudes of a desert exploration . . .

—Frederick Coville

In 1890 the population center for the United States was twenty miles east of Columbus, Indiana. Some nineteen hundred miles to the southwest lay Death Valley.

On January 20, 1891, a small party of men, reported Frederick Coville, emerged "into Death Valley near the south end of its alkali-flat" (figure 40). In the bottom of this California desert playa, more than two hundred feet below sea level, extended a "snow-white stretch of salt and

Figure 40. Death Valley, looking east from the slope of Telescope Peak, 1996. (Photo by R. G. Beidleman.)

alkali, and to the northward, perhaps fifty miles away, mountains, valleys, and salt-flat vanished in haze." Over the next three months of fieldwork, members of the Death Valley Expedition, the first biological survey in California authorized by an act of Congress to study the geographic distribution of animals and plants in the West, explored little-known parts of the Golden State.

Dr. C. Hart Merriam was this expedition's driving force. In 1881 the Division of Entomology was created in the U.S. Department of Agriculture, and five years later came the first appropriation for a branch of economic ornithology. Merriam was a main objector to the focus on economics, feeling that geographic distribution was more significant. With encouragement from the new American Ornithologists' Union, a section of ornithology was created within the Department of Agriculture, with Merriam as its head. He then intensified his efforts on behalf of biological surveys, with the result that the USDA finally embarked on comprehensive studies of western natural history, including the 1889 life zone study by Merriam at Arizona's San Francisco Peaks, followed by the Death Valley Expedition.

Under Merriam, the acting chief of the Death Valley Expedition was zoologist Theodore Palmer of Agriculture's new Division of Ornithology and Mammalogy, who was well familiar with California, having grown up in Oakland. In 1895 he would give his extensive collection of bird specimens and eggs to the California Academy of Sciences, only to have it destroyed in the earthquake of 1906. The expedition's zoologist, Vernon Bailey, had served as chief field naturalist under Merriam in 1888 during a Wyoming survey and was Merriam's assistant for the 1889 Arizona expedition. Eventually Bailey became Merriam's right-hand man in the Bureau of Biological Survey and married Merriam's sister Florence, a field naturalist in her own right. Ornithologist Albert K. Fisher had been a classmate of Merriam's at New York's College of Physicians and Surgeons and was now with the USDA. Edward Nelson, a naturalist whose field experiences dated back to an expedition in the West with paleontologist E. D. Cope in 1872, was with John Muir in Alaska, and had been living and exploring in the Southwest before he joined Merriam. Albert Koebele, famous for having introduced Australian Lady Beetles into California in 1888 to control citrus scale insects, joined the expedition in Death Valley for a brief period. Frederick Coville, a botanist with the USDA, was responsible for the expedition's plant studies. His assistant was Frederick Funston, a botanist from Kansas now with the U.S. Department of Agriculture. In this party were among the nation's most important natural scientists of the day and into the next century.

Despite its name, the Death Valley Expedition investigated other southwestern regions as well. But especially in Death Valley the naturalists came to appreciate that the "line of frontier" was definitely not passé. After a month without mail, some of the party set out from the Amargosa Valley on February 19 for the closest post office, the old silver-mining town of Panamint City. The mail run took five days one way, including mushing through what Coville recalled as a "terrific snowstorm" in the Panamint Mountains.

The naturalists found the greatest diversity of animals and plants in Death Valley around the desert "oases" of Bennett and Mesquite wells, Saratoga Springs, and especially Furnace Creek and the adjacent Pacific Borax Company ranch. Altogether, seventy-eight species of birds were encountered, with two-thirds at Furnace Creek and the ranch. The Western Meadowlark occurred most widely throughout the valley, while attracted to Furnace Creek, especially during migration and winter, were such species as the Ruby-crowned Kinglet, Yellow-rumped Warbler, Lazuli Bunting, a few American Robins, Killdeer (noisy during moonlit nights), and one April Day a solitary Eared Grebe. The Prairie Falcon that paused atop a haystack to survey a covey of Gambel's Quail was quickly accessioned. The zoologists collected eleven species of reptiles, including nine lizards, the Side-blotched Lizard being very common, as well as a whipsnake and Sidewinder. There were two amphibians, the Pacific Tree-frog and Red-spotted Toad, with an unidentified tadpole. At Saratoga Springs the Desert Pupfish, now endangered, was obtained. The collections even included two freshwater snails, several hundred of the Variable Paludestrina being found at Saratoga Springs. When the stomach of a pupfish was dissected, it contained a small snail as well as insect remains. Koebele obtained over sixty different insects, half of them beetles. The zoological findings were published in North American Fauna no. 7. Coville's extensive botany report, including Death Valley, was published by the U.S. National Herbarium in 1893. In Charles Hunt's 1966 *Plant Ecology of Death Valley, California,* he wrote of Coville's publication: "It is one of the classics in scientific literature—one that contributed greatly to the foundation for the field of study that later came to be known as plant ecology."

THE LINGERING CALIFORNIA FRONTIER

The first rough conquest of the wilderness is
accomplished.
                          —Frederick Jackson Turner

For many sections of California the frontier prevailed for at least a quarter of a century more, with intervening events affecting both society and natural science. In 1892 Stanford University's Hopkins Seaside Laboratory (now Hopkins Marine Station) became the first marine field station on North America's West Coast. In 1894 Thaddeus Lowe, of balloon fame, established Southern California's first astronomical observatory. When the Harriman Alaska Expedition of 1899 sailed with its shipload of American scientists, half had strong California ties: William Brewer; Merriam and two of his Death Valley Expedition colleagues, Coville and Fisher; Harvard's Charles Palache, who had mapped San Francisco's earthquake faults; William Ritter, head of the University of California's zoology department; Charles Keeler, director of the California Academy of Sciences museum; William Dall, who had been curator of ichthyology there; zoologist Robert Ridgway, who served with the King survey; geologist Grove Karl Gilbert, former member of the Wheeler survey, soon to head up studies on the effects of hydraulic mining in California; and

John Muir. This year Merriam's North American Fauna no. 16, the biological survey of Mount Shasta, appeared.

In 1908, through the energetic efforts of Annie Alexander, the University of California's Museum of Vertebrate Zoology was established, and two years later the Department of Paleontology. During 1909 on the Colorado River the Laguna Dam was completed, about fifteen miles north of old Fort Yuma, the first of many dams on the lower Colorado, which changed the region's ecology. In 1913 California hosted the West Coast's first international scientific entourage, the International Phytogeographical Excursion, guided in the field by University of California botanist Willis Jepson and California Academy of Sciences botany curator Alice Eastwood. In 1914 the completion of the Panama Canal put California six thousand miles closer by sea to the East Coast. And in January 1915 the first transcontinental telephone call was made between New York and California.

In a sense the 1915 celebrations of San Francisco's Panama-Pacific International Exposition and San Diego's Panama California Exposition signaled the end of the historic frontier for California. Yet for those with a naturalist's bent, there remained many frontiers ahead. Indeed, also in 1915, on the east side of the Golden State, geographer Ellsworth Huntington and his desert botanist companion Samuel Parish were continuing to explore in the vast solitude of Death Valley. And the population center of the United States was still in Indiana.

Selected References

GENERAL REFERENCES

Adler, Kraig. 1979. *A brief history of herpetology in North America before 1900*. Soc. Study Amph. and Rept. Herpetological Circular 8.

Alden, Roland H., and John D. Ifft. 1943. *Early naturalists in the Far West*. San Francisco: Calif. Acad. Sci.

Alt, David, and Donald W. Hyndman. 2000. *Roadside geology of northern and central California*. Missoula, MT: Mountain Press.

Badè, William F. 1924. *The life and letters of John Muir*. 2 vols. Boston: Houghton Mifflin.

Badger, Geoffrey. 1996. *The explorers of the Pacific*. 2nd ed. Kenthurst, Australia: Kangaroo Press.

Bancroft, Hubert H. 1886–90. The works of Hubert H. Bancroft: A history of California. 7 vols. San Francisco: History Company.

Barnhart, John H. 1965. *Biographical notes on botanists*. 3 vols. Boston: G. K. Hall.

Beauchamp, R. Mitchel. 1986. *A flora of San Diego County, California*. National City, CA: Sweetwater River Press. Includes type localities and collectors for new San Diego County species.

Bebee, Rose M., and Robert M. Senkewicz. 2001. *Lands of promise and despair: Chronicles of early California, 1535–1846*. Berkeley: Heyday Books.

Beidleman, Richard G. 1960. European naturalists on the early American frontiers. *Actes du IXe Congrès international d'histoire des sciences*, pp. 338–39.

———. 1964. Biology and the North American wilderness. *BioScience* 14(12): 22–27.

———. 1969. Field biology in frontier America. *The Science Teacher* 36(9): 52–55.

———. 1970. George Engelmann, botanical gatekeeper for the West. *Horticulture* 48(4):42–43, 52–53, 56–58.

———. 1973. John Bidwell. *Horticulture* 51(11):50–53.

Brendel, Frederick. 1879–80. Historical sketch of the science of botany in North America from 1635 to 1840. *American Naturalist* 13:754–71; 14:25–38.

Brewer, William H. 1880. List of persons who have made botanical collections in California. In *Geological survey of California: Botany,* by Sereno Watson, 2:553–59. Boston: Little, Brown.

Britten, James, and G. S. Boulger. 1893. *A biographical index of British and Irish botanists.* London: West, Newman.

California Division of Beaches and Parks. 1956. *Fort Ross State Historical Monument.* Sacramento: Calif. State Printing Office.

Coats, Alice M. 1969. *The plant hunters.* New York: McGraw-Hill.

Commite, Anne, ed. 2001. *Women in world history: Biographical encyclopedia.* Waterford, CT: Yorkin Publications.

Dakin, Susanna B. 1949. *The lives of William E. P. Hartnell.* Stanford, CA: Stanford Univ. Press.

Dance, S. Peter. 1966. *Shell collecting: An illustrated history.* Berkeley: Univ. Calif. Press.

Desmond, Ray. 1994. *Dictionary of British and Irish botanists and horticulturists.* London: Taylor & Francis.

Drake, Ellen T., and William M. Jordan. 1985. *Geologists and ideas: A history of North American geology.* Boulder, CO: Geol. Soc. Amer.

Drury, Aubrey. 1939. *California: An intimate guide.* New York: Harper & Brothers.

Dupree, A. Hunter. 1957. *Science in the federal government.* Cambridge, MA: Harvard Univ. Press.

———. 1959. *Asa Gray.* Cambridge, MA: Harvard Univ. Press.

Eastwood, Alice. 1939. Early botanical explorers on the Pacific Coast and the trees they found there. *Calif. Hist. Soc. Quart.* 18(4):335–46.

Engstrand, Iris H. W. 1981. *Spanish scientists in the New World: The eighteenth-century expeditions.* Seattle: Univ. Wash. Press.

Ertter, Barbara. 1995a. The changing face of California botany. *Madroño* 42(2):114–22.

———. 1995b. The changing face of western botany. *Kalmiopsis* 1995:18–25.

———. 1998. The California Geological Survey: Botanical impact and current implications. Draft article.

———. 2000. People, plants, and politics: The development of institution-based botany in California. In *Cultures and institutions of natural history,* ed. Michael T. Ghiselin and Alan E. Leviton. Calif. Acad. Sci. Memoir 25: 203–48. San Francisco: Calif. Acad. Sci.

———. 2001 [orig. 2000]. Our undiscovered heritage: Past and future prospects for species-level botanical inventory. *Madroño* 47(4):237–52.

———. 2004. The flowering of natural history institutions in California. *Proc. Calif. Acad. Sci.* 55(suppl. 1, no. 4):58–87.

Essig, E. O. 1931. *A history of entomology.* New York: Macmillan.

———. 1933. *The Russians in California.* San Francisco: Calif. Hist. Soc.

Essig, E. O., Adele Ogden, and Clarence J. DuFour. 1991. *Fort Ross: California outpost of Russian Alaska 1812–1841*. Kingston, Ontario: Limestone Press.

Evans, Howard E. 1993. *Pioneer naturalists*. New York: Henry Holt.

Ewan, Joseph. 1950. *Rocky Mountain naturalists*. Denver: Univ. Denver Press.

———. 1955. San Francisco as a mecca for nineteenth century naturalists. In *A century of progress in the natural sciences: 1853–1953*, pp. 1–63. San Francisco: Calif. Acad. Sci.

———, ed. 1969. *A short history of botany in the United States*. New York: Hafner.

Ewan, Joseph, and Nesta Ewan. 1981. *Biographical dictionary of Rocky Mountain naturalists*. The Hague: Dr. W. Junk.

Farquhar, Francis P. 1965. *History of the Sierra Nevada*. Berkeley: Univ. Calif. Press.

Fenton, Carroll L., and Mildred A. Fenton. 1945. *The story of the great geologists*. Garden City, NY: Doubleday, Doran.

Fischer, Dan L. 2001. *Early southwest ornithologists, 1528–1900*. Tucson: Univ. Ariz. Press.

Garraty, John A., and M. C. Carnes, eds. 1999. *American national biography*. New York: Oxford Univ. Press.

Geiser, Samuel W. 1948. *Naturalists on the frontier*. Dallas: Southern Methodist Univ. Press.

Goetzmann, William H. 1979. *Army exploration in the American West: 1803–1863*. Lincoln: Univ. Nebr. Press.

———. 1986. *New lands, new men*. New York: Viking Press.

———. 1993. *Exploration and empire: The explorer and the scientist in the winning of the American West*. New York: History Book Club.

Goetzmann, William H., and William N. Goetzmann. 1986. *The West of the imagination*. New York: W. W. Norton.

Graustein, Jeannette E. 1967. *Thomas Nuttall: Naturalist*. Cambridge, MA: Harvard Univ. Press.

Gray, Jane Loring, ed. 1893. *Letters of Asa Gray*. 2 vols. Boston: Houghton Mifflin.

Grinnell, Joseph. 1932. Type localities of birds described from California. *Univ. Calif. Publ. Zoology* 38(3):243–324.

Grinnell, Joseph, J. S. Dixon, and Jean M. Linsdale. 1937. *Fur-bearing mammals of California*. Berkeley: Univ. Calif. Press.

Gutierrez, Ramón, and Richard J. Orsi, eds. 1998. *Contested Eden*. Berkeley: Univ. Calif. Press.

Harlow, Neal. 1989. *California conquered: The annexation of a Mexican province 1846–1850*. Berkeley: Univ. Calif. Press.

Hedrick, U. P. 1950. *A history of horticulture in America to 1860*. New York: Oxford Univ. Press.

Heidler, David S., and Jeanne T. Heidler. 2000. *Encyclopedia of the American Civil War*. 4 vols. Santa Barbara, CA: ABC-CLIO.

Henry, John F. 1984. *Early maritime artists of the Pacific Northwest coast, 1741–1841*. Seattle: Univ. Wash. Press.

Higgins, Edith B. 1959. Type localities of vascular plants in San Diego County, California. *Trans. San Diego Soc. Nat. Hist.* 12:347–406.

Hill, Joseph J. 1927. *The history of Warner's Ranch and its environs.* Los Angeles: privately printed.

Hoobler, Dorothy, and Thomas Hoobler. 1980. *Photographing the frontier.* New York: G. P. Putnam's Sons.

Hoover, Mildred B., Hero E. Rensch, Ethel G. Rensch, and William N. Abaloe. 1990. *Historic spots in California.* 4th ed., rev. by Douglas E. Kyle. Stanford, CA: Stanford Univ. Press.

Howard, Thomas F. 1998. *Sierra crossing: First roads to California.* Berkeley: Univ. Calif. Press.

Hume, Edgar E. 1942. *Ornithologists of the United States Army Medical Corps.* Baltimore: Johns Hopkins Univ. Press.

Humphrey, Harry B. 1961. *Makers of North American botany.* New York: Ronald Press.

Jepson, Willis L. Field books, Jepson correspondence volumes, and Jepson collections. Berkeley: Univ. Calif. Jepson Herbarium Archives.

———. 1942. Early botanical ascents of Mount Shasta. *Sierra Club Bull.* 27(4):23–30.

Johnson, Robert, and Clarence Buel, eds. 1888. *Battles and leaders of the Civil War,* vol. 3. New York: Century Company.

Keay, John. 1991. *History of world exploration.* New York: Mallard Press.

Kelly, Howard A. 1914. *Some American medical botanists.* Troy, NY: Southworth.

Kemp, Peter, ed. 1988. *The Oxford companion to ships and the sea.* New York: Oxford Univ. Press.

Kibbe, Alice L. 1953. *Afield with plant lovers and collectors.* Carthage, IL: by the author.

Lavender, David. 1982. *Colorado River country.* New York: E. P. Dutton.

Lemmon Collection (John G. and Sarah Plummer Lemmon). Berkeley: Univ. Calif. Herbarium Archives.

Leviton, Alan E., and M. L. Aldrich, eds. 1997. *Theodore Henry Hittell's The California Academy of Sciences 1853–1906.* San Francisco: Calif. Acad. Sci.

Lewis, Donovan. 1993. *Pioneers of California.* San Francisco: Scottwall Associates.

Mallis, Arnold. 1971. *American entomologists.* New Brunswick, NJ: Rutgers Univ. Press.

Malone, Dumas, ed. 1936. *Dictionary of American biography.* 20 vols. New York: Charles Scribner's Sons.

Marschner, Janice. 2001. *California: A snapshot in time, 1850.* Sacramento: Coleman Ranch Press.

Mathias, Mildred E. 1989. The fascinating history of the early botanical exploration and investigations in Southern California. *Aliso* 12(3):407–33.

McElroy, Walter, ed. 1946. *Monterey Peninsula.* 2nd ed. American Guide Series. San Francisco: Hooper Printing.

McKelvey, Susan. 1955. *Botanical exploration of the Trans-Mississippi West: 1790–1850.* Jamaica Plain, MA: Arnold Arboretum.

Meade, Gordon M. 1987. Surgeons, birds, and the U.S. Army, part 1. *Birding* 19(1):33–40.

Mearns, Barbara, and Richard Mearns. 1992. *Audubon to Xantus.* New York: Academic Press.

Meisel, Max. 1929. *A bibliography of American natural history: The pioneer century, 1769–1865.* 3 vols. Brooklyn: Premier Publ.

Morning, John. 2002. *Early American naturalists exploring the American West, 1804–1900.* New York: Cooper Square Press.

Morwood, William. 1973. *Traveler in a vanished landscape: The life & times of David Douglas, botanical explorer.* New York: Clarkson N. Potter.

Nilsson, Karen B. 1994. *A wild flower by any other name.* Yosemite, CA: Yosemite Nature Assn.

Norris, L. David, J. C. Milligan, and O. B. Faulk. 1998. *William H. Emory, soldier-scientist.* Tucson: Univ. Ariz. Press.

Norris, Robert M., and Robert W. Webb. 1990. *Geology of California.* 2nd ed. New York: John Wiley & Sons.

Ornduff, Robert. 2000. Piss and vinegar: Skeletons in our botanical closet. *Fremontia* 28(2–4):18–20.

Paddison, Joshua, ed. 1999. *A world transformed: Firsthand accounts of California before the gold rush.* Berkeley: Heyday Books.

Palmer, Theodore S. 1928. Notes on persons whose names appear in the nomenclature of California birds. *Condor* 30(5):261–307.

Parry, Joan. 1956. Who first discovered these western flowers? *Pacific Discovery* 9(1):12–20.

Perry, Claire. 1999. *Pacific Arcadia: Images of California 1600–1915.* New York: Oxford Univ. Press.

Pittman, Ruth. 1995. *Roadside history of California.* Missoula, MT: Mountain Press.

Rawls, James J. and Richard J. Orsi, eds. 1999. *A golden state.* Berkeley: Univ. Calif. Press.

Reveal, James L. 1992. *Gentle conquest.* Washington, DC: Starwood Publishing.

Reveal, James L., and J. S. Pringle. 1993. Taxonomic botany and floristics. In *Flora of North America north of Mexico,* 1:157–92. New York: Oxford Univ. Press.

Roberts, David. 2000. *A newer world: Kit Carson, John C. Frémont, and the claiming of the American West.* New York: Simon & Schuster.

Roderick, Wayne. 2001. Early plant explorers of the Pacific Coast. *Manzanita* 5(4):1–7.

Rodgers, Andrew Denny III. 1942. *John Torrey: A story of North American botany.* Princeton, NJ: Princeton Univ. Press.

———. 1968. *American botany 1873–1892.* New York: Hafner.

Ronda, James P. 2005. *Beyond Lewis & Clark: The army explores the West.* Tacoma: Wash. State Hist. Soc.

Rothenberg, Marc, ed. 1996. *The papers of Joseph Henry,* vol. 7. Washington, DC: Smithsonian Inst. Press.

Royce, Josiah. 2002 [orig. 1856]. *California: A study of American character.* Berkeley: Heyday Books.

Russell, Carl P. 1947. *One hundred years in Yosemite.* Berkeley: Univ. Calif. Press.

Saunders, Charles F. 1933. *Western wild flowers and their stories.* Garden City: Doubleday, Doran.

Schoenherr, Allan A. 1992. *A natural history of California.* Berkeley: Univ. Calif. Press.

Schubert, Frank N. 1980. *Vanguard of expansion: Army engineers in the Trans-Mississippi West, 1819–1879.* Washington, DC: Office of the Chief of Engineers, Hist. Div.

Slack, Nancy G. 1993. Botanical exploration of California: From Menzies to Muir. In *John Muir, life and work,* ed. Sally M. Miller, pp. 194–242. Albuquerque: Univ. New Mexico Press.

Smith, Michael L. 1987. *Pacific visions: California scientists and the environment, 1850–1915.* New Haven, CT: Yale Univ. Press.

Spongberg, Stephen A. 1990. *A reunion of trees.* Cambridge, MA: Harvard Univ. Press.

Stadtman, Verne A., ed. 1968. *The centennial record of the University of California.* Berkeley: Univ. Calif. Printing Dept.

Stanton, William. 1991. *American scientific expeditions 1803–1860.* Philadelphia: Amer. Philosophical Soc. Library.

Starr, Kevin. 1973. *Americans and the California dream, 1850–1915.* New York: Oxford Univ. Press.

Stephen, Leslie, and Sidney Lee, eds. 1908–09. *Dictionary of national biography.* 22 vols. London: Macmillan.

Stillman, F. D. B. 1869. Footprints of early California discoveries. *Overland Monthly* 2:256–63.

Taft, Robert. 1953. *Artists and illustrators of the old West: 1850–1900.* Princeton, NJ: Princeton Univ. Press.

Thayer, Christopher. 1995. Botanical exploration in the East Bay: Then and now. *Four Seasons* 10(1):4–21.

Thomas, John H. 1969. Botanical explorations in Washington, Oregon, California and adjacent regions. *Huntia* 3:5–62.

Thompson, Gerald. 1983. *Edward F. Beale & the American West.* Albuquerque: Univ. New Mexico Press.

United States Works Progress Administration. 1943. *California.* New York: Hastings House.

Utley, Robert M. 1997. *A life wild and perilous: Mountain men and the paths to the Pacific.* New York: Henry Holt.

Viola, Herman J. 1987. *Exploring the West.* Washington, DC: Smithsonian Books.

Walker, Dale L. 1999. *Bear Flag Rising: The conquest of California, 1846.* New York: Tom Doherty Assoc.

Wallace, Edward S. 1995. *The great reconnaissance.* Boston: Little, Brown.

Watson, Sereno. 1880. *Geological Survey of California: Botany,* vol. 2. Boston: Little, Brown.

White, David A., ed. 1998. *News of the plains and Rockies 1803–1865,* vol. 5: *Later explorers, 1847–1865.* Spokane, WA: Arthur H. Clark.

Wilson, James G., and John Fiske. 1887–89. *Appletons' cyclopaedia of American biography*. New York: Appleton.

MAPS AND GAZETTEERS

California Division of Mines and Geology. 1966. *Geologic map of California*. Sacramento: Dept. of Conservation.
California State Parks. Office of Historic Preservation. 1996. *California historical landmarks*. Sacramento: Calif. State Parks.
DeLorme. 1988a. *Northern California atlas & gazetteer*. 2nd ed. Freeport, ME: DeLorme Mapping.
———. 1988b. *Southern and Central California atlas & gazetteer*. 4th ed. Yarmouth, ME: DeLorme Mapping.
Durham, David L. 1998. *California's geographic names*. Clovis, CA: Word Dancer Press.
Gudde, Erwin G., and William Bright. 1998. *California place names*. 4th ed. Berkeley: Univ. Calif. Press.
Hart, James D. 1987. *A companion to California*. Rev. ed. Berkeley: Univ. Calif. Press.

PRIMARY TAXONOMIC REFERENCES
(FOR SCIENTIFIC AND COMMON NAMES)

Birds

American Ornithologists' Union. 1957. *Check-list of North American birds*. 5th ed. Washington, DC: AOU.
———. 1998. *Check-list of North American birds*. 7th ed. Washington, DC: AOU.
Small, Arnold. 1994. *California birds: Their status and distribution*. Vista, CA: Ibis Publ.

Fishes

Eschmeyer, William N., and Earl S. Herald. 1983. *A field guide to Pacific Coast fishes of North America*. Boston: Houghton Mifflin.
Gilbert, Carter R., and J. D. Williams. 2002. *National Audubon Society field guide to fishes*. Rev. ed. New York: Alfred A. Knopf.
McGinnis, Samuel M. 1984. *Freshwater fishes of California*. Berkeley: Univ. Calif. Press.

Invertebrates

Essig, E. O. 1926. *Insects of western North America*. New York: Macmillan.
Morris, Percy A. 1980. *Pacific Coast shells*. 2nd ed. Boston: Houghton Mifflin.
Powell, Jerry A., and C. L. Hogue. 1979. *California insects*. Berkeley: Univ. Calif. Press.

Mammals

Jameson, E. W., Jr., and H. J. Peeters. 1988. *California mammals.* Berkeley: Univ. Calif. Press.

Miller, Gerrit S., Jr., and Remington Kellogg. 1955. *List of North American recent mammals.* U.S. Nat. Mus. Bull. 205. Washington, DC: Smithsonian Inst.

Plants

Abrams, Leroy. 1940–60. *Illustrated flora of the Pacific states.* Stanford, CA: Stanford Univ. Press. Dated, but good reference for common names.

Beauchamp, R. Mitchel. 1986. *A flora of San Diego County, California.* National City, CA: Sweetwater River Press. Regional common names, some type localities.

Beidleman, Linda H., and E. N. Kozloff. 2003. *Plants of the San Francisco Bay Region.* Berkeley: Univ. Calif. Press. Regional common names.

Hickman, James C., ed. 1993. *The Jepson manual: Higher plants of California.* Berkeley: Univ. Calif. Press. Primary reference for current scientific and many common names.

Reptiles and Amphibians

Stebbins, Robert C. 1985. *Western reptiles and amphibians.* 2nd ed. Boston: Houghton Mifflin.

OTHER REFERENCES: PRELUDE AND PARTS ONE TO SEVEN

Prelude

Conner, Daniel, and Lorraine Miller. 1999. *Master mariner: Capt. James Cook and the peoples of the Pacific.* Vancouver, B.C.: Douglas & McIntyre.

Hough, Richard. 1979. *The last voyage of Captain James Cook.* New York: William Morrow.

Rickman, John. 1966 [orig. 1781]. *Journal of Captain Cook's last voyage to the Pacific Ocean.* Readex Microprint Corp. [London: E. Newberry].

Townsend, John Kirk. 1839. Narrative of a journey across the Rocky Mountains to the Columbia River. In Reuben G. Thwaites, ed., *Early Western travels,* vol. 21. Cleveland, OH: Arthur H. Clark.

Withey, Lynne. 1987. *Voyages of discovery: Captain Cook and the exploration of the Pacific.* Berkeley: Univ. Calif. Press.

Part One: The Oceanic Expeditions

The French La Pérouse Expedition at Monterey, 1786

Badger, Geoffrey. 1996. *The explorers of the Pacific.* Kenthurst, Australia: Kangaroo Press.

Discombe, Reece, and Pierre Anthonioz. 1960. Voyage to Vanikoro. *Pacific Discovery* 13(1):4–17.

Dunmore, John, ed. 1994. *The journal of Jean-François de Galaup de La Pérouse,* vol. 1. London: Hakluyt Society.

Fisher, Robin, and Hugh Johnston. 1979. *Captain James Cook and his times.* Seattle: Univ. Wash. Press.

La Pérouse, J.-F. G. de. 1798. *A voyage round the world.* 3 vols. London: J. Johnson. English translation.

Margolin, Malcolm. 1989. *Monterey in 1786: The journals of Jean François de La Pérouse.* Berkeley: Heyday Books.

Phillip, Arthur. 1950. *The voyage of Governor Phillip to Botany Bay.* Melbourne, Australia: Advertiser Printing Office.

Schmidt, Earl F. 1987. In recognition of a nearly forgotten American bicentennial. *San Francisco Westerners Argonaut* 10:2, 5–10.

Silkirk, Henry. 1918. La Pérouse and the French monuments at Botany Bay. *Jour. and Proc. Royal Australian Hist. Soc.* 4(7):339–42.

Spary, E. C. 2000. *Utopia's garden: French natural history from Old Regime to revolution.* Chicago: Univ. Chicago Press.

The Spanish Malaspina Expedition with Naturalist Haenke, 1791

Beddall, Barbara G. 1979. Scientific books and instruments for an eighteenth-century voyage around the world: Antonio Pineda and the Malaspina expedition. *Jour. Soc. Biblio. Nat. Hist.* 9(2):95–107.

Ceska, A., ed. 2002. Tadeo Haenke. *Botanical Electronic News* nos. 287, 288.

Cutter, Donald C. 1960. *Malaspina in California.* San Francisco: John Howell Books.

——, ed. 1980. *Journal of Tomás de Suría of his voyage with Malaspina to the Northwest Coast of America in 1791.* Fairfield, WA: Ye Galleon Press.

——. 1990. *California in 1792: A Spanish naval visit.* Norman: Univ. Okla. Press.

Galbraith, Edith C., ed. 1924. Malaspina's voyage around the world. *Calif. Hist. Soc. Quart.* 3(3):215–37.

Ibáñez, Victoria, and Robert J. King. 1996. A letter from Thaddeus Haenke to Sir Joseph Banks. *Archives Nat. Hist.* 23(2):255–60.

Jepson, Willis L. 1899. Early scientific expeditions to California: II. *Erythea* 7(11):129–34.

Madulid, Domingo A. 1989. The life and work of Luis Née, botanist of the Malaspina expedition. *Archives Nat. Hist.* 16(1):33–48.

Silva, Paul C. 1996. California seaweeds collected by the Malaspina expedition, especially *Pelvetia. Madroño* 43(3):345–54.

Vaughan, Thomas, E. A. P. Crownhart-Vaughan, and M. P. de Iglesias. 1977. *Voyages of enlightenment: Malaspina on the Northwest coast, 1791–1792.* Portland: Oregon Hist. Soc.

The British Vancouver Expedition with Menzies, 1792–1795

Anderson, Bern. 1960. *Surveyor of the sea.* Seattle: Univ. Wash. Press.

Balfour, F. R. S. 1945. Archibald Menzies, 1754–1842: Botanist, zoologist, medico and explorer. *Proc. Linnean Soc. London,* sess. 156(3):170–83.

David, Andrew. 1991. *The artists on Vancouver's voyage to the northwest coast of America.* Burnaby, B.C.: Simon Fraser Univ.

Eastwood, Alice, ed. 1924. Menzies' California journal. *Calif. Hist. Soc. Quart.* 2(4):265–340.

Galloway, David, and Eric W. Grove. 1987. Archibald Menzies MD, FLS (1754–1842): Aspects of his life, travels and collections. *Archives Nat. Hist.* 14(1):3–43.

Grinnell, Joseph. 1932. Archibald Menzies, first collector of California birds. *Condor* 34(6):243–52.

Groves, Eric W. 1992a. Archibald Menzies: An early botanist on the West Coast of North America. Paper presented at the Vancouver Conference on Exploration & Discovery, April 1992.

———. 1992b. Letter, Groves to Richard Beidleman from Coulsdon, England, July 14, 1992.

———. 2001. Archibald Menzies (1754–1842): An early botanist on the northwestern seaboard of North America, 1792–1794, with further notes on his life and work. *Archives Nat. Hist.* 28(1):71–122.

Howell, John Thomas. 1938. A botanical visit to the Vancouver Pinnacles. *Leaflets Western Bot.* 2(6):97–102.

Jepson, Willis L. 1929. The botanical explorers of California, VI: Archibald Menzies. *Madroño* 1:262–66.

Johnson, Marion H. 1976. George Vancouver: The last search for the Northwest Passage. *Oceans* 9(5):30–37.

McMillan, Ian I. 1966. The California quail. *Pacific Discovery* 19(6):9–13.

Morwood, William. 1973. *Traveler in a vanished landscape: The life & times of David Douglas, botanical explorer.* New York: Clarkson N. Potter.

Newcombe, C. F., ed. 1923. Menzies' journal of Vancouver's voyage, April to October, 1792. Archives Brit. Columbia Memoir 5. Victoria, B.C.: Archives Brit. Columbia.

Vancouver, George. 1967. *A voyage of discovery to the North Pacific Ocean, and round the world.* 3 vols. Bibliotheca Australiana 30. New York: Da Capo Press.

Wilbur, Marguerite E., ed. 1953. *Vancouver in California 1792–1794.* Los Angeles: Glen Dawson.

Young, Peggy. 1977. The seagoing botanist. *Pacific Discovery* 30(2):14–23.

The Russian Rezanov Expedition with Langsdorff, 1806

Atherton, Gertrude. 1906. *Rezanov.* New York: Authors and Newspapers Association.

Nickerson, Thomas, et al. 2000. *The loss of the ship* Essex, *sunk by a whale.* New York: Penguin Books.

Russell, Thomas C., ed. 1926. *The Rezanov voyage to Nueva California in 1806.* San Francisco: Private Press of Thomas C. Russell.

Vaughan, Thomas, and Bill Holm. 1990. *Soft gold: The fur trade & cultural exchange on the northwest coast of America.* Portland: Oregon Hist. Soc. Press.

Russia Visits Again

Blomkvist, E. E. 1972. A Russian scientific expedition to California and Alaska, 1839–1849. *Oregon Hist. Quart.* 73(2):101–70.

Hohbach, Ann J. 1976. The Russian Kotzebue expedition of 1816. Manuscript.

Jepson, Willis L. 1929. Johann Friederich Eschscholtz. *Madroño* 1:253.

Kotzebue, Otto von. 1821. *A voyage of discovery into the South Sea and Bering's Straits.* 3 vols. Trans. by Hannibal E. Lloyd of Weimar ed. N.p.

———. 1830. *A new voyage round the world in the years 1823, 24, 25, and 26,* vol. 1. London: Henry Colburn and Richard Bentley.

Lewis, Oscar. 1936. *A sojourn at San Francisco Bay 1816.* San Francisco: Book Club of Calif. Chamisso account.

Lincoln, A. 1966. Discoveries of the "Rurik" expedition. *Pacific Discovery* 19(6):14–19.

Mahr, August C. 1932. The visit of the "Rurik" to San Francisco in 1816. *Stanford Univ. Publ. History, Economics & Political Sci.* 2:1–194, 267–460.

Mornin, Edward. 1999. Adelbert von Chamisso: A German poet-naturalist and his visit to California. *Calif. History* 78(1):2–13.

Schlechtendal, Diedrich G. L. von. 1943. Dem Andenken as Adelbert von Chamisso als Botaniker. *London Jour. Bot.* 2:483–91. Translation from *Linnaea* 13:93–106 (1839).

HMS Blossom's *Visit with Beechey, Collie, and Lay, 1826–1827*

Beechey, Frederick W. 1831. *Narrative of a voyage to the Pacific and Beering's Strait.* 2 vols. London: n.p.

———. 1832. Narrative of a voyage to the Pacific and Beering's Straits. Philadelphia: Carey and Lea.

Bretschneider, E. 1898. *History of European botanial discoveries in China,* pp. 290–94. London: Sampson Low, Marston.

Hooker, William Jackson, and G. A. Walker-Arnott. 1830–41. *The botany of Captain Beechey's voyage.* London: Henry Bohn.

Lincoln, A. 1969. The Beechey expedition visits San Francisco. *Pacific Discovery* 22(1):1–8.

Richardson, John, N. A. Vigors, et al. 1839. *The zoology of Captain Beechey's voyage.* London: Henry G. Bohn.

HMS Sulphur's *Visits with Belcher, Hinds, and Barclay, 1837 and 1839*

Belcher, Edward. 1843. *Narrative of a voyage round the world, performed in Her Majesty's Ship Sulphur, during the years 1836–1842.* 2 vols. London: Henry Colburn.

Bentham, George. 1844. *The botany of the voyage of H.M.S. Sulphur: The botanical descriptions.* London: Smith, Elder.

Hinds, R. B. 1843. The regions of vegetation; being an analysis of the distribution of vegetable forms over the surface of the globe in connexion with climate and physical agents. In Belcher 1843 (see above), 2:323–460.

———, ed. 1844. *The zoology of the voyage of H.M.S. Sulphur, under the com-*

mand of Captain Sir Edward Belcher during the years 1836–1842. 2 vols. London: Lords Commissioners of the Admiralty.

Hinds, R. B., and George Bentham. 1844. *The botany of the voyage of H.M.S. Sulphur, under the command of Captain Sir Edward Belcher, during the years 1836–1842.* London: Lords Commissioners of the Admiralty.

Pierce, Richard A., and John H. Winslow, eds. 1969. *H.M.S. Sulphur at California, 1837 and 1839.* San Francisco: Book Club of Calif. Includes Midshipman Simpkinson's account, as well as Belcher's.

Raven, Peter H. 1964. George Barclay and the "California" portion of the *Botany of the Sulphur. Aliso* 5(4):469–77.

The French Visit Again

Anonymous. 1973. Carlo Giuseppe Guglielmo Botta. *Encyc. Brit.* 4:5.

Bancroft, Hubert H. 1886. *The works of Hubert H. Bancroft.* Vol. 4: *California,* pp. 147–49. San Francisco: History Company.

Carter, Charles F., ed. 1929. Duhaut-Cilly's account of California in the years 1827–1828. *Calif. Hist. Soc. Quart.* 8:130–66, 214–50, 306–36.

Chamberlin, J. Lockwood. 1960. Voyage of the Venus. *Nautilus* 74(2): 65–68.

Coues, Elliott. 1900. The "churca" *(Geococcyx californianus). Auk* 17(1):66.

Delougaz, Pinhas P. 1973. Mesopotamia archaeology. *Encyc. Brit.* 2:243.

Fagan, Brian M., ed. 1996. *The Oxford companion to archaeology,* p. 94. New York: Oxford Univ. Press.

Frugé, August, and Neal Harlow, eds. 1999. *A voyage to California, the Sandwich Islands, and around the world in the years 1826–1829: Auguste Duhaut-Cilly.* Berkeley: Univ. Calif. Press.

Howell, John Thomas. 1935. Sea-gulls and tarweeds: A distributional mix-up. *Leaflets Western Bot.* 1(16):189–201.

Kastner, Joseph. 1977. *A world of naturalists,* p. 295. London: John Murray.

Roberson, Don. 1985. *Monterey birds,* pp. 112, 138. Monterey, CA: Monterey Peninsula Audubon Society. Includes a photograph of the Swallow-tailed Gull at Moss Landing and Pacific Grove.

Rudkin, Charles N. 1956. *Voyage of the Venus: Sojourn in California.* Los Angeles: Glen Dawson.

Sherratt, Andrew, ed. 1980. *The Cambridge encyclopedia of archaeology,* p. 21. New York: Crown Publishers/Cambridge Univ. Press.

Streshinsky, Shirley. 1993. *Audubon: Life and art in the American wilderness,* p. 235. New York: Villard Books.

The United States Exploring Expedition in California, 1841

Barnhart, John H. 1919. Brackenridge and his book on ferns. *Jour. New York Botanical Garden* 2:117–24.

Bartlett, Harley H. 1940. The reports of the Wilkes expedition, and the work of the specialists in science. *Proc. Amer. Phil. Soc.* 82:(5):601–705.

Bryan, G. S. 1940. The purpose, equipment and personnel of the Wilkes expedition. *Proc. Amer. Philosophical Soc.* 82(5):551–60.

Dana, James D. 1849. Notes on Upper California. *Amer. Jour. Science,* 2nd ser., 7:247–64.

Eastwood, Alice. 1945. An account and list of the plants in the Brackenridge Journal. *Calif. Hist. Soc. Quart.* 24(4):337–42.

Eyde, Richard H. 1986. William Rich of the great U.S. Exploring Expedition and how his shortcomings helped botany become a calling. *Huntia* 6(2):165–96.

Gilman, Daniel C. 1899. *The life of James Dwight Dana.* New York: Harper & Brothers.

Maloney, Alice B., ed. 1945. A botanist on the road to Yerba Buena: Journal of William Dunlop Brackenridge October 1–28, 1841. *Calif. Hist. Soc. Quart.* 24(4):320–36.

Morgan, W. J., D. B. Tyler, J. L. Leonhart, and M. F. Loughlin, eds. 1978. *Autobiography of Rear Admiral Charles Wilkes, U.S. Navy 1798–1877.* Washington, DC: Department of the Navy.

Philbrick, Nathaniel. 2003. *Sea of Glory.* New York: Viking.

Poesch, Jessie. 1961. *Titian Ramsey Peale 1799–1885 and his journals of the Wilkes expedition.* Philadelphia: Amer. Philosophical Soc.

Sperlin, O. B., ed. 1930–31. Our first horticulturist: The Brackenridge journal. *Wash. Hist. Quart.* 21(3):218–29; 22:129–45, 218–29.

Stanton, William. 1975. *The great United States Exploring Expedition of 1838–1842.* Berkeley: Univ. Calif. Press.

Tyler, David B. 1968. *The Wilkes expedition: The first United States Exploring Expedition (1838–1842).* Philadelphia: Amer. Philosophical Soc.

Viola, Herman J., and Carolyn Margolis, eds. 1985. *Magnificent voyagers.* Washington, DC: Smithsonian Inst. Press.

Wilkes, Charles. 1845. *Narrative of the United States Exploring Expedition,* vol. 5. Philadelphia: C. Sherman.

Part Two: The Early Peripatetic Naturalists

David Douglas: Don David El Botánico

Beidleman, Richard G. 1969. Douglas in pursuit of plants. *Horticulture* 47(2):30–31, 49, 59.

Davies, John. 1981. *Douglas of the forests.* Seattle: Univ. Wash. Press.

Harris, Harry. 1941. The annals of *Gymnogyps* to 1900. *Condor* 43(1):23–27.

Harvey, Athelstan G. 1947. *Douglas of the fir: A biography of David Douglas botanist.* Cambridge, MA: Harvard Univ. Press.

Hooker, William Jackson, and G. A. Walker-Arnott. 1841. *The botany of Captain Beechey's voyage.* London: n.p. Includes West Coast plants collected by David Douglas in "California—Supplement," 2:316–409.

Howell, John Thomas. 1942. Concerning David Douglas. *Leaflets Western Bot.* 3(7):160–62.

Jepson, Willis Linn. 1933. David Douglas in California. *Madroño* 2:97–100.

Morwood, William. 1973. *Traveler in a vanished landscape: The life & times of David Douglas, botanical explorer.* New York: Clarkson N. Potter.

Netboy, Anthony. 1971. Plant collector in the wilderness. *Pacific Discovery* 24(2):18–25.

Shepherd, Steven L. 2001. Trials of an intrepid botanist: David Douglas. *California Wild* 54(2):8–14, 53.

Thomas Coulter: First to the California Desert
Coulter, Thomas. 1835. Notes on Upper California. *Jour. Royal Geogr. Soc. London* 5:59–70. Contains map of Alta California.
Coville, Frederick V. 1895. The botanical exlorations of Thomas Coulter in Mexico and California. *Bot. Gaz.* 20(12):519–31. Plate 34 is a reproduction of Coulter's map.
Don, David. [1834–37.] Descriptions of five new species of the genus Pinus; discovered by Dr. Coulter in California. *Trans. Linnean Soc. London* 17: 439–44.
Harvey, William H. 1845. Description of a new genus of Papaveraceae, detected by the late Dr. Coulter in California. *London Jour. Bot.* 4:73–76, pl. 3.
———. 1869. *Memoir of W. H. Harvey.* London: n.p.
McCaugh, Rogers. 1943. The travels of Thomas Coulter, 1824–1827. *Jour. Wash. Acad. Sci.* 33:65–70.
Nelson, E. Charles. 1983. Thomas Coulter (1793–1843) in North America: Some bibliographic problems and some solutions. *Soc. Biblio. Nat. Hist. Special Publ.* 2:59–71.
Parry, Charles C. 1883. Early botanical explorers of the Pacific coast. *Overland Monthly* 2nd ser. 2: 414.
Stillman, Jacob. 1869. Footprints of early California discoverers. *Overland Monthly* 2:263. ·

A Frontier Naturalist and His Protégé Visit California Separately
Beidleman, Richard G. 1957. John Kirk Townsend on the Oregon Trail. *Audubon Magazine* 59(2):64–65, 83, 88.
———. 1958. William Gambel: Frontier naturalist. *Pacific Discovery* 11(6): 10–14.
———. 1960. Some biographical sidelights on Thomas Nuttall, 1780–1859. *Proc. Amer. Philosophical Soc.* 104(1):86–100.
———. 1967. Thomas Nuttall: Botanist of the American wilderness. *Horticulture* 45(6):36–37, 45.
Coville, Frederick V. 1899. The botanical explorations of Thomas Nuttall in California. *Proc. Biol. Soc. Wash.* 13:109–21.
Dana, Richard H. 1840. *Two years before the mast.* New York: Harper.
Gambel, William. 1843–47. Descriptions of some new and rare birds of the Rocky Mountains and California. *Proc. Acad. Nat. Sci. Phila.* 1:258–62; 2:263–72; 3:44–48, 110–15, 154–58, 200–205.
———. 1847–49. Remarks on the birds observed in upper California, with descriptions of new species. *Jour. Acad. Nat. Sci. Phila.* 2nd ser., 1:25–56, 215–29.
Graustein, Jeannette. 1967. *Thomas Nuttall: Naturalist.* Cambridge, MA: Harvard Univ. Press.
Jepson, Willis L. 1934. The overland journey of Thomas Nuttall. *Madroño* 2:143–47.

———. 1935. Centennial of Thomas Nuttall's visit to California. Program for the ceremony at the Jepson Laurel, Spring Valley Lakes, San Mateo County, October 20, 1935. Jepson collections, Berkeley: Univ. Calif. Jepson Herbarium Archives.

Jervis, Thomas. Correspondence with the author relative to William Gambel.

Keen, A. Myra. 1966. West American mollusk types at the British Museum (Natural History) I. T. A. Conrad and the Nuttall collection. *Veliger* 8(3):167–72.

Maxwell, Richard T. 1955. *Visit to Monterey in 1842*. Los Angeles: Glen Dawson.

Middleton, William J. 1914. Life and personal letters of William Gamble M.D. of Philadelphia, Pa. Scrapbook of Gambel letters at Pennsylvania Historical Society. Gambel's name is misspelled.

Nuttall, Thomas. 1848. Descriptions of plants collected by William Gambel. *Jour. Acad. Nat. Sci. Phila.* 2nd ser., 1:149–89.

———. 1850. Descriptions of plants collected by Mr. William Gambel in the Rocky Mountains and upper California. *Proc. Acad. Nat. Sci. Phila.* 2nd ser., 4:7–26.

———. 1852. *The North American sylva*. Philadelphia: by the author.

Pennel, Francis W. 1936. Travels and scientific collections of Thomas Nuttall. *Bartonia* 18:1–51.

Smith, Gene A. 2000. *Thomas ap Catesby Jones*. Annapolis, MD: Naval Institute Press.

Stone, Witmer. 1910. William Gambel. *Cassinia* 14:1–8.

Wistar, Isaac. 1914. *Autobiography of Isaac Jones Wistar*. 2 vols. Philadelphia: Wistar Institute of Anatomy and Biology.

Woodward, Arthur. 1934. Benjamin Davis Wilson's observations on early days in California and New Mexico. *Hist. Soc. Southern Calif. Quart.* 16:74–150.

Karl Theodor Hartweg

Anonymous. 1970. Hartweg in Monterey area. *Noticias del Puerto de Monterey* 14(1):2–4, 11–13.

Bentham, George. 1848–57. *Plantae Hartwegianae*. 3 vols. London: n.p.

Davis, William H. 1929. *Seventy-five years in California*. San Francisco: John Howell.

Gudde, Erwin G., ed. 1933. The memoirs of Theodor Cordua, the pioneer of New Mecklenburg in the Sacramento Valley. *Quart. Calif. Hist. Soc.* 12: 279–306.

Hartweg, Theodore. 1846–48. Journal of a mission to California in search of plants. *Jour. Hort. Soc. London* 1:180–85; 2:121–25, 187–91; 3:217–28.

Jepson, Willis Linn. 1897. The explorations of Hartweg in America. *Erythea* 5:31–35, 51–56.

Part Three: The Overland Expeditions and Their Naturalists

John Charles Frémont: In Pursuit of the California Flora

Beidleman, Richard G. 2002. John Charles Frémont and his floral forays into Oregon. *Kalmiopsis* 9:4–10.

Egan, Ferol. 1977. *Frémont: Explorer for a restless nation.* Garden City, NY: Doubleday.

Frémont, John Charles. 1845. *Report of the exploring expedition to the Rocky Mountains in the year 1842, and to Oregon and north California, in the years 1843–'44, by Brevet Captain J. C. Fremont of the Topographical Engineers.* U.S. 28th Congr., 2nd sess., Sen. doc. 11, no. 174.

———. 2001 [orig. 1887]. *Memoirs of my life.* New York: Cooper Square Press.

Gray, Asa. 1845. Characters of some new genera and species of plants of the natural order Compositae, from the Rocky Mountains and upper California. *Boston Jour. Nat. Hist.* 5:104–11.

Herr, Pamela, and Mary Lee Spence. 1993. *The letters of Jessie Benton Frémont.* Urbana: Univ. Illinois Press.

Jackson, Donald, and Mary Lee Spence, eds. 1970a. *The expeditions of John Charles Frémont,* vol. 1: *Travels from 1838 to 1844.* Urbana: Univ. Illinois Press.

———, eds. 1970b. *The expeditions of John Charles Frémont.* Map Portfolio. Urbana: Univ. Illinois Press.

Nevins, Allan. 1955. *Fremont: Pathmarker of the West.* New York: Longmans, Green.

Parish, S. B. 1908. Frémont in Southern California. *Muhlenbergia* 4:57–62.

Richmond, Patricia Joy. 1989. *Trail to disaster: The route of John C. Frémont's fourth expedition from Big Timbers, Colorado, through the San Luis Valley, to Taos, New Mexico.* Col. Hist. Soc. Monograph 4. Denver: Col. Hist. Soc.

Roberts, David. 2000. *A newer world: Kit Carson, John C. Frémont, and the claiming of the American West.* New York: Simon & Schuster.

Spence, Mary Lee, ed. 1984. *The expeditions of John Charles Frémont,* vol. 3: *Travels from 1848 to 1854.* Urbana: Univ. Illinois Press.

Spence, Mary Lee, and Donald Jackson, eds. 1973. *The expeditions of John Charles Frémont,* vol. 2: *The Bear Flag revolt and the Court-martial.* Urbana: Univ. Illinois Press.

Torrey, John. 1845. Note concerning the plants collected in the second expedition of Captain Frémont. In Frémont 1845, appendix C, 311–19.

———. 1853. Plantae Frémontianae; or, descriptions of plants collected by Col. J. C. Frémont in California. *Smithsonian Contr. Knowledge* 6(1):1–24.

Torrey, John, and John Charles Frémont. 1845. Descriptions of some new genera and species of plants, collected in Captain J. C. Frémont's exploring expedition to Oregon and north California, in the years 1843–'44. In Frémont 1845, appendix C, 311–19.

Welsh, Stanley L. 1998. *John Charles Frémont: Botanical explorer.* St. Louis: Missouri Bot. Garden Press.

Battles and Botany: Emory and the Army of the West, 1846–1847

Calvin, Ross, ed. 1968. *Lieutenant Emory reports: Notes of a military reconnaissance.* Albuquerque: Univ. New Mexico Press.

DeVoto, Bernard. 1984. *The year of decision, 1846.* New York: Book-of-the-Month Club.

Emory, William Hemsley. 1848a. *Notes of a military reconnaissance, from Fort*

Leavenworth, in Missouri, to San Diego, in California, including part of the Arkansas, Del Norte, and Gila Rivers. U.S. 30th Congr., 1st sess., exec. doc. no. 41. Includes journal by Captain A. R. Johnston and reports on the flora by John Torrey and George Engelmann.

———. 1848b. *Notes of a military reconnaissance, from Fort Leavenworth in Missouri, to San Diego, in California.* Washington, DC: Wendell and Van Benthuysen.

Norris, L. David, J. C. Milligan, and O. B. Faulk. 1998. *William H. Emory: Soldier-scientist.* Tucson: Univ. Ariz. Press.

The Mexican Boundary Surveys

Bancroft, Peter. 1979. Tourmaline. *Pacific Discovery* 32(3):10–14.

Baptista, Luis F. 1987. Andrew Jackson Grayson and the "Solitary Dove." *Pacific Discovery* 40(2):30–37.

Bartlett, John R. 1965 [orig. 1854]. *A personal narrative of explorations and incidents in Texas, New Mexico, California, Sonora, and Chihuahua: 1850–1853*, vol. 2. Chicago: Rio Grande Press.

Edwards, E. I., ed. 1955. *The Whipple report.* Los Angeles: Western Lore Press.

Emory, William H. 1857–59. *Report on the United States and Mexican boundary survey.* 3 vols. in 2. U.S. 34th Cong., 1st sess., Senate exec. doc. no. 108. Washington, DC: Government Printing Office.

Fireman, Janet R., ed. 2004. R. P. Effinger's excellent adventure: The unknown letters of a young Ohio lawyer. *California History* 82(1):2–75.

Hine, Robert V. 1968. *Bartlett's West: Drawing the Mexican boundary.* New Haven, CT: Yale Univ. Press.

Hodder, Frank H. 1984. *Audubon's western journal, 1849–1850.* Tucson: Univ. Ariz. Press.

Lesley, Lewis B. 1930. The international boundary survey from San Diego to the Gila River, 1849–1850. *Quart. Calif. Hist. Soc.* 9(1):3–15.

McPherson, William, ed. 1932. *From San Diego to the Colorado in 1849: The journal and maps of Cave J. Couts.* Los Angeles: William M. Ellis.

Norris, L. David, J. C. Milligan, and O. B. Faulk. 1998. *William H. Emory: Soldier-scientist.* Tucson: Univ. Ariz. Press.

Parish, S. B. 1909. Parry and Southern California botany. *Plant World* 12(7):158–62.

Rusby, H. H. 1890. A biographical sketch of Dr. George Thurber. *Torrey Bot. Club Bull.* 17:204–10.

Wallace, Edward S. 1955. *The great reconnaissance.* Boston: Little, Brown.

Part Four: Iron Horses and River Steamers: The 1850s Surveys

Albright, George L. 1921. *Official explorations for Pacific railroads.* Univ. Calif. Publ. Hist. 11. Berkeley: Univ. Calif. Press.

Alt, David, and Donald W. Hyndman. 2000. *Roadside geology of Northern and Central California.* Missoula, MT: Mountain Press.

Beckwith, Edward G. 1853–54. *Journals for the Gunnison/Beckwith expedition.* Beckwith/Fox Collection, box 8. San Marino, CA: Huntington Library.

Bettelheim, Matthew. 2005. Mormorata: The famed mud turtle of the San Francisco Market. *Calif. Hist.* 82(4):26–42.

California Office of Historic Preservation. 1996. *California historical landmarks*. Sacramento: Calif. State Parks.

Cassin, John. 1991. *Birds of California, Texas, Oregon, British and Russian America*. Austin: Texas State Hist. Assn.

Cutright, Paul R. 1969. *Lewis and Clark: Pioneering naturalists*. Urbana: Univ. Illinois Press.

Dawson, William L. 1923. *The birds of California*, vol. 3. San Diego: South Moulton.

Foreman, Grant, ed. 1941. *A pathfinder in the West: The itinerary of Lieutenant A. W. Whipple during his explorations for a railroad route from Fort Smith to Los Angeles in the years 1853 & 1854*. Norman: Univ. Okla. Press.

Fowler, Harlan D. 1950. *Camels to California*. Stanford, CA: Stanford Univ. Press.

Gordon, Mary M., ed. 1988. Through Indian country to California: John P. Sherburne's diary of the Whipple expedition, 1853–1854. Stanford, CA: Stanford Univ. Press.

Gross, Thomas L. T. 2000. Volcanoes in the Susanville region, Lassen, Modoc, Plumas Counties, northeastern California. *California Geology* 53(5):4–23.

Heermann, Adolphus L. 1853. Notes on the birds of California, observed during a residence of three years in that country. *Jour. Acad. Nat. Sci. Phila.* 2nd ser., 2:259–72.

Huseman, Ben W. 1995. *Wild river, timeless canyons: Balduin Mölhausen's watercolors of the Colorado*. Fort Worth, TX: Amon Carter Museum.

Ives, Joseph C. 1861. *Report upon the Colorado River of the West, explored in 1857 and 1858*. 36th Congr., 1st sess., Sen. exec. doc., unnumbered. Washington, DC: Government Printing Office. Includes geology, botany, and zoology reports.

Jepson, Willis L. 1962. John Milton Bigelow. *Madroño* 16(6):179–83.

Johnston, Ivan M. 1928. The botanical activities of Thomas Bridges. *Contr. Gray Herbarium* 81:98–106.

Lesley, Lewis B. 1929. *Uncle Sam's camels*. Cambridge, MA: Harvard Univ. Press.

Madden, Henry M. 1949. *Xántus: Hungarian naturalist in the pioneer West*. Palo Alto, CA: Books of the West.

Möllhausen, Heinrich B. 1858. *Diary of a journey from the Mississippi to the coasts of the Pacific, with a government expedition*. 2 vols. London: n.p.

Rosenberg, Kenneth V., et al. 1991. *Birds of the lower Colorado River valley*. Tucson: Univ. Ariz. Press.

Schaffer, Jeffrey P. 1999. *Lassen Volcanic National Park & vicinity*. Berkeley: Wilderness Press.

Schiel, Jacob H. 1959. *Journey through the Rocky Mountains and the Humboldt Mountains to the Pacific Ocean*. Norman: Univ. Okla. Press.

Stewart, George R. 1962. *The California trail*. New York: McGraw Hill.

Sturhahn, Joan. 1976. *Carvalho: Artist-photographer-adventurer-patriot*. Merrick, NY: Richwood Publ.

Thompson, Gerald. 1983. *Edward F. Beale & the American West*. Albuquerque: Univ. New Mexico Press.

United States War Department. Pacific Railroad Surveys. 1855–1859. *Reports of explorations and surveys to ascertain the most practicable and economic route for a railroad from the Mississippi River to the Pacific Ocean*. 12 vols. (vol. 12 in two parts).

Utley, Robert M. 1997. *A life wild and perilous: Mountain men and the paths to the Pacific*. New York: Henry Holt.

Waller, A. E. 1942. Dr. John Milton Bigelow, 1804–1878. *Ohio Arch. and Hist. Quart*. 51:313–31.

White, David A., ed. 1999. *News of the plains and Rockies 1803–1865*, vol. 6: *Gold seekers, California, 1849–1856; Railroad forerunners, 1850–1865*. Spokane, WA: Arthur H. Clark.

Zwinger, Ann. 1986. *John Xántus: The Fort Tejon letters, 1857–1859*. Tucson: Univ. Ariz. Press.

Part Five: The California Geological Survey

Brewer, William H. 1974. *Up and down California*. 3rd ed. Berkeley: Univ. Calif. Press.

Brewer, William, Sereno Watson, and Asa Gray. 1880. *Geological survey of California: Botany*, vol. 1. 2nd ed. Boston: Little, Brown.

Brewster, Edwin T. 1909. *Life and letters of Josiah Dwight Whitney*. Boston: Houghton Mifflin.

Fahy, N. E. 1985. Joseph D. Whitney, the California state geological survey and Yosemite Valley. *Calif. Geol*. 38(6):129–33.

Fulton, John F., and Elizabeth H. Thompson. 1947. *Benjamin Silliman: Pathfinder in American science*. New York: Henry Schuman.

Gabb, W. M. 1864. *Paleontology of California*, vol. 1: *Triassic and Cretaceous fossils*. Geological survey of California. Philadelphia: Caxton Press of Sherman & Co.

———. 1869. *Paleontology of California*, vol. 2: *Cretaceous and Tertiary fossils*. Geological survey of California. Philadelphia: Caxton Press of Sherman & Co.

Hodder, Frank H., ed. 1984. *Audubon's western journal, 1849–1850*. Tucson: Univ. Ariz. Press.

Jepson, Willis Linn. 1933. The botanical explorers of California, VIII: Thomas Bridges. *Madroño* 2(10):84–88.

Johnston, Ivan M. 1928. The botanical activities of Thomas Bridges. *Contr. Gray Herbarium* 81:98–106.

Kay, E. Alison. 1976. Hawaiian natural history: 1778–1900. In *Natural history of the Hawaiian Islands: Selected readings*, ed. E. Alison Kay, pp. 626–27. Honolulu: Univ. Hawaii Press.

Oakeshott, Gordon B. 1985. Contributions of the state geological surveys: California as a case history. In *Geologists and ideas: A history of North American geology*, ed. Ellen T. Drake and William M. Jordan, pp. 323–35. Boulder: Geol. Soc. Amer.

Vaughan, Thomas, and Bill Holm. 1990. *Soft gold*. Portland: Oregon Hist. Soc. Press.

Watson, Sereno. 1880. *Geological survey of California: Botany*, vol. 2. Boston: Little, Brown.

White, Gerald T. 1968. *Scientists in conflict*. San Marino, CA: Huntington Library.

Part Six: Institutions and Naturalists

Barber, Jack. 1945. Edward Lee Greene. *The Catholic World* 160:444–49.

Bartlett, Harley H. 1916. The botanical work of Edward Lee Greene. *Torreya* 16(7):151–75.

Brandegee, Katharine. 1893. The botanical writings of Edward L. Greene. *Zoe* 4:63–103.

Egerton, Frank N., ed. 1983. *Landmarks of botanical history: Edward Lee Greene*. 2 vols. Stanford, CA: Stanford Univ. Press.

Gingerich, Owen. 1978. James Lick's observatory. *Pacific Discovery* 31(1):1–11.

Greene, Edward L. 1929. Why the Eschscholtzia? *Madroño* 1(14):195–97.

Gudde, Erwin G. 1955. A century of astronomy and geodesy. In *A century of progress in the natural sciences, 1853–1953*, ed. Edward L. Kessel, pp. 65–74. San Francisco: Calif. Acad. Sci.

Holden, Edward S. 1936. A visit to the Lick Observatory. In *West of the Rocky Mountains*, ed. John Muir, pp. 119–28. Philadelphia: Running Press.

Jepson, Willis L. 1918. Edward Lee Greene, the man and botanist. *Newman Hall Review* 1(2):24–29.

———. 1943. Edward Lee Greene the individual. *Amer. Midland Naturalist* 30(1):3–5.

Kirby-Smith, H. T. 1976. *U.S. observatories: A directory and travel guide*. New York: Van Nostrand Reinhold.

Kistler, Ellen D. 1936. Bibliography of the botanical writings of Edward Lee Greene. *Madroño* 3(8):328–48.

LeConte, Joseph. 1930 [orig. 1875]. *A Journal of ramblings through the High Sierra of California by the University Excursion Party*. San Francisco: Sierra Club.

Main, Angie K. 1929. Life and letters of Edward Lee Greene. *Trans. Wisc. Acad. Sci., Arts and Letters* 24:147–85.

McIntosh, Robert P. 1961. The Greene-Nieuland Herbarium at the University of Notre Dame. *Indiana Acad. Sci. Proc.* 71:235–38.

———. 1983. Edward Lee Greene: The man. In Egerton 1983, 1:18–53.

McVaugh, Rogers. 1983. Edward Lee Greene: An appraisal of his contributions to botany. In Egerton 1983, 1:54–92.

Osterbrock, Donald E., J. R. Gustafson, and W. J. S. Unruh. 1988. *Eye on the sky: Lick Observatory's first century*. Berkeley: Univ. Calif. Press.

Ridout, Lionel U. 1958. Priests, pistols and polemics. *Calif. Hist. Soc. Quart.* 37(1):51–61.

Rush, Elizabeth. 2004. The irascible Reverend Greene: Botanist and contrarian. *California Wild* 57(2):35–39.

Soulé, Frank. 1902. Joseph LeConte in the Sierra. *Sierra Club Bull.* 4(1):1–11.

Stephens, Lester D. 1982. *Joseph LeConte: Gentle prophet of evolution.* Baton Rouge: Louisiana State Univ. Press.

Wright, Helen. 1987. *James Lick's monument.* Cambridge, MA: Cambridge Univ. Press.

Part Seven: The Postwar Naturalists

The Latter-Day Peripatetic Naturalists

Beaty, Janice J. 1864. Plantsman of the West. *Pacific Discovery* 17(6):10–15. About Edward Palmer.

Bell, William A. 1869. *New tracks in North America.* London: Chapman and Hall. About Charles Parry.

Davis, Helen B. 1936. Life and work of Cyrus Guernsey Pringle. Burlington: Univ. Vermont Press.

Jones, Marcus. 1930. C. C. Parry. *Contr. Western Bot.* 17:3–6.

———. 1933–35. Botanists I have known. *Contr. Western Bot.* 18:1–18.

Kibbe, Alice L. 1953. *Afield with plant lovers and collectors.* Quincy, IL: Carthage College.

Lenz, Lee W. 1986. *Marcus E. Jones.* Claremont, CA: Rancho Santa Ana Botanic Garden.

McVaugh, Rogers. 1956. *Edward Palmer: Plant explorer of the American West.* Norman: Univ. Okla. Press.

Orcutt, C. R. 1890. Charles Christopher Parry. *West Amer. Scientist* 7(50):1–5.

Parish, S. B. 1909. Parry and Southern California botany. *Plant World* 12(7): 158–62.

Preston, C. H. 1893. Biographical sketch of Charles Christopher Parry. *Proc. Davenport Acad. Sciences* 6:35–52.

Weber, William A. 1997. *King of Colorado botany: Charles Christopher Parry, 1823–1890.* Boulder: Univ. Press Col.

Welsh, Stanley L. 1988. Utah botanical explorer Charles Christopher Parry. *Great Basin Naturalist* 48(1):9–18.

Frontier Naturalists Calling California Home

Badè, William F. 1924. *The life and letters of John Muir.* 2 vols. Boston: Houghton Mifflin.

Beidleman, Richard G. 2004. Lemmons and poppies. *The Jepson Globe* 14(2): 1–2.

Bonta, Marcia M. 1992. *Women in the field: America's pioneering women naturalists.* College Station: Texas A. & M. Univ. Press.

Brewster, Edwin T. 1909. *Life and letters of Josiah Dwight Whitney.* Boston: Houghton Mifflin.

Coan, Eugene. 1982. *James Graham Cooper: Pioneer western naturalist.* Boise: Univ. Press Idaho.

Crosswhite, Frank S. "J. G. Lemmon & wife," plant explorers in Arizona, California, and Mexico. *Desert Plants* 1(1):12–21.

Crosswhite, Frank S., and Carol D. Crosswhite. 1985. The plant collecting Brandegees, with emphasis on Katharine Brandegee as a liberated woman scientist of early California. *Desert Plants* 7(3):128–39, 158–62.

Ehrlich, Gretel. 2000. *John Muir: Nature's visionary.* Washington, DC: Nat. Geogr. Soc.

Ewan, Joseph. 1944. Sara Allen Plummer Lemmon and her "Ferns of the Pacific Coast." *Amer. Midland Naturalist* 32(Jul.–Nov.):513–18.

Guinn, James M. 1907. *A history of the state of California and biographical record of Oakland and environs,* vol. 2. Los Angeles: Historic Record Co.

Jaeger, Edmund C. 1953. Bold Kate Brandegee: Pioneer California woman botanist. *Calico Print* 9(2):8–9, 33.

James, Edward T., ed. 1974. *Notable American women, 1607–1950.* Cambridge, MA: Harvard Univ. Press.

Jones, Marcus. 1929. The Brandegees. *Contr. Western Bot.* 15:15–18.

Lemmon, John G. 1903. Kings River Outing, 1902: Botanical notes, including an important discovery. *Sierra Club Bull.* 4(4):292–300.

———. 1908. Notes by a pioneer botanist. *Muhlenbergia* 4(2):17–21.

Lemmon Collection (John G. and Sarah Plummer Lemmon). Berkeley: Univ. Calif. Herbarium Archives.

MacFarlane, James. 1890. *An American geological railway guide.* 2nd ed. rev. by James H. MacFarlane. New York: D. Appleton.

Main, Angie K. 1929. Life and letters of Edward Lee Greene. *Trans. Wisc. Acad. Sci., Arts and Letters* 24:147–85.

Marshall, Robert B. 1916–19. John Muir. *Sierra Club Bull.* 10:23–24.

Muir, John. 1916. *The mountains of California.* 2 vols. Boston: Houghton Mifflin.

Rush, Elizabeth. 1997. On her terms: Katharine Brandegee, first woman of western botany. *Pacific Discovery* 50(1):22–27.

San Francisco Bulletin, July 8, 10, 17, 1904. Lemmon and the California Promotion Commitee.

Setchell, William B. 1926. Townshend Stith Brandegee and Mary Katharine (Layne) (Curran) Brandegee. *Univ. Calif. Publ. Bot.* 13(9):155–78.

Smith, Emory E. 1902. *The golden poppy.* San Francisco: Murdock Press.

Tompkins, Walker A. 1983. *Santa Barbara history makers.* Santa Barbara, CA: McNally & Loftin.

Wolfe, Linnie M. 1945. *Son of the wilderness: The life of John Muir.* New York: Alfred A. Knopf.

Postlude: The End of California's "Frontier"

Coville, Frederick V. 1893a. *Botany of the Death Valley expedition.* U.S. Nat. Herbarium Contr. 4. Washington, DC: Government Printing Office.

———. 1893b. Descriptions of new plants from Southern California, Nevada, Utah, and Arizona. *Proc. Biol. Soc. Wash.* 7(May):78.

Fisher, A. K., et al. 1893. *The Death Valley expedition: A biological survey of parts of California, Nevada, Arizona, and Utah,* part II. North American Fauna 7. Washington, DC: Government Printing Office.

Goetzmann, William H., and Jay Sloan. 1982. *Looking far north: The Harriman expedition to Alaska, 1899*. New York: Viking Press.

Palmer, Theodore S. 1989. *Chronology and names of the Death Valley region in California, 1849–1949*. Hollywood, CA: Borgo Press.

Turner, Frederick Jackson. 1893. The significance of the frontier in American history. *Proc. State Hist. Soc. Wisc.* Dec. 14:79–112.

———. 1920. *The frontier in American history*. New York: Henry Holt. The 1893 article is included.

Index

Text:	10/13 Sabon
Display:	Sabon
Cartographer:	Bill Nelson
Compositor:	Binghamton Valley Composition, LLC
Printer and Binder:	Thomson-Shore, Inc.

Text: 10/13 Sabon
Display: Sabon
Cartographer: Bill Nelson
Compositor: Binghamton Valley Composition, LLC
Printer and Binder: Thomson-Shore, Inc.